T0215415

Water Resources Monograph 18

LANDSLIDES
Processes, Prediction, and Land Use

Roy C. Sidle
Hirotaka Ochiai

American Geophysical Union
Washington, DC

Library of Congress Cataloging-in-Publication Data

Sidle, Roy C.
  Landslides : processes, prediction, and land use / Roy C. Sidle, Hirotaka Ochiai.
    p. cm. -- (Water resources monograph ; 18)
  Includes bibliographical references and index.
  ISBN-13: 978-0-87590-322-4
  ISBN-10: 0-87590-322-3
  1. Landslides. 2. Landslide hazard analysis.   I. Ochiai, Hirotaka.
  II. Title. III. Series.
  QE599.A2.S56 2006
  551.3'07--dc22

                                        2006013210

ISBN-13: 978-0-87590-322-4
ISBN-10: 0-87590-322-3
ISSN 0170-9600

Cover: Debris flow/landslide of 17 February 2006 that buried Philippine village of Guinsahugon; photograph taken on 19 February 2006 by Photographer's Mate 1st Class Michael D. Kennedy and used courtesy of Navy Visual News (U.S. Navy imagery used in illustration without endorsement expressed or implied).

Table of Contents

Preface

This book marks the culmination of 5 years of writing as well as the cumulative experience gained from working in various regions of North America, New Zealand, Southeast Asia, Japan, Europe, and China over the past 30 years. About 20 years ago I wrote the Water Resources Monograph *Hillslope Stability and Land Use* for American Geophysical Union with two colleagues from New Zealand. Whilst still relevant to the issues of land management and landslides, it is obvious that many advances have been made related to landslide process understanding, prediction methods, and management implications; thus, a totally new book was warranted on this topic. This time I have included as co-author my very good friend Dr. Hirotaka Ochiai, who not only wrote most of Chapter 4 and much of the "Seismicity" section of Chapter 3, but also introduced me to the incredible breadth of research and management activity related to landslides in Japan and was very influential in my eventual placement as Professor at Disaster Prevention Research Institute (DPRI), Kyoto University. Dr. Ochiai is one of the leading members of the Japanese landslide community, not only as a researcher, but also as a liaison with management agencies and a mentor for young scientists. He opened doors for me that most foreign scientists will never experience or appreciate, and for this I am forever indebted to him. Much of the knowledge that I gained from working in Japan is imbedded in this book.

As the title implies, the book focuses on landslide processes, prediction methods, and effects of land use. It is meant to serve as a benchmark reference for researchers, engineers, land managers, educators, planners, and policy makers dealing with landslide issues, and would be appropriate as a university textbook for an upper division or graduate-level course. The timeliness of the text is underscored by the recent landside tragedy in southern Leyte Island in the Philippines which killed an estimated 1800 people. Both the detailed descriptions of landslide processes (Chapter 3) and the effects of land use on landslides (Chapter 6) could stand alone as significant parts of university courses. The text has been written in one "voice" rather than a compilation of chapters by individuals. These points represent the forte and uniqueness of this book amongst other texts that have focused on engineering aspects of landslides and correction measures, as well as collections of process-based studies or case investigations. Throughout the book, the terms *landslide, mass wasting*, and *soil mass movement* are used largely synonymously. Strictly speaking, the processes of soil creep and dry ravel would not

Landslides: Processes, Prediction, and Land Use
Water Resources Monograph 18
Copyright 2006 American Geophysical Union
10.1029/18WM01

be considered landslides but can be classified as soil mass movements or mass wasting processes. The term *slope stability* is used to refer to all types of potential soil mass movement processes that occur on hillslopes. While I have tried to be as geographically inclusive as possible, the contents are necessarily biased towards areas in which I have actively worked. Of the approximately 1300 references included herein, refereed scientific journal papers were cited where possible. Nevertheless, due to the nature of certain examples and management implications, it was necessary to cite a wide array of symposium proceedings, agency reports, and other non-refereed materials. Because of the numerous citations, portions of long reference strings in the text have been placed in footnotes at the bottom of respective pages to improve readability.

The writing of this book has been influenced by a career of collaborations and discussions with colleagues from around the world, far too numerous to mention here, but greatly valued. In reflecting on such discourses, it is often through points of difference and disagreement where the most notable achievements have been made. The opinions expressed in the book, particularly related to agency policies, are mine; where these are critical, I do so in the hope of implementing new discussions on these topics that will benefit our environment. To remain totally uncommitted and unbiased, I accepted no support from any environmental groups, industries, international organizations, special interest programs or any other institutes related to the writing of this book. I thank my employer, Kyoto University, DPRI, for allowing me the time and total freedom to write the bulk of this text. Many people at DPRI contributed useful references to the book. I am grateful to Toshitaka Kamai, Robert Olshansky, Aurelian Trandafir, Takashi Gomi, and Tewodros Ayele Taddese for reviewing portions of the manuscript and giving me valuable suggestions. Special appreciation is extended to Dr. Walt Megahan and Professor Nelson Fernandes for reviewing the entire text and providing insightful comments. Finally, I wish to thank Dr. Karin Laursen for exemplary work and patience in preparing the references and figures as well as providing useful advice on clay mineralogy and petrology related to landslides.

Roy C. Sidle
Professor and Head of the Slope Conservation Section
Geohazards Division
Disaster Prevention Research Institute
Kyoto University
Uji, Kyoto 611-0011, Japan

Introduction and Overview of Landslide Problems

Landslides are important natural geomorphic agents that shape mountainous areas and redistribute sediment in gentler terrain. Much of the present Earth's landscape has been extensively sculpted by episodic large landslides; more subtle, but significant modifications have also been made by frequent, smaller-scale mass movements. Moreover, streams and rivers receive a large portion of their natural sediment supply from landslides and related debris flows, both active and historic. With the development and settlement of unstable terrain, landslides heretofore considered as natural processes have become natural disasters. Additionally, human activities have strongly influenced the extent and timing of landslide occurrence, especially the more frequent, smaller-scale soil mass movements. The interactions of such anthropogenic influences with landslide processes are a major focus of this book.

Herein, landslides are defined as a variety of processes that result in the downward and outward movement of slope-forming materials composed of natural rocks, soil, artificial fill, or combinations of these materials. The displaced mass may move in a number of ways: falling, toppling, sliding, spreading, flowing, or by their combinations. Gravity is always the primary driving factor, but it may be supplemented by water. Landslides differ from surface erosion processes, where water is the driving/transport mechanism. Most of the discussion related to landslide processes and land use effects focuses on landslides within the soil mantle or weathered regolith. A few references are made to failures in bedrock, but this is not the primary focus. The process of dry ravel/dry creep is included because it is gravity driven (i.e., a mass wasting process), albeit a surface phenomenon [e.g., *Krammes*, 1965; *Kirkby and Statham*, 1975], and is affected by land management activities [e.g., *Mersereau and Dyrness*, 1972; *Sidle et al.*, 1993]. Debris flows are discussed because they represent a special type of failure that can initiate either on hillslopes or within channels when augmented by an accumulation of pore water [*Innes*, 1983; *Coussot and Meunier*, 1996; *Iverson et al.*, 1997; *Imaizumi et al.*, 2005]. Herein, we focus on debris flows that initiate on hillslopes or in steep headwater channels, as opposed to the more commonly studied debris flows in low-gradient channels [see e.g., *Takahashi*, 1991; *Lorenzini and Mazza*, 2003; *Rickenmann and Chen*, 2003; *Jakob and Hungr*, 2005].

Several excellent summaries and books have been published on prediction, analysis, and control of landslides, particularly from an engineering or geotechnical perspective [*Turner and Schuster*, 1996; *Abramson et al.*, 2002; *Cornforth*, 2005]. Additionally, several significant books have focused on specific landslide processes and environments

Landslides: Processes, Prediction, and Land Use
Water Resources Monograph 18
Copyright 2006 American Geophysical Union
10.1029/18WM02

[*Brunsden and Prior*, 1984; *Eisbacher and Clague*, 1984; *Dikau et al.*, 1996]. However, relatively little attention has been paid to the influence of natural factors and extensive land management practices on slope stability, as well as the effects of landslides on natural and human-affected environments. This book follows an earlier work by *Sidle et al.* [1985] that addressed interactions of land use and slope stability. Incorporated herein are new concepts and findings related to hydrogeomorphic process, seismicity, weathering, and vegetation influences; land use interactions and amelioration measures; landslide hazard analysis, prediction techniques, and warning systems; damage assessments; global landslide coverage, including many examples from Japan and Southeast Asia; climate change scenarios; and interrelations among various hazard types.

OVERALL SIGNIFICANCE OF LANDSLIDES

Most parts of the world experience some level of landslide activity or more chronic mass wasting. For example, landslides have occurred in all 50 states in the USA, and almost all nations in the Western Hemisphere have experienced major socioeconomic impacts related to landslides [*Swanston and Schuster*, 1989; *Schuster and Highland*, 2001]. Because landslides only attract widespread attention when people are killed or property is damaged, it is difficult to ascertain the level of, or any increases in, landslide activity during the past few centuries. Even the most devastating landslide in recent history, the December 1920 dry loess flow triggered by an earthquake in central China, was not known to the outside world for nearly a year after its occurrence [*Close and McCormick*, 1922; *Derbyshire et al.*, 2000]. The relatively high level of landslide damage recorded in Japan over the past century may somewhat reflect the better documentation of natural hazards compared with developing nations [*Takahashi*, 1994; *Nunamoto et al.*, 1999] (Table 1.1).

Very large landslides can sometimes be recognized in geologic archives [*Hewitt*, 1998; *Loope et al.*, 1999; *Sukhija et al.*, 1999; *Weidinger et al.*, 2002], however, smaller landslides are usually obscured in sediment records [*DeRose et al.*, 1993; *Sidle et al.*, 2004a]. Although it is difficult to sort out specific sequences of smaller landslide events from sediment deposits, it is clear that widespread land cover changes (particularly forest clearance and conversion) have accelerated landslide erosion in the past few centuries [e.g., *Kuruppuarachchi and Wyrwoll*, 1992; *Luckman et al.*, 1999; *Slaymaker*, 2000]. Notable examples are New Zealand, India, Nepal, and Western Australia. Additionally, residential development and engineering earthworks in cities such as Hong Kong; San Francisco; San Paulo and Rio de Janeiro, Brazil; Basilicata, Italy; Guatemala City; and Kobe, Sendai, and Niigata, Japan, have contributed to numerous landslide occurrences [e.g., *Nilsen et al.*, 1979; *Gostelow et al.*, 1997; *Smyth and Royle*, 2000; *Yoshida et al.*, 2001[1]]. Where human intervention has accelerated landslide occurrence and hazards, the prior environmental conditions controlling slope stability may have been in a state of tenuous equilibrium. By reducing vegetation rooting strength, oversteepening and overloading natural slopes, altering hydrologic regime, and modifying soil properties, land use can disrupt this tenuous equilibrium and increase landslides in both the long- and short-term, as well as impacting humans, property, and the environment.

---

[1] *Kamai et al.*, 2004; *Kwong et al.*, 2004; *Sidle et al.*, 2005.

TABLE 1.1. Examples of the most destructive landslides in the past century.

Location	Date	Failure type/ description	Cause	Failure volume (m³)	Major losses and damage
Usoy, Tajikistan Murgab River	1911	massive rock and debris avalanche	earthquake (M = 7.4)	2.0 billion	town of Usoy destroyed; 54 people killed; unstable sediment dam on Murjab River now affects 5 million people downstream
Kedri, east Java, Indonesia Kelud Volcano	1919	lahars	draining of Crater Lake	not reported	5160 deaths; 150 km² of arable land devastated; 104 villages destroyed
Gansu Province, China; also affected Ningxia & Shaanxi Provinces	16 December 1920	dry loess flows	earthquake (M = 8.5)	huge, but unknown	≈180,000 people killed; 40 lakes formed by 675 dry loess flows; 50,000 km² of land affected
Alma-Ata, Kazakhstan Malaya Almaatinka River	1921	huge debris flow	snowmelt	not known	500 people killed
Sichuan, China Min River	1933	several large landslides & landslide dam failed	Deixi earthquake (M = 7.5)	not reported	6800 deaths from landslides; 2500 people drowned when landslide dam failed
Khait, Tajikistan	1949	rock slide transformed to debris avalanche; many other landslides	earthquake (M = 7.5)	not reported	about 18,000 people killed as boulders buried Khait and other villages and farms to depths up to 150 m
Kanogawa River Valley, Shizuoka, Japan	September 1958	landslide and debris flows	typhoon storm	not reported	1094 deaths; 19,754 homes destroyed

TABLE 1.1. Examples of the most destructive landslides in the past century. *cont.*

Location	Date	Failure type/description	Cause	Failure volume (m^3)	Major losses and damage
Rupanco region, Chile	May 1960	widespread landslides	M = 7.5 earthquake preceded by heavy rain	not reported	210 deaths; many buildings, port facilities, roads and agricultural fields destroyed
Mt. Huascaran, Peru	10 January 1962	debris and rock avalanche	ice and rock avalanche triggered debris avalanche	13 million	4000–5000 casualties; town of Ranrahirca destroyed; US$200 million in damages
Vajont Gorge near Longarone, Italy (Vajont Dam disaster)	9 October 1963	rockslide in limestone weakened by fractures, old erosion & faulting	groundwater accretion during heavy seasonal rains & rising of lake; slide entered reservoir	300 million	1899 deaths by downstream flood that destroyed or heavily damaged towns of Castellavazzo, Pirago, Villanova, Forance, Faè, & Codissago; cost US$200 million
Anchorage, Alaska, USA	27 March 1964	liquefaction failures & landslides in quick, glacial clay deposit	major earthquake (M = 9.2)	not reported	9 deaths; 215 homes destroyed, 157 commercial properties damaged; lifelines damaged; cost ≈ US $200 million
Rio de Janero, Brazil	January 1966	widespread debris slides, avalanches, and flows	heavy rainfall in areas excavated & cleared of vegetation	not reported	about 1000 casualties from landslides and resultant flooding in and around Rio de Janeiro
SW of Rio de Janeiro, Brazil (Serra das Araras)	23 January 1967	thousands debris slides, avalanches and flows	heavy and intense rain	>10 million	about 1200 deaths attributed to landslides; >100 homes destroyed; damage to lifelines
Los Angeles area, California, USA	Jan.–Feb. 1969	shallow landslides & resultant debris flows	heavy rain in mid Jan. followed by rain in Feb.	not reported	18 deaths; 175 homes and major highway damaged: cost >$40 million

TABLE 1.1. Examples of the most destructive landslides in the past century. *cont.*

Location	Date	Failure type/ description	Cause	Failure volume (m³)	Major losses and damage
Mt. Huascaran, Peru	31 May 1970	debris avalanche/debris flow	earthquake (M = 7.7)	30–50 million	destroyed Yungay and partly destroyed Ranrahirca; 18,000 dead; many homeless
Mantaro Valley, Peru Mayunmarca landslide	25 April 1974	rock/debris avalanche dammed Mantaro River; dam breached suddenly	probably rainfall and fluvial undercutting	1.6 billion	resulting flooding and debris from breach of landslide dam killed 450 downstream; many farms, roads, and bridges destroyed
Guatemala City region, Guatemala	4 February 1976	>10,000 debris slides, rock falls, slumps & debris avalanches	earthquake (M = 7.5)	unknown	≈250 deaths; >500 homes damaged; rail lines damaged; several landslide-dammed lakes breached
Yungjing County, Gansu Province, China	August 1979	numerous landslides in loess deposits	heavy rainfall	unknown	≈800 people killed or injured; >200 ha of farmland damaged
Mount St. Helens, Washington, USA	18 May 1980	debris avalanche/rockslide	volcanic blast preceded by M = 5.1 earthquake	2.8 billion	53 deaths; US$1.1 billion losses to civil works, agriculture/ forests, plus personal property loss, clean-up cost, and loss of tourism and wildlife
Gansu Province, China (Saleshan Landslide)	August 1983	deep landslide in loess deposits	prolonged rainfall followed by large storm	35 million	227 deaths; 22 injuries; 2 km² of farmland overwhelmed; 4 villages destroyed
Spanish Fork Canyon, Utah, USA (Thistle)	13–17 April 1983	translation slide in old earthflow and debris flow deposits	rain-on-snow (El Niño event)	21 million	dammed river; flooded Thistle; destroyed railroad and highway; 1330 jobs lost; total cost >US$400 million (direct + indirect)

TABLE 1.1. Examples of the most destructive landslides in the past century. *cont.*

Location	Date	Failure type/ description	Cause	Failure volume (m³)	Major losses and damage
Jizukiyama, Nagano City, Japan	26 July 1985	deep-seated slump-earthflow	>1 month of high rain; intense rainfall prior to failure	3.5 million	25 deaths; 4 injuries; 64 homes damaged; nursing home, roads, farms, forests, & waterworks damaged; cost US$40 million
Nevado del Ruiz/ Armero, Colombia	13 November 1985	lahars	rapid melting of snow & ice caused by volcanic eruption & pyroclastic flow	60 million	23,000 people killed; 5000 injured; >5000 homes destroyed; destroyed Amero town
Napo Province, Ecuador (Reventador Volcano)	5 March 1987	rock & debris slides, avalanches & flows in wet soils	two earthquakes (M = 6.1 & M = 6.9)	75-110 million	≈1000 people killed; 70 km of oil pipe line and major highway destroyed; much damage to hydroelectric development and agriculture; direct losses ≈ US$1 billion
Kaiapit, Morobe Province, Papua New Guinea	6 September 1988	large landslides-debris flows	likely moderate earthquake that destabilized rather dry regolith	1.8 billion	74 people killed; 800 people affected; 3 villages destroyed; 4 landslide lakes formed, 3 of which breached
Gilan Province, north Iran	20 June 1990	>76 large landslides, many types, including liquefaction	Manjil earthquake (M = 7.7)	huge, but unknown	≈300 deaths; roads blocked, preventing rescuers from accessing victims; buildings, infrastructure & farms damaged
Toluk, Suusamyr Valley, Kyrgyzstan	19 August 1992	many landslides; large rock avalanche	Suusamyr earthquake (M = 7.3)	≈1 million (rock slide)	≈35 deaths; extensive damage to the Bishkek-Osh highway

TABLE 1.1. Examples of the most destructive landslides in the past century. *cont.*

Location	Date	Failure type/ description	Cause	Failure volume (m³)	Major losses and damage
Kagoshima Bay area, Japan	July– September 1993	many debris avalanches & flows	seasonal heavy rain (0.8–1.2 m in July)	not reported	105 deaths; heavy damage to housing, roads and public facilities
Harihara River, Izumi, Kagoshima, Japan	10 June 1997	deep landslide that triggered a debris flow	4 days of heavy rain	160,000	21 deaths; 13 injuries; 15 homes destroyed
Malpa, northern India	17 August 1998	large rockfall/debris avalanche	4 days of heavy rain	"large"	207 deaths; 5.2 million rupees in other direct costs & 0.5 million rupees indirect costs
Nishigo, Shirakawa, and Nasu, Japan	26–31 August 1998	>1000 landslides and debris flows	5 days of heavy rain (max. ≈ 1.2 m)	not reported	9 people killed; many homes and buildings destroyed
North coast of Venezuela near Caracas	14–16 December 1999	widespread shallow landslides and debris flows along a 40-km coastal strip	nearly 1 m of rain in less than a 3-day period	unknown	≈30,000 deaths, 8000 residences & 700 apartments destroyed; extensive infrastructure damage; all losses include flood damages
Bajo Caliente, Costa Rica (Arancibia landslide)	27 June 2000	debris avalanche & subsequent block slide	possibly fracturing & faulting of bedrock	12 million	9 deaths; 2.5 km² of cropland & 9 homes destroyed; US$0.4 million damage
Kara Taryk (Osh region), Kyrgyzstan	20 April 2003	large landslide in Soviet-era uranium mining area	rain-on-snow	1.5 million	38 deaths; 13 homes destroyed; potential pollution of a river
Southwest Guizhou Province, P.R. China	11 May 2003	road-related landslide	heavy rain and road construction	not reported	35 road workers killed and 2 buildings and road destroyed

TABLE 1.1. Examples of the most destructive landslides in the past century. *cont.*

Location	Date	Failure type/ description	Cause	Failure volume (m³)	Major losses and damage
Ratnapura and Hambantota Districts, south-central Sri Lanka	May 2003	many landslides and debris flows	continual heavy rains (extensive associated flooding)	not reported	>260 deaths (some due to floods) >24,000 homes & schools destroyed; ≈180,000 families homeless
Minamata and Hishikari, southern Kyushu, Japan	20 July 2003	debris avalanches & debris flows; many small landslides	heavy and intense rainfall	37,000 at Minamata landslide	25 deaths (23 at Minamata debris flow; 2 at Hishikari landslide; 7 homes destroyed; roads, power & hot spring lines damaged
Miyagawa area, Mie Prefecture, Japan	October 2004	numerous landslides and debris flows	heavy and intense rainfall	unknown	17 deaths, 9 injuries; 87 homes damaged/destroyed; extensive forest road damage
Niigata Prefecture, Japan	23 October 2004	thousands of landslides on natural slopes, roads, paddy fields, and fills	three successive and large earthquakes	unknown	2 deaths; Yamakoshi village isolated; 2 landslide dams caused flooding; many homes and roads damaged by fill failures
La Conchita, California, USA	10 January 2005	landslide-debris flow in a slump-earthflow deposit from 1995	378 mm of rain in the preceding 2 weeks	0.2 million	10 deaths; 30 homes seriously damaged or destroyed
Guinsahugon village, Leyte, The Philippines	17 February 2006	large, deep-seated landslide, debris flow	685 mm of rain in about 2 weeks; possibly affected by a very small (M = 2.3) earthquake	"large"	1800 deaths; 375 homes and a school destroyed; 40 ha of land buried

DISTRIBUTION OF LANDSLIDE HAZARDS WORLDWIDE

Because of the variable reliability of reporting and documentation of landslides in different nations, it is difficult to accurately quantify the global distribution of landslide hazards and related damages. While recent advances in remote sensing and geographic information systems techniques can provide contemporary estimates of large landslides in remote and poorer regions of the world [e.g., *Leroi et al.*, 1992; *Dhakal et al.*, 2000; *Perotto-Baldiviezo et al.*, 2004], comprehensive longer term landslide data from such areas, as well as damage estimates, remain elusive. Nevertheless, certain generalizations related to the regional extent of landslide hazards can be drawn based on geomorphic, tectonic, and climatic environments together with land use patterns and past records of landslides.

The circum-Pacific region is naturally susceptible to landslides because of a combination of high and intense rainfall, mountainous terrain, active tectonics, volcanism, geological history, and lithology/pedology conditions. During the past several hundred years, the widespread land cover change throughout most areas around the Pacific Rim has contributed to increased rates of landslide activity. Additionally, landslide risk in the region has substantially increased due to population growth and the concentration of people in high hazard areas; such demographic patterns have been most evident in the past 30 yr [e.g., *Harp et al.*, 1981; *Elmhirst*, 1999; *Templeton and Scherr*, 1999]. The circum-Pacific region is characterized by actively converging margins of lithosphere plates; thus, volcanism and intense seismic activity at shallow, intermediate, and deep focal levels are common features. This general convergent tectonic pattern has caused major compression and uplift of deep crustal metamorphic and igneous rocks as well as soft clay-rich sedimentary rocks. Large vertical (tens of kilometers) and horizontal (hundreds of kilometers) displacements along major faults, in combination with crustal compression, have produced extensive mélange areas of extreme crustal deformation. Triple junctions among tectonic plates near Japan, Papua New Guinea, and Taiwan make these regions particularly susceptible to earthquake-triggered landslides. Many of the coastal areas along the margins of the Pacific Rim experience high annual rainfall with distinct seasonal distributions. Typhoons or monsoons, which often deliver both large amounts and intense rainfall, are common along the Asian continental and equatorial margins of the Pacific. In contrast to western North America, South America, and New Zealand, where recent settlement and land cover change has exacerbated landslide occurrence, much of the Asian margins of the Pacific Rim have been inhabited for millennia; however, recent demographic shifts to mega-cities and exploitation of natural forests have increased landslide risk in Asia. Particularly in Southeast Asia and some developing parts of East Asia, land cover changes promoted by overpopulation, internal politics, transmigration, the drug trade, and economic investments from industrialized nations whose natural resources have already been depleted or are protected from such exploitation have contributed to landslide hazards and risk [*Templeton and Scherr*, 1999; *Verburg et al.*, 1999; *Derbyshire et al.*, 2000; *Sidle et al.*, 2004a].

Many other regions of the world experience significant landslide activity, including the European Alps; Intermountain region, Rocky Mountains, and Appalachian Mountains of USA and Canada; parts of the Middle East; mountains of the Caribbean

Islands; Brazil; Himalayas; and central Asia. China, India, and Nepal have long, but incomplete histories of landslide occurrence and damage. Landslide initiation in East Asia is exacerbated by previous and contemporary glaciation, tectonic uplift, frequent earthquakes, high-intensity rainfall, and episodic snowmelt [*Li*, 1989; *Froehlich et al.*, 1990; *Sukhija et al.*, 1999; *Gupta and Virdi*, 2000]. Deep-seated landslides and slope movements (i.e., slumps, earthflows, soil creep) are abundant in some areas where highly weathered regoliths and certain clayey soils occur. Characteristics of these deep-seated slope movements as well as other types of landslides are discussed in Chapter 2 along with their geographic occurrence.

LOSS OF HUMAN LIFE

Statistics on loss of human lives from landslide disasters is complicated by highly variable reporting amongst nations of differing economic status as well as a contemporary bias due to the general trend of improving data collection in many nations. Additionally, natural disaster statistics often combine landslide casualties and damages with those of related hazards (e.g., floods, earthquakes) when they occur during the same disaster [e.g., *Yoon*, 1991; *Water Induced Disaster Prevention Technical Centre*, 1993]. One of the earliest landslide fatality records was the Wudu landslide in Wushan County, Gansu Province, central China, which killed 760 people in 186 B.C. [*Li*, 1989]. The greatest loss of life during a landslide disaster was triggered by the large (M = 8.5) Haiyuan earthquake in Gansu, Shaanxi, and Ningxia provinces of China on 16 December 1920. Thousands of landslides in mostly dry loess material occurred throughout a region of 50,000 km^2, including more than 650 landslides that were >0.5 km wide; the subsequent dry loess flows killed an estimated 180,000 people, dammed many streams and rivers, and significantly changed the geomorphology of the region [*Derbyshire et al.*, 2000].

Based on incomplete records compiled by UNESCO from 1971 to 1974, *Varnes* [1981] estimated that 2378 people worldwide were killed by landslides, about 600 per year. Of these casualties, 2126 (89%) occurred within the circum-Pacific region. Average annual estimates of deaths attributed to landslides and debris flows for various nations are shown in Table 1.2; these figures are derived from multiple sources of varying quality and do not represent the same aggregate time periods; nevertheless, they give a general idea of casualties in different geographic and socioeconomic regions of the world. The high death rates in Nepal and China reflect the vulnerability of people in hazardous mountain areas where few structural countermeasures exist and where little attention is paid to structures related to potential landslide hazards. Italy, by far, has the highest number of landslide casualties of all European nations, albeit much less than many Asian countries. While Japan represents an extreme case for severe landslide hazards due to its topography, climate, tectonic activity, and lithology, the high number of deaths may also be related to vulnerabilities of people who have a false sense of security in protective measures [*Sidle and Chigira*, 2004]. In Japan, many residences, infrastructures, and businesses are located in potential landslide hazard areas, in some cases because these sites are virtually unavoidable, but in other cases because there is a perception that the government will provide the necessary protective measures, either via structures or warning systems. Oftentimes, the citizens would be better served by

TABLE 1.2. Approximate average number of human deaths per year attributed to landslides and debris flows in various nations.

Nation	Average deaths yr^{-1}
Canada	5
USA	25-50
Hong Kong	11
China	140-150
Japan	170
Nepal	186
Brazil	88
Bulgaria	0.5
Italy	18
Portugal	1
Spain	2.3
Norway	5-7
New Zealand	1-2
Australia	0.2-0.25

a combination of improved warning systems and prudent land use zoning that would restrict certain types of development in potentially hazardous sites, rather than relying too much on expensive structural control measures. Additionally, the more accurate reporting of landslide casualties in Japan compared with most other nations, particularly in developing countries of Asia, may account for some of the higher loss of life statistics.

Nevertheless, average annual casualties (e.g., Table 1.2) are poor indicators of loss of life within any given year, as shown by the temporal distribution of deaths from landslides in Japan, Italy, Nepal, and Hong Kong (Figure 1.1). The episodic occurrence of landslide casualties is apparent in all four countries, with the densely populated territory of Hong Kong and Italy showing the greatest year to year variability. Hong Kong also has the highest number of casualties per unit land area (due to its small size and dense population located near steep hillsides), followed by Nepal. Japan and Italy have much lower casualties per unit area (Figure 1.1).

An inspection of the major damaging landslides in the past century shows that the majority of these disasters occurred within the Pacific Rim (Table 1.1). Similarly, more than half of the very large (>10^6 m^3) landslides summarized by *Coates* [1977] also occurred in the circum-Pacific region. Of all the Pacific Rim nations, Japan has the most comprehensive data for landslide damage and deaths. In the 15-yr period from 1945 to 1959, about 69% of the deaths due to flood and sediment disasters in Japan were caused by landslides and debris flows [*Takahashi*, 1994]. During the 10-yr period from 1966 to 1975, 1714 of 3096 deaths (55%) attributed to such natural disasters in Japan were caused by landslides and debris flows [*National Research Center for Disaster Prevention*, 1980]. After about 1978, the percentage of casualties related to rainfall-induced landslides/debris flows compared to all rainfall-induced disasters continued to decline to about 43% in the 10-yr period from 1987 to 1996 [*Nunamoto et al.*, 1999]. Thus, it appears that the advanced warning systems and high level of natural hazard awareness in Japan, together with the very expensive

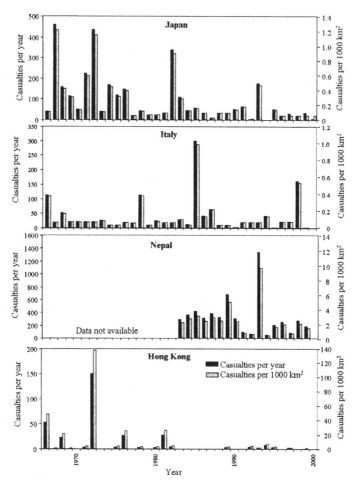

Figure 1.1. Trends in deaths caused by landslides; comparison amongst Japan, Italy, Nepal, and Hong Kong.

structural landslide control measures have significantly reduced landslide casualties. Nevertheless, almost every year Japan suffers more deaths from landslides compared to any other industrialized nation because of its vulnerable setting.

Landslides triggered by earthquakes and lahars typically cause the largest loss of life due to their unexpected nature. Of the 14 destructive landslide episodes that killed more than about 1000 people shown in Table 1.1, two major disasters were related to lahars and five were triggered by major earthquakes. Rainfall and typhoons were responsible for initiating slightly more than half of the landslide disasters described in Table 1.1; however, the average number of casualties in these disasters was more than 3-fold lower (about 1100; skewed upwards by uncertain figures for the 1999 Venezuela landslides; *Wieczorek et al.*, 2001) compared to earthquake-triggered landslides (about 3700, even with the huge death toll from the 1920 central China dry loess flows removed). Notably, 9 of the 11 rainfall-triggered disasters that killed more than 200 people occurred in developing nations, thus emphasizing the greater exposure and vulnerability of people in poor regions to landslide hazards. As this book goes to press, the rapid, deep-seated landslide/debris flow that buried the village of Guinsahugon in the Philippines on 17 February 2006, killing an estimated 1800 people, including some 250 occupants of an elementary school, is a grave reminder of the vulnerability of residents in poor mountain regions to such primarily rainfall-triggered disasters (Table 1.1). All seven of the earth-quake-triggered landslide and lahar disasters that individually claimed about 1000 or more lives were in developing nations. For these mass movements that occurred without any obvious precursor, the lack of advanced warning systems significantly increased the death toll. Additionally, it should be noted that none of the landslides that caused extensive loss of life were deep-seated, slow-moving earthflows or slumps. While such deep-seated landslides may generate huge costs (see examples later in this section), the rate of movement generally affords sufficient time for people to safely evacuate the area of impact.

ECONOMIC CONSEQUENCES

At national and even regional scales, it is difficult to accurately assess economic dam-age from landslides because most countries and provinces do not compile comprehensive records of landslides and related damage, and data collection methods vary from area to area. Localized areas with extensive or intensive landslide problems have been studied, but economic data are typically collected only for selected disasters. Thus, the find-ings presented here are heavily biased to regions or nations where more comprehensive records of landslide occurrence and damage have been documented. For other regions where economic losses are reported, they are likely to be underestimated.

Both direct and indirect costs arise from landslide damages. Direct costs relate to the actual physical damage incurred and can be defined as the costs of replac-ing, repairing, rebuilding, or maintenance resulting from direct landslide damage or destruction to property and land within the affected area. All other costs are indi-rect. Although economic assessments of landslide damages are extremely difficult to compile, especially on a national basis, some estimates are presented in Table 1.3 to provide very general comparisons amongst various nations. These statistics are merely annual averages and are based on different time frames of variable quality data. Generally, indirect costs are absent, but where these have been estimated, the values are included. Overall, the figures (especially for indirect costs) are likely con-servative. Of the industrialized nations, Japan and Italy sustain the greatest economic damage from landslides. Given the lower property values in the developing nations of China and India, the landslide costs in these mountainous countries are very high. Countries like New Zealand, Canada, and Nepal generally incur landslide damage in rural areas; thus, costs are proportionally lower. Landslide damages in Scandinavia are minor compared to mountainous terrain in the rest of Europe; the collective landslide damage in Europe is less than in the USA.

TABLE 1.3. Estimated average annual costs (in US$) of landslides in various nations.

	Average annual direct costs[1]	Average annual total costs[1]	Comments[1]
Canada		$70 million	A more recent estimate of total costs is up to $1.4 billion annually
Japan	$1.5 billion	$4 billion	
Korea	$60 million	--	Based on poor records
United States	$1.2 billion	$1.6–3.2 billion	Direct costs only include damage to private dwellings
Italy	--	$2.6–5 billion	Rough estimate
Sweden	$10–20 million		
Spain	$0.2 billion		
former USSR	$0.5 billion		
China	$0.5 billion	--	Costs based on valuations in 1989
India	$1.3 billion		
Nepal	$19.6 million	--	Includes flood damage, but likely incomplete
New Zealand	--	$26.3 million	90% of costs are sustained in rural areas

[1]Based on data from the following: *Cruden et al.* [1989]; *Li* [1989]; *Swanston and Schuster* [1989]; *Yoon* [1991]; *Water Induced Disaster Prevention Technical Centre* [1993]; *Schuster and Highland* [2001]; and various personal communications.

Direct Costs

Examples of property damage resulting from landslides near urban or residential areas include damage to homes, offices, and industrial complexes; displaced infrastructure, such as roads, railways, power lines, gas lines, communication lines, and sidewalks; severed water lines, sewage systems, stormwater drains, and irrigation systems; and damage to other urban facilities or personal possessions, such as automobiles, parks, street lights, and other personal belongings. In rural areas, property damage from landslides is typically less because of more dispersed investment; however, losses are generally more difficult to quantify [*DeGraff*, 1991]. Examples of direct costs associated with landslides in agricultural lands include damage to rural buildings, loss of crops, inundation of farm ponds, severance of irrigation and drainage ditches, damage to roads and trails, loss of animals, and clean-up costs. Direct property losses on forest and rangelands from landslides include loss of timber, blockage of stream channels that must be cleared, and damage to low-volume roads, rail lines, and drainage systems. Other property damage that could be classified as direct costs includes loss of storage capacity of a reservoir due to siltation, blockage of navigation in rivers, and damage to underground and surface mining operations [*Li*, 1989]. For urban and rural environments alike, significant direct costs are associated with the repair and maintenance of damaged property [*Schuster and Highland*, 2001]. While large rapid landslides and debris flows generate extensive property damage as well as loss of life, the costs related to deep-seated, slower moving landslides are also very significant (even though these failures rarely cause casualties). For example, Hamilton County, Ohio

(USA), which includes the city of Cincinnati, experiences widespread, slow-moving earthflows in glacial and colluvial deposits. The average annual cost of landslide damage from 1973 to 1978 was estimated at about US$31 million or $5.80 per capita [*Fleming and Taylor*, 1980]. Other examples of such costly, but not life-threatening, landslides are discussed later in this chapter.

Indirect Costs

Indirect costs resulting from landslides are difficult to assess and typically ignored or overlooked in national, regional, or site-specific economic analyses. In some cases, indirect costs may exceed direct costs [e.g., *Schuster and Highland*, 2001; *Burke et al.*, 2002]. Examples of indirect costs include (1) loss of industrial productivity, resulting from damage to facilities or interruption of transportation; (2) loss of human productivity, due to injuries, death, displacement, or trauma; (3) loss of site productivity of agricultural and forest lands; (4) costs of preventive measures to reduce future damage (e.g., rerouting roads, improving drainage systems, constructing structures to prevent future landslides); (5) depreciated real estate values in the vicinity of damage and in areas threatened by future landslides; (6) costs incurred through the obligation of the public sector to obtain title to property that has recurring landslide problems; (7) loss of tax revenue on devalued property or in the mandatory conversion of private property to public property; (8) loss of tourism revenues; (9) costs to individuals who sustained damage to personal property, including time lost from work, legal fees, moving expenses, and social costs; (10) expenses for emergency services (e.g., rescue and evacuation efforts, temporary housing, and emergency supplies) and public health workers; (11) perishable losses due to infrastructure breakdown; and (12) impacts on water quality and aquatic habitat. Secondary physical effects of landslides, such as induced flooding and resulting downstream sedimentation, can incur both direct and indirect losses. Sometimes these secondary effects, as in the case of the Vaiont, Italy, reservoir disaster [*Kiersch*, 1964], constitute the principal economic loss. Other examples of significant indirect costs associated with landslides are discussed in the next section.

EXAMPLES OF DETAILED ECONOMIC ANALYSES OF LANDSLIDE DAMAGES

Four examples are presented of economic assessments of landslide costs for very different cases. Firstly, a long-term assessment of costs associated with conversion of forests to pasture in parts of rural New Zealand is detailed. Secondly, a case study of the largest recorded economic damage by a single landslide is given—the Thistle landslide in Utah, USA. Thirdly, a detailed assessment of urban damages due to the 2001 Nisqually Earthquake near Seattle, Washington, USA, is presented. Finally, an example from a remote mountain region in a developing country (India) is provided in which both direct and indirect landslide costs were assessed. Only in this last case were any deaths attributed to the landslides; the first three cases each experienced extensive monetary damage, but no loss of life.

Widespread Landslide Damage From Forest Conversion, New Zealand

An important large-scale example where indirect costs dominate the damages incurred by landslides is the continued loss of site productivity of agricultural and forest lands in New Zealand related to the historic forest conversion to pasture and recent conversion of native forests to plantations. Based on pollen and diatom analysis of Holocene lake bed sediments together with tephra chronology and historical evidence from a landslide-prone region of North Island, New Zealand, *Page and Trustrum* [1997] deduced that erosion (both landslide and surface erosion) losses were low prior to Polynesian settlement, when native forests covered the region. In the early 15th century, Polynesians began clearing indigenous forests by fire; with the subsequent fern/shrub cover, a 5- to 6-fold increase in erosion occurred. The arrival of Europeans in 1873 prompted a shift in land cover to pasture with a corresponding 8- to 17-fold increase in erosion compared to indigenous forests [*Page and Trustrum*, 1997]. In spite of this well-documented chronology, it was not possible to separate periodic landslides from extensive gully erosion that occurred during this period. Other studies have documented the effects of widespread conversion of forests to pasture on severe erosion and land productivity declines following initial European settlement (up through 1920) in North Island [*Ministry of Works and Development*, 1970; *Trustrum et al.*, 1983; *Sidle et al.*, 1985]. Moderate to extreme mass movement erosion affected a total of 9280 km^2 of forest lands converted to pasture on North Island.

Soil removal by landslides has severely reduced the productivity of pastoral lands. In the mid-1960s a major economic appraisal of land use and development options was undertaken for 6500 km^2 of eastern Raukumara Peninsula, where severe landslide erosion after the 1930s caused agricultural declines as well as river channel aggradation and flooding, requiring replacement or repair of many bridges and roads [*Ministry of Works and Development*, 1970]. Of the various options for conservation practices along with the continuation of pastoral farming, all had negative present values and rates of monetary return regardless of discount rate. Measurements of pasture production on landslide scars of various ages indicate that sites disturbed 2 yr previously produce only 20% as much pasture as uneroded sites; pasture production on sites disturbed 15 to 40 yr previously gradually increases to 75% [*Lambert*, 1980; *Trustrum et al.*, 1983]. In summer (dry season), the 40-yr-old landslide sites yield only half of the pasture production of uneroded sites. Studies on similar terrain in another district note that 7- to 8-yr-old disturbed sites produce half the forage compared to uneroded sites [*Garrett*, 1980]. Together, these data indicate that the average pasture production during the 15 yr after landslide erosion is less than half of the production on uneroded sites and, for several decades thereafter, will only be about 75% of the production on uneroded areas. In the Wairarapa district alone, during the wet winter of 1977, landslides eroded about 4% of the 1400 km^2 land area. This immediate loss of pasture equates to ≈$600,000 (1980 NZ dollars) in the first year [*Hawley*, 1980; *Trustrum and Stephens*, 1981]. The total cost of pasture lost from this area due to landslides during the subsequent 40 yr is estimated at more than NZ$9 million. In the Wairarapa study sites, long-term stripping of the soil mantle by landslides has eroded 41–56% of the total area. Such extensive mass erosion could reduce New Zealand's export revenue as much as 5%, or several hundred million dollars [*Sidle et al.*, 1985]. The extent of such an erosion problem in a country with a high investment in water, soil, and pasture management provides a stark warning for other countries intending to convert hillslope forests to agriculture, particularly developing nations with few resources for conservation. The economic significance of New Zealand's landslide problems, albeit largely indirect costs, begs the implementation of long-term monitoring and assessment to accurately evaluate the effects of agricultural development on site productivity in unstable terrain.

The Costly Thistle Landslide, Central Utah (USA)

An El Niño–driven storm in April 1983 on a record snowpack triggered thousands of landslides and debris flows in the Wasatch and Manti LaSal Mountains of Utah. By far the most damaging landslide occurred near the outlet of Spanish Fork Canyon by the village of Thistle, Utah (Figure 1.2), and led to the first Presidential Disaster Declaration in Utah's history. This large, deep (22×10^6 m^3, average depth $\approx$ 40 m) slide in old earthflow and debris flow deposits dammed the Spanish Fork River, creating a 60-m-deep lake that submerged the village of Thistle; severed US Highways 6/50 and 89; and destroyed the Denver and Rio Grande Western Railroad that ran through the area [*Kaliser and Fleming*, 1986]. Record levels of rain fell in autumn 1982; thus, antecedent moisture in the unstable mass was very high. Isolated slumping, creep, and movement of the unstable landform were periodically recognized during several decades prior to the 1983 landslide; the first record of movement at the toe of the Thistle landslide was in the early 1900s [*Slosson et al.*, 1992]. In the 1950s, the railroad company removed several thousand cubic meters of landslide debris from the tracks; thus, the railroad cut could have contributed to the activation of the 1983 landslide. The railway served several large coal mines in central Utah, as well as being the primary corridor for national shipping in the region. The railway, connecting Denver, Colorado, to Salt Lake City, Utah, follows the Soldier Creek and Spanish Fork drainages because of the relatively gentle gradient through the Wasatch Mountains in a passage where older landslides have eroded unstable landforms. This route has been used long before the railways by Native American and Spanish explorers and later trappers and pioneers [*Sumsion*, 1983].

The railroad company spent $\approx$US$69 million (in 2000 currency) to rebuild the line, including the construction of a 900-m tunnel that bypassed the landslide and dammed lake. Direct costs associated with the Thistle landslide were estimated at US$344 million. Indirect costs were numerous and were estimated at about the same level (or possibly higher) [*Schuster and Highland*, 2001], making the total losses from this landslide about US$688 million. Indirect costs included the loss of revenue from local coal, petroleum, and uranium industries; tourism; and small businesses due to closure of railways and highways [*Schuster and Highland*, 2001]. Additionally, about 1330 jobs were lost as the result of the landslide. The largest indirect cost (estimated at US$139 million) was the revenue lost by the railroad company. Because of the forewarning during the early stages of movement and the relatively slow progression of the landslide, no lives were lost; nonetheless, the Thistle landslide is the largest documented loss of revenue from any single slope failure worldwide.

Landslide Damage From the February 2001 Nisqually, Washington (USA),
Earthquake

An earthquake (M = 6.8) on 28 February 2001 triggered numerous landslides and ground failures in the region near Seattle, Washington. Strong ground shaking caused several significant lateral spreads, rock falls, embankment failures, and landslides on hillslopes, as well as significant liquefaction; these slope failures were spread over a wide region without any particular areal concentration [*Highland*, 2003]. The Federal Emergency Management Agency (FEMA) conducted a detailed investigation of losses incurred in the eight most damaging landslide complexes. By far the most costly landslides were the lateral spreads in fill material underlying the Deschutes Parkway (causing extended closure of the road) and around the margins of Capitol Lake (the dike along the southern portion of the lake failed) [*Highland*, 2003]. Numerous water and sewer lines were severed and the estimated landslide damages were more than US$22 million. The second highest damage occurred in the Maplewild Avenue area, where an uncompacted fill supporting five homes formed a translational slide. Total damage sustained to homes and road right-of-ways was about US$7.6 million. The Salmon

Figure 1.2. The Thistle landslide (outlined in right side of photo) in April 1983 at the outlet of Spanish Fork Canyon, Utah, caused the largest economic losses of any single landslide despite having no casualties. The previous locations of the two highways are shown; the former railroad was located essentially parallel to Route 6. Route 6 has now been reconstructed along the mountainside and a tunnel has been built for the new railway.

Beach landslide demolished two homes along the shore of Puget Sound and damaged sewer, water, and electrical lines; the estimated damage of US$1.5 million does not include the high risk of failure (including the exposure of eight additional homes) of an adjacent land mass that was destabilized during the earthquake [*Highland*, 2003]. A landslide in glacial till along the Cedar River (in Renton) destroyed an erosion control structure and temporarily dammed the river. Just upstream, a debris flow destroyed a home. Damage from these two landslides in Renton totaled US$1.7 million. Landslide damage, largely from lateral spreads, totaled more than US$0.5 million for the areas of Tomlin State Park and a mobile home park (both in Thurston County). The park incurred broken water and sewer lines, as well as damage to bridges, trails, and a shelter [*Highland*, 2003]. A slump/flow in the roadfill of US Highway 101 removed one lane of the four-lane corridor, causing an estimated US$0.9 million in damages. Clearly, landslides in unstable fill materials were responsible for the greatest proportion of the costs incurred during the earthquake.

Highland [2003] estimated preliminary losses incurred by landslides resulting from the Nisqually Earthquake at US$34.3 million; however, this figure does not include many of the losses from smaller landslides, and a better estimate of total landslide losses may exceed US$40 million. Furthermore, extensive damage to the Alaskan Way Viaduct (a major transportation artery with a seawall in downtown Seattle) during the earthquake may have been due in part to lateral spreading and liquefaction [*Highland*, 2003]. A recent feasibility study (costing at least US$0.5 million) conducted by Washington State Department of Transportation, the Federal Highway Administration, and the City of Seattle recommended replacing the viaduct with a tunnel (thus no need to repair the sea wall) at the estimated cost of US$4.2 billion. Even if only 10–15% of these projected incurred costs are attributable to the earthquake-induced landslides, the total landslide damage from the Nisqually Earthquake would rival that of the 1983 Thistle Landslide in Utah.

Damage From the 1998 Malpa Landslide, Kumaun Himalaya, India

Just after midnight on 17 August 1998, following 4 days of steady rainfall, a large rockfall/debris avalanche occurred near a fault zone in the Malpadhur Range of the Great Himalaya in northern India (adjacent to western Nepal and southern China). This general area, referred to as Kumaun Himalaya, has a history of destructive landslides and debris flows. Part of the landslide deposited in the Malpa stream, creating a temporary lake that burst on the night of 17 August, initiating a flash flood that destroyed the mountain village of Malpa. Some warning of the impending disaster during the preceding daytime would have prevented the deaths of many of the people who had taken shelter in a floodplain area downstream of the landslide—an area that was inundated by the flash flood when the stream channel shifted about 14 m to the west [*Kumar and Satyal*, 1999]. At least 207 people were killed by the landslide and resulting flood, including 118 inhabitants in the area, as well as a number of Hindu pilgrims, police, and Nepalese citizens. The massive landslide cut off Malpa and surrounding areas from the rest of the world, hampering rescue efforts. Several days later when the first assistance finally arrived, only 10 survivors were rescued from the debris and thousands of stranded villagers had to be evacuated. While some farming was practiced at this high elevation (altitude 2065–2125 m) site, much of the land was either not suitable for farming or had been degraded by past farming practices to the extent that it was no longer arable. The area was used by trekkers and numerous mountain footpaths existed to support this recreational use as well as to sustain local livelihoods, transit, and policing in the border areas.

In one of the most comprehensive economic assessments of landslide damages ever conducted in a remote developing region, *Kumar and Satyal* [1999] analyzed most direct costs and many of the indirect costs related to the Malpa landslide. Monetary values were placed on loss of lives based on compensation paid by the government to families of victims (total assessment 20.1 million rupees). Other direct costs assessed included the destruction of 10 homes and huts, a bridge, 81 cattle and horses, 106 other smaller animals, and crops/vegetation (Table 1.4). Cumulatively, these direct costs totaled ≈5.2 million rupees, slightly more than 25% of the compensation paid for casualties. The indirect costs related to rescue and evacuation efforts were not estimated; however, most other indirect costs were assessed, including lost future production on arable land covered by landslide debris, unusable barren land, and lost recreational revenues due to inaccessible trekking paths (Table 1.4). These indirect costs totaled ≈0.5 million rupees. This analysis clearly shows that direct costs associated with compensation for loss of life, structures, and livestock were most important, in descending order. Although the assessed indirect costs accounted for <10% of total costs, less tangible, but important damages, such as interruptions in local transit, decreased quality of life, and impaired water quality, as well as cost of rescue efforts, were not considered. Additionally, *Kumar and Satyal* [1999] attempted to partition costs borne by the private and public sectors and found that damage to structures was largely reimbursed through the public sector, whereas most other costs were borne by the private sector, including all tabulated indirect costs. Overall, this analysis is quite unique for a landslide disaster in a remote mountainous area, especially in a developing nation.

OTHER ENVIRONMENTAL IMPACTS

In some cases, such as the preceding New Zealand example, it is the direct and indirect impacts of landslides on the natural environment that cause the greatest concern, rather than property and infrastructure damages. Aside from the indirect environmental cost of lost site productivity following landslides, other environmental damages can occur that are significant, but difficult to quantify. The transport of landslide sediment into stream and river channels is a significant environmental consequence that needs to be assessed from the perspectives of fisheries and aquatic habitat, water quality, recreation, and aesthetics.

In western North America, a major concern related to in-stream increases in landslide sediment from forest roads and harvesting in steep terrain is the deterioration of aquatic habitat [e.g., *Platts and Megahan*, 1975; *Swanson et al.*, 1987; *Allison et al.*, 2004]. High levels of suspended sediment in streams following landslides can damage the gills of fish

TABLE 1.4. Detailed cost estimates (in US$) associated with the 17 August 1998 Malpa landslide, Kumaun Himalaya, India [based on data from *Kumar and Satyal*, 1999].

Specific types of losses	Public costs	Private costs	Total costs
Direct costs			
Compensation for casualties	471,400	--	471,400
Livestock	3100	21,000	24,100
Structures (homes, huts, bridge)	51,300	11,200	62,500
Crops/vegetation	--	650	650
Other direct costs	13,700	17,400	31,100
Indirect costs			
Decline in future agricultural productivity	--	1400	1400
Unusable barren land	--	2500	2500
Lost recreational revenues (trekking)	--	7400	7400

and cause mortality [*Herbert and Merkens*, 1961; *Phillips*, 1971]. Although juvenile coho salmon can acclimate to moderate sediment increases when background levels are low, they do not acclimate well when suspended sediment increases from high background levels [*Bisson and Bilby*, 1982]. Suspended sediment can also block the transmission of light in water and thus reduce the depth to which photosynthesis occurs. Fine sediment that is not carried in suspension often becomes entrapped in spawning gravels, reducing the transport of oxygen to incubating eggs and inhibiting the removal of waste products that accrete as embryos develop [*Meehan*, 1974; *Lisle*, 1989]. Entrapped sediment also forms a barrier to emergence of fry by blocking pore space [*Phillips*, 1971]. Excess sediment can damage rearing habitat, areas where young fish feed and grow, by creating unfavorable conditions for growth of aquatic insects [*Meehan*, 1974].

Landslides and debris flows can inflict direct damage or drastic change on stream channels and aquatic habitat by scouring and depositing large quantities of sediment and wood. Debris flows, either triggered by landslides or mobilized by high flows in headwater streams, may disturb channels in steep terrain, scouring long reaches down to bedrock and inflicting at least short-term damage to aquatic habitat through changes in primary and secondary production [*Swanson et al.*, 1987; *Lamberti et al.*, 1991; *Gomi et al.*, 2002; *Roghair et al.*, 2002]. Recovery time in these systems depends on the condition of aquatic habitat following the disturbance, as well as the dynamics of woody debris recruitment and the routing of sediment through adjusting channels [*Lamberti et al.*, 1991; *Benda and Dunne*, 1997a; *Nakamura et al.*, 2000; *Roghair et al.*, 2002]. At channel gradient breaks, abrupt tributary junctions, channel constrictions, and large roughness elements, debris flows may deposit enough sediment and wood to block channels and prevent fish migration [e.g., *Pearce and Watson*, 1983]. Debris flows also alter the course and morphology of channels, as well as riparian vegetation and the nature and distribution of woody debris jams [*Swanson et al.*, 1982a; *Benda and Dunne*, 1997a; *Benda et al.*, 2003a; *Gomi et al.*, 2003, 2004; *Chen*, 2006].

Sediment generated by mass movements can adversely affect downstream municipalities, industries, agriculture, recreational users, and domestic water systems. Many communities and individuals rely on diversion of surface water for domestic water supplies, and water laden with sediment may require additional treatment or at best is aesthetically unpleasing. The city of Vancouver, British Columbia, is quite sensitive to this issue because its domestic water supply is fed by steep, albeit unmanaged, catchments that frequently experience landslides and debris flows. In the Slocan Valley of southeastern British Columbia, the issue of forest management and domestic water supplies (mainly untreated diversions of surface runoff from headwater streams) has become a major battleground between the forest industry and environmental activists. Headwater streams in landslide-prone mountain areas may feed sediment-laden water to farmlands downstream or down-river. When this water is diverted for agricultural use, the life of irrigation pumps can be shortened by turbid water and the infiltration capacity of land continually irrigated with such water may be reduced. Water with high concentrations of sediment is aesthetically undesirable for recreational users, and many industries incur additional treatment or production costs when using turbid water. Whilst these damages are significant, they are difficult to quantify in an economic analysis of landslide impacts, and thus are usually ignored.

Landslides and debris flows can indirectly affect flooding in both headwaters and larger river systems. The magnitude and timing of peak flows emanating from headwater streams may be affected by landslides and debris flows due to scouring of channels to bedrock, thus removing frictional elements that resist flow [*MacFarlane and Wohl*, 2003]. The consequent higher flow velocities may lead to reduced times of concentration and increased peak flows that could persist in channels until roughness recovers [*Thomas and Megahan*, 1998]. Other research found that both woody debris and sediment supplies were important to the formation of roughness in headwater forested, clearcut, and debris flow channels [*Gomi et al.*, 2003]. In riverine systems with active landslide erosion, cumulative storage of sediment has decreased channel conveyance capacity and contributed to flooding [*Lu and Higgitt*, 1998; *Sidle et al.*, 2004a].

Besides their negative consequences, landslides may benefit the environment in a number of ways as natural sources and supplies of sediment to streams. In temperate forested catchments, landslides scour sediments and wood from headwater channels, which form debris fans at confluences and in higher order reaches [e.g., *Benda et al.*, 2004; *Gomi et al.*, 2004; *Wilford et al.*, 2004]. Such debris flow deposits in low-energy environments may enhance aquatic habitat by forming ponds, increasing channel complexity via large wood, releasing nutrients from buried organics, augmenting off-channel habitat, replenishing spawning gravels, and altering the composition of riparian forests [e.g., *Nakamura et al.*, 2000; *Benda et al.*, 2004]; nevertheless, long-term benefits of such large deposits remain unclear. It is also unclear how sporadic increases in sediment from mass wasting induced by land management may affect the long-term recovery processes in channels [*Gomi et al.*, 2002].

2

Characteristics of Various Types of Landslides

CLASSIFICATION SYSTEMS

Slope movements have been classified in many ways, with each method having some particular usefulness or applicability related to the recognition, avoidance, control, or correction of the hazard. *Sharpe* [1938] proposed a comprehensive scheme for classifying mass wasting based on geomorphology in which processes were divided into four categories: slow-flowage types (creep and solifluction), rapid-flowage types (earthflows, mudflows, and debris avalanches), landslides (slumps, debris slides, debris falls, rockslides, and rockfalls), and subsidence. The widely used classification scheme developed by *Varnes* [1978] distinguishes five types of mass movement (falls, topples, slides, spreads, and flows) plus combinations of these principal types along with the type of material (bedrock, coarse soils and predominately fine soils) (Table 2.1). Further subdivision is based on speed of movement. A single landslide may pass through several phases as it progresses downslope; such slope movements, particularly those of debris or earth, are considered *complex* (Table 2.1), but one type usually predominates in different parts of the moving mass or at different times during the period of displacement. In this classification, *Varnes* [1978] avoids the specific use of the term *creep*; later modifications [*Cruden and Varnes*, 1996] note that creep processes can be included in various landslide categories by using either *very slow* or *extremely slow* descriptors for rate of movement. Other modifications of the original *Varnes* [1978] classification have been made, including an eight-category system developed by the European Community Program *EPOCH* [1993] and others [*Brunsden*, 1985; *Hutchinson*, 1988]; however, in general, changes to the original system have been minimal. *Sassa* [1985, 1989] developed a geotechnical classification of landslides based on the initiation mechanism; landslides are categorized into the following general classes: slides, liquefaction, and creep. Slides are further classified based on their stress state as peak-strength or residual-state slides in reference to initial and reactivated slope failures, respectively. Liquefaction is classified as either mass liquefaction or sliding surface liquefaction; mass liquefaction generally occurs in loose soils, whereas sliding surface liquefaction can occur even in moderately dense to dense regoliths in response to grain crushing and subsequent pore pressure accretion within the shear zone [e.g., *Sassa*, 1996]. In Sassa's classification, creep occurs in regoliths prior to stress paths reaching the failure line.

Keefer [1984a] employed similar principles and terminology as *Varnes* [1978] to classify earthquake-induced landslides based on material type (soil or rock), type of

Landslides: Processes, Prediction, and Land Use
Water Resources Monograph 18
Copyright 2006 American Geophysical Union
10.1029/18WM03

TABLE 2.1. An abbreviated and modified version of the landslide classification scheme developed by *Varnes* [1978].

Type of movement		Type of material		
		Bedrock	**Engineering soils**	
			Coarse	*Fine*
Falls		Rock fall	Debris fall	Earth fall
Topples		Rock topple	Debris topple	Earth topple
Slides	**Rotational**	Rock slump	Debris slump	Earth slump
	Translational	Rock block slide; rock slide	Debris block slide; debris slide	Earth block slide; earth slide
Lateral spreads		Rock spread	Debris spread	Earth spread
Flows		Rock flow (deep creep)	Debris flow (soil creep)	Earth flow (soil creep)
Complex slope movements (i.e., combinations of two or more types)				

movement (disrupted or coherent), and other characteristics (e.g., velocity, depth, water content). This classification includes three main categories of earthquake-induced landslides: (1) disrupted slides and falls; (2) coherent slides (e.g., slumps, earthflows, block glides); and (3) lateral spreads and flows. Based on the frequency of occurrence during 40 earthquakes (M 5.2 to 9.2), *Keefer* [1984a] further categorized landslides as very abundant (>2500 event^{-1}), abundant (250–2500 event^{-1}), and uncommon.

Limitations exist for any type of classification, and landslides are certainly no exception. This is especially true for older landslides where much of the slide mass has been reworked over the years. *Cruden and Varnes* [1996] use the prehistoric Blackhawk landslide in the San Bernadino Mountains of southern California as an example of a significant mass movement that is difficult to classify according to their system. Additionally, the *Cruden and Varnes* [1996] system, among others, employs an elaborate set of descriptors that, while informative for engineering geologists, may be confusing for land managers dealing with slope stability. For the purposes of assessing the effects of land management in potentially unstable terrain as well as the degree of landslide risk, the most important aspects to consider are the size of the landslide (typically represented by depth to the failure plane), the rate of movement, response to climate and earthquakes, and sensitivity to various anthropogenic disturbances (e.g., slope undercutting and overloading, vegetation modification, climate change).

BROAD FUNCTIONAL CATEGORIZATION OF MASS WASTING PROCESSES

The classification systems for landslides and slope movements described in the previous section have been developed from either geomorphological or geotechnical perspectives. As such, these systems provide detailed descriptions of the mode of failure, materials, velocity, and failure mechanism of landslides. A rather simple categorization of soil mass movements that is useful for land managers was proposed by *Sidle and Dhakal* [2002] based on earlier work that showed strong interrelations of processes within each of these functional groupings [e.g., *Swanston and Dyrness*, 1973; *Burroughs et al.*, 1976; *Sidle et al.*, 1985; *Massari and Atkinson*, 1999]. The role of climate, which is the dominant trigger mechanism, varies among different failure types. This categorization recognizes the importance of combinations of mass movements and follows the terminology employed by *Varnes* [1978] as much as possible. However, herein, a broader definition of soil mass movements is employed that includes the specific and unique processes of soil creep and surficial dry ravel due to their important interactions with land management. Five functional categories of mass movements are described: (i) shallow, rapid landslides (debris slides, avalanches, and flows); (ii) rapid, deep slides and flows; (iii) slower, deep-seated landslides (e.g., slumps, earthflows, lateral spreads); (iv) slow flows and deformations (soil creep and solifluction); and (v) surficial mass wasting (dry ravel). This last category, surficial mass wasting, is generally not considered together with landslides; however, it is a gravity-driven process that is technically a mass movement (distinct from surface erosion by water), whose movement rates can be exacerbated by certain management practices in many environments. Precipitation or other climatic triggering factors associated with each of the five mass movement categories are shown in Table 2.2. It is apparent that each mass movement type requires

TABLE 2.2. Rainfall and other climatic triggering factors associated with each of the five general mass movement categories.

Types of soil mass movements	Rainfall and other climate conditions that contribute to initiation of slope failures	Timing of the landslide
Shallow, rapid landslides	A large total amount of rain during an individual storm, often with a period of high intensity; exacerbated by wet antecedent conditions. Also, during prolonged and rapid periods of snowmelt.	During the storm event or at the peak of rapid snowmelt.
Rapid, deep slides and flows	An extended rainy period (or snowmelt period) followed by a large to moderately large storm; if preferential flow paths exist to route water quickly to the failure plane then these failures can occur during an isolated high-intensity rain event.	Usually following extended periods of rain, but towards the latter part of a moderate to large storm.
Slower, deep-seated landslides	Typically requires extended rainfall (or snowmelt) to initiate movement; movement continues through the rainy (or snowmelt) season and then subsides when rainfall is infrequent.	The lag time between the start of a rainy season (or snowmelt season) and the onset of slope movement is usually at least several days, but can be several weeks or more.
Slow flows and deformations	Same as for slower, deep-seated landslides (solifluction responds to snowmelt).	Same as for slower, deep-seated landslides.
Surficial mass wasting	Initiated by freeze–thaw (frost heave), wetting–drying, and wind under natural conditions.	May occur year-round, but most active when frost heave and wetting–drying processes predominate.

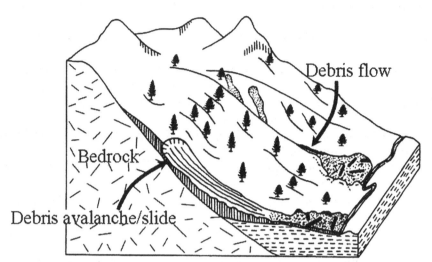

Figure 2.1. Schematic of a shallow, rapid landslide (debris slide–debris avalanche) on a steep hillslope where thin soils are underlain by relatively impermeable bedrock. On the right side is the lower extent of a debris flow that was triggered by landslides; the debris flow extends into a headwater channel.

different precipitation and climatic inputs, with temporal sensitivities ranging from seasonal (or even annual) to intervals of hours or minutes. These broad categories of mass movement tend to be associated with certain types of damage, and with particular avoidance, remediation, and control measures [*Erley and Kockelman*, 1981; *Malik and Farooq*, 1996; *Bromhead*, 1997]. Thus, this categorization of soil mass movement is useful to assess the impacts of land use, as well as climate change [*Bailey*, 1971; *Sidle*, 1980; *Swanston and Howes*, 1994; *Sidle and Dhakal*, 2002].

Shallow, Rapid Landslides

Debris slides, debris avalanches, and debris flows are typical shallow movement types in steep (>25° gradient) terrain [e.g., *Swanston*, 1970; *Bogucki*, 1976; *O'Loughlin and Pearce*, 1976; *Crozier et al.*, 1980[1]]. Soils are characteristically <2 m deep and have low cohesive properties. The shallow soil mantle overlies either bedrock or another low permeability layer (e.g., glacial or marine till) that acts as a failure plane (Figure 2.1). The failure plane is oriented approximately parallel to the soil surface, allowing for assessment using the infinite slope model. These translational landslides range in size from very small to rather large, are typically longer than their width, and have length to depth ratios <0.1 [*Wentworth*, 1943; *Skempton and Hutchinson*, 1969; *Crozier et al.*, 1980; *Cruden and Varnes*, 1996]. In many sites, these translational failures initiate as slower moving debris slides (<1 m yr^{-1} to 0.3 m min^{-1}). Water may then be incorporated

---

[1] *Rapp and Nyberg*, 1981; *VanDine*, 1985; *Maharaj*, 1993; *Massari and Atkinson*, 1999.

into the failure mass transforming it into a more rapidly moving debris avalanche (0.3 m min^{-1} to > 3 m s^{-1}). With increasing liquefaction and subsequent channelization, the failure may then become a very rapid (>3 m s^{-1}) debris flow [*Temple and Rapp*, 1972; *Sidle et al.*, 1985; *Jordan*, 1994; *Cruden and Varnes*, 1996; *Chen*, 2006].

Debris slides and avalanches, as well as debris flows, initiate on slopes that are typically either concave or linear in plan form [*Swanston*, 1969; *Gao*, 1993; *Massari and Atkinson*, 1999; *Palacios, et al.*, 2003] (Figure 2.2a), whereas channelized debris flows follow and augment the plan form of the channel (Figure 2.2b) [*Cruden and Varnes*, 1996; *Palacios et al.*, 2003; *Gomi et al.*, 2004; *Chen*, 2006]. These linked sequences of debris slides, avalanches, and flows are major sources of sedimentation in mountain streams. Sediment and large woody debris movement associated with landslides and debris flows can strongly affect stream channel substrate and form [*Harvey*, 1991; *Lisle and Madej*, 1992; *Rice and Church*, 1996] and related aquatic habitat [*Keller and Swanson*, 1979; *Lamberti et al.*, 1991; *Reeves et al.*, 1995; *Gomi et al.*, 2002]. In particular, debris flows exert the largest fluvial disturbances due to their magnitude and energy [e.g., *Wieczorek et al.*, 1989; *Iverson et al.*, 1997; *Chen*, 2006]. Debris flows that enter or initiate in steep headwater streams are sometimes called debris torrents [e.g., *Swanston and Swanson*, 1976; *VanDine*, 1985; *Lu et al.*, 2001; *VanDine and Bovis*, 2002].

Lahars are a special type of debris flow or hyperconcentrated flow that initiate on volcanic slopes in recently deposited ash or debris (Figure 2.3; see Chapter 3). Debris flows and lahars undergo bulking as they move downslope and through channel systems; thus, the depth of transported material can increase significantly compared to

Figure 2.2. (a) Debris avalanche on a steep, forested hillslope near Juneau, Alaska. (b) Channelized debris flow near Vancouver, British Columbia.

Figure 2.3. A lahar deposit from Mount Unzen, Kyushu, Japan.

depth of the initial failure, depending on the channel configuration [e.g., *Pierson and Scott*, 1985; *Lewkowicz and Hartshorn*, 1998]. Slushflows, resulting from wet snow avalanches within hillslope gullies and valleys [e.g., *Rapp*, 1960; *Clark and Seppälä*, 1988; *Nyberg*, 1989], have similar erosional consequences as debris flows, but are not covered here since they represent fluvial, not mass wasting processes. Debris flows produce characteristic debris fans in the deposition zone (Figure 2.4), which are also hazardous sites [*Kellerhals and Church*, 1990; *Wilford et al.*, 2004]. Hazards on debris fans and the linkages between hillslope failures and in-channel debris flows are both discussed in Chapter 3 (see Geomorphic Factors).

Figure 2.4. A large debris fan formed by the Akakuzure landslide in Ooigawa basin, Shizuoka, Japan.

Shallow, rapid landslides are typically triggered by rainstorms, rapid snowmelt, earthquakes, or combinations of these factors. Many shallow, rapid landslides can be triggered by moderate to large magnitude earthquakes [*Harp and Jibson*, 1996; *Keefer*, 2002; *Khazai and Sitar*, 2003]; nevertheless, although not well documented, shallow mass movements triggered by rainstorms likely comprise the largest volume of mass wasting worldwide and certainly the greatest number of slope movements. Additionally, these failures create the largest amount of environmental damage. Damage to property and loss of lives associated with shallow landslides varies greatly, but generally is lower compared to rapid, deep-seated landslides.

During major rainstorms or snowmelt events, pore water pressure usually builds up in the soil mantle just above (or sometimes from below) the lithic contact or other hydrologic impeding layer [e.g., *Sidle and Swanston*, 1982; *Onda et al.*, 1992; *Harp et al.*, 1990; *Dai et al.*, 1999]. A common hydrologic sequence for shallow landsliding involves wet antecedent conditions followed by a prolonged period of rainfall with a burst of high intensity [e.g., *Campbell*, 1975; *Okuda et al.*, 1979; *Sidle and Swanston*, 1982; *Keefer et al.*, 1987] (Table 2.2). Hence, these failures are largely dependent on the characteristics of individual storms, particularly intensity and, to a lesser extent, total storm precipitation [*Campbell*, 1975; *Cannon and Ellen*, 1985; *Finlay et al.*, 1997]. Additionally, the extent of pore pressure response is influenced by antecedent moisture [*Sidle* 1984a, 1986; *Haneberg*, 1991a; *Sidle and Tsuboyama*, 1992; *Tsuboyama et al.*, 2000]. Although antecedent moisture may predispose hillslopes and headwater channels to shallow, rapid failures, an important distinguishing feature from deep-seated mass movements is that shallow, rapid landslides are triggered by an individual event (i.e., rainfall or snowmelt).

Rapid, Deep Slides and Flows

Deep and rapid landslides may include large debris slides and avalanches, debris flows, dry flows, bedrock slides, and certain block glides and rapid earthflows. Some of the terrain and material characteristics are similar to those of shallow, rapid landslides—i.e., steep hillslopes and generally low-cohesion material—but responses to triggering factors and management activities often differ, as does the potential disaster. Deeper slides (>5 m) often include a significant proportion of weathered or fractured bedrock within the slide mass (Figure 2.5); however, the regolith is typically not clay-rich, which would favor the development of slower, deep-seated mass movements. Exceptions include dry flows initiated during earthquakes [*Derbyshire et al.*, 2000]. For deep slides in weathered bedrock that are triggered by rainfall, they may or may not respond directly to individual rainfall events, depending on the hydraulic properties of the soil and bedrock, as well as the existence of tension cracks that extend from the soil surface to some depth in the regolith. Thus, for some cases where fracturing of bedrock and tension cracks are largely absent, these deep slides may require substantial antecedent rainfall for failure to occur [e.g., *Govi*, 1999; *Rogers et al.*, 1999]. In other cases, where bedrock is highly fractured and/or tension cracks exist in the regolith, the landslide may initiate in direct response to storm inputs [e.g., *Rodrigues and Ayala-Carcedo*, 2003; *Sidle and Chigira*, 2004] (Figure 2.5). Many studies have shown that liquefaction

Figure 2.5. A rapid, deep-seated landslide that occurred during an intense storm at Hogawachi in Minamata, Japan. The landslide triggered a rather large debris flow that killed 15 people downstream. Fractures in the weathered andesite bedrock appeared to contribute to the rapid buildup of pore water pressure during this storm.

occurs along the shearing (sliding) surface in saturated cohesionless materials just after the initial failure as the result of excess pore water pressure generation [e.g., *Iverson and Major*, 1986; *Anderson and Sitar*, 1995; *Iverson et al.*, 1997; 2000[1]]. *Sassa* [1996] proposed that grain crushing along the failure plain is partly responsible for this increase in pore pressure and consequently a factor in the initiation of many large, long-runout landslides, especially when regoliths are dense. This explanation seems reasonable for shear surfaces of some deep-seated landslides where the magnitude of effective stresses is large enough to cause particle breakage during shearing. However, the dominant factor related to excess pore pressure generation during progressive shear displacement appears to be the tendency of the soil to undergo volume changes (i.e., contraction) within the shear zone as a consequence of particle rearrangement (i.e., sliding and rolling amongst particles) during unidirectional shearing [*Wafid et al.*, 2004].

Often these larger landslides are triggered by earthquakes; such strong ground motion is sometimes required to initiate failures in deeper regoliths [e.g., *Eisbacher and Clague*, 1984; *Yanase et al.*, 1985; *Shoaei and Ghayoumian*, 2000[2]] (Figure 2.6). Earthquakes can exacerbate pore pressures when sites are wet [*Wright and Mella*, 1963; *Ochiai et al.*, 1985; *Ayonghe et al.*, 1999], as well as mobilize regoliths during the shaking process [e.g.,

---

[1]*Dai et al.*, 1999; *Wang and Sassa*, 2002.
[2]*Duman et al.*, 2005; *Sidle et al.*, 2005.

Fukuda and Ochiai, 1993; *Harp and Jibson*, 1996; *Havenith et al.*, 2000]. Block glides are deep, intact masses of regolith that move along a rather gentle discontinuity in the bedding sequence [*Ibsen et al.*, 1996]. While most reported movement rates are rather slow [*Ibsen et al.*, 1996], notable exceptions have been observed in response to earthquakes and high pore water pressures. A large, rapid block glide triggered during the October 2004 Chuetsu Earthquake in Niigata, Japan, dammed the confluence of the Inokawa and Maesawa Rivers, causing much damage to Yamakoshi village [*Sidle et al.*, 2005] (Figure 2.6).

Given the size and unexpected occurrence of these earthquake-triggered failures, they represent very formidable hazards, albeit much less frequent than shallow, rapid landslides. Areas of the world that experience significant rapid, deep landslides include Japan; Peruvian Andes; Anatolia, Turkey; Italian Alps; western Iran; Kyrgyzstan; Papua, New Guinea; Central America; the Himalaya; Gansu Province, China; South Island, New Zealand; and western North America, to name a few [*Plafker et al.*, 1971; *Harp et al.*, 1981; *Eisbacher and Clague*, 1984[1]]. They are partly distinguished from shallow, rapid failures in this text due to their relative insensitivity to land management practices, with the possible exception of roads or other practices that may concentrate water onto particularly susceptible sites (i.e., those where tension cracks or other preferential flow paths exist that will generate concentrated zones of pore water pressure accretion).

Figure 2.6. A deep-seated block glide that failed rapidly during the 24 October 2004 Chuetsu earthquake, Yamokoshi village, Japan. The block glide traveled relatively undisrupted across the Imokawa River, where it dammed the Imokawa and Maesawa rivers.

---

[1] *Yanase et al.*, 1985; *King et al.*, 1989; *Froehlich et al.*, 1990; *Fukuda and Ochiai*, 1993; *Shroder and Bishop*, 1998; *McSaveney et al.*, 1999; *Rogers et al.*, 1999; *Derbyshire et al.*, 2000; *Havenith et al.*, 2000; *Shoaei and Ghayoumian*, 2000; *Duman et al.*, 2005; *Sidle and Chigira*, 2004.

Figure 2.7. Hummocky topography with immature drainages due to ground movement by creep and earthflows, north coast of California.

Slower, Deep-Seated Landslides

Deep-seated landslides that move at rates generally <1 m day^{-1} include slumps, earthflows, and lateral spreads. Both active and dormant slow, deep-seated landslides characteristically occur in somewhat gently sloping topography that is typically hummocky with immature drainage systems [*Swanson and Swanston*, 1977; *Bechini*, 1993; *Massari and Atkinson*, 1999; *Ocakoglu et al.*, 2002] (Figure 2.7). Failure occurs in deep, heavily weathered clay-rich soils or regoliths that exhibit plastic behavior over a range of water contents. Examples include the Franciscan mélange in northwestern California [*Kelsey et al.*, 1981; *Lewis*, 2002]; the Cascade Range in western Oregon [*Swanson and Swanston*, 1977; *Ambers*, 2001]; the Columbia River Valley of northeastern Washington [*Jones et al.*, 1961]; south-central British Columbia [*Bovis and Jones*, 1992]; the upper Ohio Valley [*Sharpe and Dosch*, 1942]; southern Rwanda [*Moeyersons*, 1989]; the Raukumara and Otago Peninsulas of New Zealand [*Crozier*, 1968; *Pearce et al.*, 1981]; the Flysch Zone adjacent to the Austrian Alps [*Rohn et al.*, 2003]; east and southeast France [*Julian and Anthony*, 1996; *Campy et al.*, 1998; *Malet et al.*, 2002]; the central Apennines of Italy [*Massari and Atkinson*, 1999]; the West Black Sea Region of Turkey [*Ocakoglu et al.*, 2002]; Shikoku Island [*Furuya et al.*, 1999] and Hyogo Prefecture [*Okunishi*, 1982], Japan; and the mountains of Uzbekistan [*Niyazou*, 1982].

Slumps and earthflows often occur in combination; the initial movement is typically a rotational slump, and subsequent downslope movement of the remoulded material is by earthflow [e.g., *Swanson and Swanston*, 1977; *Okunishi*, 1982; *Bechini*, 1993; *Rohn et al.*, 2003; Figures 2.8 and 2.9]. Within deep-seated slump–earthflow complexes, relatively intact blocks of regolith have been displaced with upright woody vegetation [e.g., *Ocakoglu et al.*, 2002]—so-called block slides or block glides (Figure 2.9). Movement rates of deep-seated slumps and earthflows

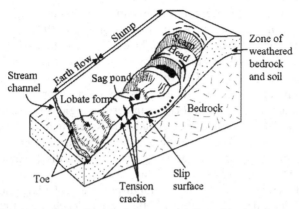

Figure 2.8. Schematic of a combination slump–earthflow showing characteristic features.

typically range from millimeters per year (extremely slow) to meters per day (moderate to rapid), although relatively slower rates in the range of centimeters to several meters per year are more common [*Campbell*, 1966; *Ker*, 1970; *Yeend*, 1973; *Swanson and Swanston*, 1977[1]]. Slumps and earthflows are typically larger than shallow, rapid landslides and move in response to seasonal or at least multi-day accretions of groundwater related to inputs from rainfall or snowmelt [*Campbell*, 1966; *Haruyama*, 1974; *Swanson and Swanston*, 1977; *Wasson and Hall*, 1982[2]]; once a critical level of groundwater is present, movement accelerates rapidly. Because

Figure 2.9. A block glide as one component of a larger slump–earthflow complex, Drift Creek Slide, Oregon Coast Ranges.

---

[1] *Wasson and Hall*, 1982; *Vonder Linden*, 1989; *Julian and Anthony*, 1996; *Furuya et al.*, 1999; *Malet et al.*, 2002; *Coe et al.*, 2003; *Rohn et al.*, 2003.

[2] *Iverson and Major*, 1987; *van Asch and van Steijn*, 1991; *Julian and Anthony*, 1996; *Malet et al.*, 2002; *Coe et al.*, 2003.

of the complex network of preferential flow pathways that develop within and near earthflows [e.g., *Swanson and Swanston*, 1977; *Sidle et al.*, 1985; *Bisci et al.*, 1996], the movement of surface water to deep failure planes can be rapid (in the order of days or less). In some cases, earthflow movement has occurred in direct response to individual rainstorms [e.g., *Okunishi*, 1982; *Furuya et al.*, 2002]. Despite their slower rates, deep-seated mass movements are responsible for the transport of large volumes of sediment to streams and rivers in certain regions [*Dietrich and Dunne*, 1978; *Sasaki et al.*, 2000], such as the Eel and Mad Rivers in northern California, the Waipaoa River in New Zealand, and the Combe d'Ain basin in France, which have some of the highest contemporary or historical sediment discharge rates [*Gregory and Walling*, 1973; *Campy et al.*, 1998; *Lewis*, 2002].

Lateral spreads can be defined as the lateral displacement of a large mass of cohesive rock or soil overlying a deforming mass of soft material [*Dikau et al.*, 1996; *Cruden and Varnes*, 1996]. Lateral spreads in soils occur in both sensitive clay and sand materials, typically with gentle slope gradients. The depth of the lateral spread depends on the depth of the deposit. Movement is initiated by high internal pore water pressure from rainfall or snowmelt, by earthquakes, or by high water contents (in the case of clays) [*Buma and van Asch*, 1996]; subsequently, liquefaction occurs rapidly along the sliding surface, causing the overlying soil to break up and spread in a rather chaotic fashion [*Selby*, 1982]. Lateral spreads in sands often occur along coasts or adjacent to water bodies, particularly where slopes have already been subject to mass movement [*Buma and van Asch*, 1996]. Sensitive clay deposits (so-called quick clays) prone to lateral spreading are found predominately in the St. Lawrence lowlands of eastern Canada, southern Norway, and portions of the Alaskan coast [*Lefebvre*, 1996]. One of the most damaging lateral spreads in recent history occurred during the Great Alaska Earthquake of 1964 in the Bootlegger Cove Clay, a wet formation consisting of weak zones of sensitive silty clay, silt, and sand bounded above and below by relatively stiff clays [*Hansen*, 1966] (Figure 2.10). Large lateral spreads

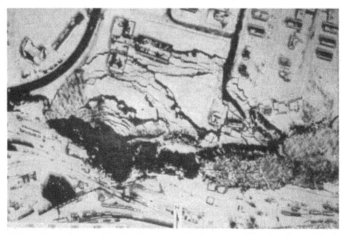

Figure 2.10. A large lateral spread in Anchorage, Alaska, caused by the Great Alaska Earthquake (M = 9.2) of 1964.

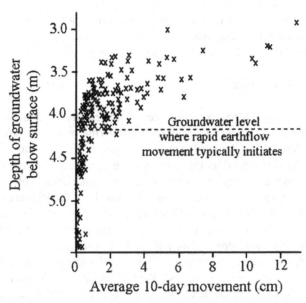

Figure 2.11. The relationship between groundwater levels and movement of the deep-seated Chausuyama landslide in Nagano Prefecture, Japan; data for 1971–1975 show a clear threshold of accelerated movement when groundwater rises above 4.2 m (below the soil surface) [based on data from *Oyagi*, 1977].

can also occur in bedrock material at the edge of escarpments and plateaus [*Cruden and Varnes*, 1996]. While not discussed at length in this book, lateral spreads are important in specific regions, particularly where quick clay deposits occur.

Deep-seated mass movements are more influenced by seasonal variations in water inputs due to the deeply weathered nature of the regolith [*Campbell*, 1966; *Swanson and Swanston*, 1977; *Wasson and Hall*, 1982; *Iverson and Major*, 1987] (Table 2.2). Typically, deep-seated movements respond non-linearly to accumulations of soil water; during the dry season little if any movement occurs in response to rainfall, but once the regolith is recharged to a critical threshold, mass movement responds to further rainfall or snowmelt inputs [*Oyagi*, 1977; *Swanson and Swanston*, 1977; *Bechini*, 1993]. An example of a clear groundwater response threshold that induces earthflow movement is given for the 9×10^6 m^3 Chausuyama landslide in Nagano Prefecture, Japan [*Oyagi*, 1977] (Figure 2.11). Once a critical threshold of groundwater level is achieved, earthflow movement generally increases dramatically with further inputs of precipitation. Such behavior attests to the seasonal surging of earthflows recorded in many areas [*Campbell*, 1966; *Wasson and Hall*, 1982; *Iverson and Major*, 1987; *Vonder Linden*, 1989; *Coe et al.*, 2003]. Due to the complex hydrologic infrastructure of deep-seated mass movements, responses to precipitation vary from site to site and generalizations are difficult to make. Localized features, such as sag ponds, terraces, and tension cracks, all serve to store and route water into the deep-seated failure [*Swanson and Swanston*, 1977; *Coe et al.*, 2003] (Figure 2.8). Minor changes in hydrologic pathways induced by land use can dramatically alter the movement of slump–earthflow combinations. Because these failures move relatively slowly, loss of life is unusual, but extensive property damage can occur.

Slow Flows and Deformations

This mass movement category is dominated by the relatively ubiquitous soil creep, and also includes the less important, regionally occurring process of solifluction. Soil creep is not a failure per se, but a plastic deformation of the soil mantle. Neither soil creep nor solifluction poses direct threats to humans, but may contribute to extensive property damage and act as catalysts for more rapid landslides. Active soil creep is often associated with slump–earthflows; thus, terrain characteristics are the same. Soil creep occurs in both shallow and deep regoliths, and movement may occur throughout the thickness of both cohesive and noncohesive soil profiles. However, movement rates typically decrease with depth [*Swanston*, 1981; *Donoghue*, 1986; *Sonoda*, 1998]. Rates of soil creep displacement are generally much slower (i.e., millimeters per year) [*Swanson and Swanston*, 1977; *Swanston*, 1981; *Jahn*, 1989; *Jungerius et al.*, 1989[1]] than annual movement of deep-seated slumps and earthflows. Soil creep responds more to the cumulative or seasonal inputs of water than to synoptic inputs [*Swanston*, 1981; *Jungerius et al.*, 1989; *Sonoda and Okunishi*, 1999].

In many locations (e.g., Franciscan mélange of northern California; southwestern Oregon; Ohio Valley region of the Appalachians; Bonne Valley of the French Alps; Shikoku Island, Japan), studies have focused on the importance of soil creep in deeply weathered and typically clay-rich soil mantles as a precursor for deep-seated earthflows, slumps, and even shallower landslides [e.g., *Sharpe and Dosch*, 1942; *Swanson and Swanston*, 1977; *Swanston*, 1981; *Moeyersons*, 1989[2]]. Soil creep has also been found to be a precursor for large failures in quick clay deposits in Norway [*Okamoto et al.*, 2004]. Strain softening of clayey regoliths during creep [e.g., *van Asch and van Stejn*, 1991] as well as subsurface erosion of less clay-rich regoliths induced by creep-like movement within the potential landslide mass [e.g., *Furuya et al.*, 1999], are likely explanations for the association of creep prior to deep-seated landslide initiation or activity. While such areas exhibit high fluxes of soil mass movement via creep, in many sites where shallow mass movements have occurred (e.g., geomorphic hollows), shallower soil creep can be a primary recharge process [*Dietrich and Dunne*, 1978; *Lehre*, 1981; *Okunishi and Iida*, 1981[3]]. Soil creep often acts together with surficial transport processes (e.g., root throw, wind snap, freeze–thaw action, dry ravel, surface wash, bioturbations) as part of a linked system that permits repeated rapid mass movement at individual failure sites [*Marron*, 1985; *Yamada*, 1999; *Sasaki et al.*, 2000]. In the long term, rates of soil creep and other recharge processes help govern the frequency at which rapid soil movements discharge material from unstable sites. Thus, the recharge rate via soil creep relative to

---

[1] *Sonoda and Okunishi*, 1999; *Sasaki et al.*, 2000.

[2] *van Asch and van Steijn*, 1991; *Bisci et al.*, 1996; *Furuya et al.*, 1999.

[3] *Sonoda and Okunishi*, 1994; *Yamada*, 1999; *Sasaki et al.*, 2000.

the unloading of rapid-failure sites is important in assessing the true long-term impact of rapid slope movements caused by land management or by land use changes [*Swanson et al.*, 1981; *Swanson and Fredriksen*, 1982; *Sonoda and Okunishi*, 1994]. In landscapes where other large or frequent soil mass movements and widespread surface wash are uncommon, soil creep may be the rate-governing process in the denudation equilibrium of whole hillslopes or complete drainage basins [*Pearce and Elson*, 1973; *Dietrich and Dunne*, 1978; *Sasaki et al.*, 2000].

Soil creep cannot be completely distinguished from the slower range of earthflow movements solely based on rate of movement. However, rates of soil creep generally decrease with depth, whereas in some slow earthflows the movement rates increase with depth for some distance below the surface [*Swanston*, 1981; *Sonoda*, 1998]. Many slow earthflow movements have well-developed zones or planes of shear failure along the flanks and on the basal surface. These bounding surfaces provide a convenient means of distinction between slow earthflow and soil creep. In the absence of such indicators, it is difficult to distinguish between these two mass movement processes because characteristic terrain features are similar, particularly for deeply weathered regoliths (Figure 2.7).

Solifluction is a specialized type of shallow earthflow found in periglacial environments that are typically underlain by permafrost. While gravity is the driving factor in solifluction [e.g., *Rohdenburg*, 1989], active solifluction has been found on slopes ranging from as gentle as 2° to as steep as 36° [see Table 1 in *Matsuoka*, 2001]. Both active and past solifluction is evidenced by conspicuous solifluction lobes, large tongue-shaped masses of soil that have well-defined bulges high at their distal margins [*Jahn*, 1967; *Easterbrook*, 1999] (Figure 2.12). Stripes and terraces can also appear as characteristic landforms as well as in association with lobes [*Matsuoka*, 2001]. During warmer seasons a perched water table develops in the plastic soil mantle above the permafrost layer, inducing slow, downslope movement along the permafrost plane [*Åkerman*, 1993; *Harris et al.*, 1997]. Solifluction can be broadly defined as slow mass wasting resulting from freeze–thaw action in fine-textured soils, although *Matsuoka's* [2001] definition includes needle ice creep, which is strictly a surface process and is included with surficial mass wasting processes herein. As described here, solifluction includes the processes of frost creep, gelifluction, and plug-like flow (also known as active-layer detachment slides). Frost creep is the downslope movement of cohesive soil material in response to frost heaving normal to the hillslope followed by nearly vertical settling after thawing [e.g., *Taber*, 1930; *Washburn*, 1979]. Frost creep is a near-surface process but can influence mass movement down to depths of 20 cm below the surface, depending on climate and site conditions [e.g., *Koaze*, 1983]. Gelifluction occurs deeper in the soil mantle and is largely associated with seasonal thawing of frozen ground together with inputs of water into the plastic soil mantle via rain and snowmelt [*Koaze*, 1983; *Harris et al.*, 1997]. These inputs promote the slow, downslope movement of saturated soil. Plug-like flows (or detachment slides) occur on slopes underlain by cold permafrost, which promote both upward and downward freezing [*Harris and Lewkowicz*, 1993]; thus, plug-like flows may include both deeper gelifluction and more rapid sliding [*Matsuoka*, 2001]. While rates of solifluction movement are generally ≤1 m yr^{-1} [*Selby*, 1982; *Matsuoka*, 2001], typically in the same range as earthflow velocities, cases of short-term, high-velocity solifluction (up to 1 m day^{-1}) have been reported when snow patches are in their final melting stage [*Jahn*, 1967].

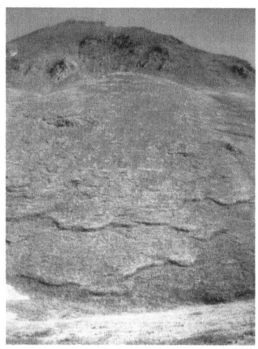

Figure 2.12. Example of active solifluction on a hillslope near Suslositna Creek, Alaska (from NOAA Geologic Hazard Photos).

Similar to velocity profiles of soil creep, annual solifluction velocities decrease rapidly with depth, with most movement occurring in the upper 1 m of soil [*Matsuoka*, 2001]. Although solifluction can be a significant denudation process as well as supplier of sediment to streams in arctic regions [e.g., *Rapp and Åkerman*, 1993], it is not discussed at length because its activity is confined to cold regions where land use is not intense. However, recent concerns over potential impacts of climate change have prompted renewed interest in this process [e.g., *Matsuoka*, 2001; *Mann et al.*, 2002].

Surficial Mass Wasting

Dry ravel and dry creep are both surficial processes, but because they are gravity driven, they are classified as mass wasting. These processes involve the downslope movement of individual soil grains, aggregates, and coarse fragments by rolling, sliding, or bounding. The main cause of dry ravel is commonly believed to be the loss of interlocking frictional resistance among soil aggregates or grains [e.g., *Hough*, 1951; *Rahn*, 1969], which may occur in relation to freezing–thawing and wetting–drying cycles [*Krammes*, 1965; *Innes*, 1983; *Asare et al.*, 1997]; however, recent research suggests that slightly deeper sliding failures in granular materials may be triggered by the successive occurrence of rolling of near-surface particles [*Onda and Matsukura*, 1997]. While dry ravel is a less perceptible type of erosion and usually transports far less sediment compared to other mass wasting processes, it

Figure 2.13. Examples of dry ravel and dry creep: (a) on heavily burned sites; (b) along road cuts; (c) trapped on the uphill side of woody debris obstructions; and (d) on sites with active freeze–thaw and wetting and drying on east-facing slopes with sparse organic horizons over mineral soils.

creates problems on steep slopes with sparse vegetative cover, especially where soils have been disturbed [*Megahan*, 1978; *Sidle et al.*, 1993; *Megahan et al.*, 2001]. In particular, fire can induce extensive dry ravel and creep in certain landscapes [*Krammes*, 1965; *Mersereau and Dyrness*, 1972; *Rice*, 1982; *Florsheim et al.*, 1991] (Figure 2.13a). Steep roadcuts and fresh roadfills, especially in low cohesion and easily weathered earth materials, are sites that experience significant ravel [*Megahan and Kidd*, 1972; *Megahan et al.*, 1983, 2001] (Figure 2.13b). Dry ravel deposits can often be observed as conical accumulations at the base of slopes and on the upslope side of trees and other obstructions (Figure 2.13c).

Naturally vegetated hillslopes generally experience low rates of dry ravel or dry creep when the surface organic horizons are undisturbed. However, the processes of needle ice formation and frost heave [e.g., *Krammes*, 1965; *Oostwoud Wijdenes and Ergenzinger*, 1998; *Matsuoka*, 2001] have the potential to dislodge surface materials and allow them to be eventually transported downslope by gravity. Such surface mass wasting processes have been observed to be seasonally significant at vegetated sites where the right combinations of temperature, moisture, and even wind conditions exist together with easterly exposure (Figure 2.13d). When mineral soils at such sites are subsequently exposed, dry ravel and creep will increase. While dry ravel and dry creep pose no threats to humans and cause only minor inconveniences to property, their main impacts are related to reductions in site productivity, sediment delivery to channels, and recharge of former landslide sites.

3

Natural Factors Influencing Landslides

GEOLOGICAL FACTORS

Rock Characteristics and Weathering

Landslides have been associated with many different geologic materials around the world. Some examples of strong associations between landslides and particular types of regolith materials include: (1) Shirasu volcanic deposits (ignimbrite) in Kyushu, Japan [*Shimokawa et al.*, 1989; *Yokota and Iwamatsu*, 1999]; (2) quick clays of Norway and Sweden [*Stal and Viberg*, 1981; *Okamoto et al.*, 2004]; (3) the Franciscan mélange in northern California [*Kelsey*, 1978]; (4) glaciolacustrine and glaciomarine sediments in British Columbia [*Lafleur and Lafebvre*, 1980; *Geertsema and Schwab*, 1996; *Fletcher et al.*, 2002]; (5) Cretaceous shales of midwestern North America [*Thomson*, 1971]; (6) the London clay [*Skempton and Delorey*, 1957]; (7) volcanic and pyroclastic regoliths in Uganda and Kenya [*Ngecu et al.*, 2004]; (8) loess deposits in China [*Billard et al.*, 1993; *Derbyshire et al.*, 2000]; (9) volcanoclastic rocks in Oregon and Washington [*Swanson and Swanston*, 1977; *Swanston*, 1978]; (10) clay-rich Cretaceous sandstones in North Island, New Zealand [*Gage and Black*, 1979; *Pearce et al.*, 1981]; (11) metamorphic sandstones and siltstones in Italy [*D'Amato Avanzi et al.*, 2004] and (12) granitic rocks in Japan, Korea, and Hong Kong [*Chigira*, 2001; *Wakatsuki et al.*, 2005; *Wen et al.*, 2004]. One older Japanese system classifies landslides into three groups based on lithology: (1) soft Tertiary rocks, (2) hydrothermally altered volcanic rocks, and (3) fractured- and crushed-zone rocks [*Takada*, 1964]. Landslide hazards in the Himalayas have also been mapped based primarily on geological attributes [*Gupta and Joshi*, 1990; *Pachauri and Pant*, 1992]. Such hazard assessments are useful to determine the general susceptibility of various lithologies to landsliding in specific regions, but are not useful for predicting the spatial and temporal extent of slope failures. Geological maps are often used to screen for landslide hazards in developing countries due to their availability. While many lithologies are strongly associated with active landsliding, the extent and nature of the weathering processes and its influence on water infiltration and interaction with the rock mass more accurately describe the susceptibility of these substrates to mass movement.

Weathering of earth surface materials involves the in situ physical and chemical alteration of rock or regolith. Biological weathering also occurs, but predominantly enhances physical and chemical effects. The processes of weathering include not only mechanical degradation and chemical decomposition but also the formation of weathering products

Landslides: Processes, Prediction, and Land Use
Water Resources Monograph 18
Copyright 2006 American Geophysical Union
10.1029/18WM04

in the regolith due to the interaction with infiltrating water and air [*Lumb*, 1962; *Yatsu*, 1988; *Chigira et al.*, 2002]. Because weathering processes affect both the regolith strength and the pathways of water, they are important determinants of slope stability in many settings [*Maharaj*, 1995; *Chigira*, 2002a; *Wakatsuki et al.*, 2005].

Mechanical or physical weathering consists of (1) frost action; (2) salt weathering, whereby salt crystals exert pressure and stresses on surface materials due to expansion and growth; (3) hydration, which causes the swelling of certain clays and adsorption of water in pore spaces; (4) thermal stress caused by temperature fluctuations; and (5) mechanical unloading, which may cause rock burst, sheeting, and exfoliation [*Yatsu*, 1988]. The most effective chemical weathering process within the regolith is dissolution, whereby water and air moving through minerals and rocks dissolve and mobilize chemical constituents based on stability/instability reactions [e.g., *Pearce et al.*, 1981; *Chigira*, 1990; 1993; *Chigira et al.*, 2002]. Additionally, oxidation–reduction (redox) reactions may be important in modifying minerals; e.g., iron reduction contributes to the conversion of smectites to illites [*Eslinger et al.*, 1979]. Crystalline hydration also contributes to chemical weathering in some environments. Biological processes modify chemical weathering by the addition of organic matter (plant tissues and micro- and macro-organisms) into the regolith and the resulting production of carbon dioxide (during decomposition) and organic acids [*Bormann et al.*, 1998]. Both mechanical and chemical weathering commonly proceeds from the surface of geological materials inward [*Oguchi and Matsukura*, 1999; *Yokota and Iwamatsu*, 1999; *Chigira et al.*, 2002; *Wakatsuki et al.*, 2005]. However, in some cases chemical weathering can occur at depth facilitated by dissolution of basic rocks by acidic water and piping or preferential flow [*Pearce et al.*, 1981; *Chigira*, 2002a; *Wen et al.*, 2004]. Mechanical weathering may also be facilitated by biological processes such as dynamic plant root systems and burrowing animals [*Sidle et al.*, 2001]. The relationship between weathering and hydraulic conductivity of regolith material may not be as simple as implied in many studies [e.g., *Yokota and Iwamatsu*, 1999; *Chigira*, 2002b; *Chigira et al.*, 2002]. *Megahan and Clayton* [1986] found that hydraulic conductivity of granite was significantly higher in slightly weathered compared to essentially unweathered conditions; however, with further weathering the increased production of clay reduces hydraulic conductivity. Such conditions are especially true for thick, weathered regoliths in the tropics [*Wolle and Carvalho*, 1994; *Vieira and Fernandes*, 2004].

Because weathering alters the mechanical, mineralogical, and hydrological attributes of regoliths, weathering profiles may sometimes determine the sliding plane for landslides. An example of such a weathering profile-induced sliding zone is a weakened and exfoliated tuff deposit that has undergone vapor-phase crystallization [*Chigira et al.*, 2002]. Soft, deformed layers of clay-rich material often form the sliding surface for large earthflows and deep-seated landslides [*Keefer and Johnson*, 1983; *Shuzui*, 2001; *Zheng et al.*, 2002]. Failure along this clay-rich sliding surface is exacerbated by the development or accumulation of smectites, slickensides, reduced iron, as well as associated groundwater. Thus, characteristics of the sliding surface may be more important determinants of deep-seated failures than parent bedrock. Slip zones in weathered granitic saprolite develop along pre-existing weak zones where silicon and organic matter have been depleted and aluminum has accumulated, suggesting that clays are formed in the slip surface by vertical leaching, preferential transport (via solution and colloids along relict sheeting joints), and deposition of aluminum-silicon [*Wen et al.*, 2004]. Surficial mechanical weathering of micro-sheeted granite has been reported to form slip surfaces for shallow landslides [*Chigira*, 2001]. *Wakatsuki et al.* [2005] found more rapid weathering in granitic slopes compared to gneiss slopes, and correspondingly shallower slip surfaces with shorter recurrence intervals on granitic slopes. *Ibetsberger* [1996] also noted the greater stability of valley slopes underlain by gneiss compared to similar slopes underlain by weathered granite in the Inner Himalaya, Nepal. In central Japan, high weathering rates in granodiorite formed thick regoliths with large capacities to store unsaturated water infiltrating during storms, thus leading to few landslides; in contrast, low weathering rates in granite formed thin regoliths with a much lower capacity for water storage, leading to frequent, small landslides [*Onda*, 1992]. Examination of weathering profiles in mudstones and sandstones indicate that oxidation of pyrite causes sulfuric acid to dissolve underlying materials (calcite, zeolite, and volcanic glass), weakening rock strength and promoting landslides [*Chigira*, 1990; *Chigira and Oyama*, 1999]. Calcite and zeolite may buffer this weakening by dissolution. Conversely, in the overlying oxidized zone, sandstone is strengthened via cementation by iron oxide or hydroxide, while mudstone is weakened due to its higher clay content and larger specific area [*Chigira and Oyama*, 1999]. Pyroclastic rocks (Shirasu) in Kyushu, Japan, weather rapidly by both mechanical and chemical processes. The strength of these deposits largely depends on the degree of welding of the weakly interlocked volcanic glass and pumice fragments [*Yokota and Iwamatsu*, 1999]. Mechanical weathering proceeds to loosen the surface materials, while chemical weathering dissolves the ferric oxide and silica cement. Clay minerals such as halloysite are formed by the weathering of volcanic glass in acidic rocks [*Pettapiece and Pawluk*, 1972; *Yokota and Iwamatsu*, 1999]. The resulting rapid weathering rates on steep slopes (0.1 mm–10 cm yr^{-1}) explain the frequent occurrence of landslides on many of these Shirasu mantled slopes in Kyushu [*Shimokawa et al.*, 1989; *Yokota and Iwamatsu*, 1999]. Rapid weathering of bedrock and soils in the wet tropics contributes to landslide initiation [*Maharaj*, 1995; *Ngecu et al.*, 2004]. In Fukushima, Japan, *Chigira* [2002b] found many landslides where the relatively shallow sliding surface was in weathered tuff and colluvium. A clearly defined weathering front in the tuff was underlain by hard, relatively impermeable tuff, which inhibited infiltration at depth and restricted root penetration.

A system for classifying various rock types (e.g., sedimentary, plutonic, pyroclastic, metamorphic) together with general characteristics related to weathering and water infiltration is being developed in Japan for use in landslide analysis [*Chigira*, 2003]. The interaction of weathering processes with water movement and pore pressure generation in unstable regoliths should more accurately describe the susceptibility of these substrates to mass movement compared to rock type alone [*Chigira*, 2002b; *Ngecu et al.*, 2004; *Wakatsuki et al.*, 2005]. Notable examples of the close association of weathering and landslides in Japan include (1) granitic bedrock; (2) igbrinite; and (3) soft mudstones and sandstones.

Bedrock Structure

Structural features of bedrock can promote landslide initiation in several ways: (1) by forming weak surfaces that are prone to sliding; (2) facilitating the exfiltration of groundwater into the overlying soil mantle; and (3) creating the opportunity for weathering that will eventually destabilize the regolith. Lithological structural features include faults, folds, bedding planes, fractures, joints, shear zones, and exfoliation. Bedrock that is fractured, jointed, faulted, or bedded in the direction parallel to the hillslope provides little support to the overlying soil mantle and facilitates the movement and concentration of subsurface water [*Burroughs et al.*, 1976; *Giraud et al.*, 1990; *Chigira*, 2001; *D'Amato Avanzi et al.*, 2004]. Joints and fracture planes provide preferential pathways for deep percolation of water in otherwise low-permeability bedrock; depending on jointing structure and topography, this subsurface water may rapidly exfiltrate into the soil mantle, creating zones of positive pore water pressure [*Harp et al.*, 1990; *Terajima and Moroto*, 1990; *Montgomery et al.*, 1997; *Sidle and Chigira*, 2004]. Downslope dipping bedrock promotes the emergence of springs and seeps on steep slopes, which are frequent sites of shallow landslide initiation [*Pierson*, 1977; *Onda et al.*, 2001]. Deep joints can route subsurface water rapidly to lower elevation sites (compared to the point of entry), resulting in massive landslides [*Swanston*, 1978]. In contrast, near-surface fracturing, jointing, and sheeting weaken the surface regolith and permit rapid percolation down to the slip surface of debris avalanches and flows [*Swanston*, 1978; *Chigira*, 2001; *Wakatsuki et al.*, 2005]. Near-surface, dense networks of stress-release fractures in diorite and graywacke caused by surcharge removal following glacial retreat are important factors contributing to the occurrence of shallow, rapid landslides and debris flows throughout southeast Alaska [*Bishop and Stevens*, 1964; *Swanston*, 1967, 1974a]. Widely spaced, low-angle joints in granite, along with a series of parallel micro-joints, produced by unloading, promote the development of a shallow surface layer of loosened and disintegrated granite that is susceptible to shallow landsliding [*Chigira*, 2001]. Unloading fractures create zones of similar levels of weathering within thick regoliths, forming abrupt transitions between impermeable bedrock and highly weathered regoliths [*Fernandes et al.*, 2004]. Highly sheared zones in metamorphic bedrock provide preferential pathways for infiltration, resulting in weathering of bedrock to clay-rich regoliths and initiating soil creep and slump-earthflows [*Swanston*, 1978]. Downslope dipping fractures, joints, and bedding planes, especially where the underlying material is more competent, serve as the lower boundary for root penetration and subsurface flow—thus, these are likely failure planes for landslides in steep terrain [*DeGraff*, 1978; *Swanston*, 1978; *Giraud et al.*, 1990]. In contrast, horizontal bedding planes and fractures or those that dip into the overlying soil mantle may actually provide localized zones of support against slope failure [*Swanston*, 1978]. However, if bedding planes are cross-cut by joints, these may exert the dominant control on slope stability. Faults are major zones of stress relief and intrusion of igneous rocks; thus, fault zones typically include rock that is fractured, crushed, or partly metamorphosed [*Burroughs et al.*, 1976]. Numerous studies have found strong empirical correlations between landslide numbers and distance to active faults [e.g., *Mathews*, 1979; *Pachauri et al.*, 1998; *Keefer*, 2000; *Yesilnacar and Topal*, 2005]. Slip surfaces can develop along pre-existing bedding faults in folded bedrock [*Wang et al.*, 2003a]. In the Western Carpathians, folding and faulting develops the slip surface for many large landslides [*Krejčí et al.*, 2002].

An example of terrain where geologic structure exerts a major control on slope stability is the Franciscan mélange in northern coastal California and southwest Oregon, USA. Lithology is comprised of a heterogeneous assemblage of highly sheared and faulted metamorphosed sedimentary rocks of mostly Late Jurassic and Cretaceous age interspersed with mafic volcanic rocks, greenstone, chert, schist, greywacke, and serpentine [*Harden et al.*, 1981]. The structural features of these rocks together with the climate promote weathering and alteration of this lithology to clay-rich soil material that is subject to extensive soil creep and earthflows, as well as a lesser amount of debris slides. The gently rolling and hummocky hillslopes of the region undergo active sculpting by earthflows and soil creep (Figure 2.7). In much of this mountainous terrain, large numbers of these landslides occur proximate to headwater streams and supply the bulk of the sediment load to these systems; once supplied, these tributary streams transport a high proportion of the landslide sediment [*Pitlick*, 1995]. These mass movement landforms are characterized by immature drainage patterns, localized springs, and sag ponds (Figures 2.7 and 2.8). *Kelsey* [1978] estimates that approximately 10% of the Franciscan mélange in the northern California Coast Range is comprised of active earthflow terrain.

Unstable Bedding Sequences

Numerous examples of bedding sequences that promote landslides exist throughout the world. The most widespread unstable bedding sequences include (1) alternating bedding of hard and soft rocks; (2) highly altered and permeable regoliths overlying relatively low permeability substrate; (3) thin soils overlying competent bedrock or till; and (4) hard caprock (with fissures and fractures) overlying deeply weathered rocks. The instability of these bedding sequences can be exacerbated by weathering, faulting, tectonic uplift, fracturing and folding; many of these bedding patterns are strongly associated with particular types of mass movements. Two characteristic examples are presented for each of these unstable sequences.

An important example of unstable alternating layers of hard and soft rocks is the flysch deposits, consisting of rhythmically alternating sandstone and clay-rich layers of marine origin. These deep (several tens of meters), interbedded deposits occupy large areas of the Carpathian Mountains, Turkey, parts of the Middle East, and other regions. Landforms are characteristically moderate to gently sloping and prone to mass movements, particularly when folded and faulted [*Krejčí et al.*, 2002]. Landslides in this lithology are complex, including deep-seated slides, slumps, and flows, as well as shallower translational slides. Mass movement is controlled by bedding planes where either pore water pressures develop at the interface between the overlying (more permeable) sandstone and the underlying claystone, or the strength of the clay deposit is weakened by water infiltrating through the sandstone [*Krejčí et al.*, 2002; *Duman et al.*, 2005]. Additionally, tectonic fractures in these deposits influence the pathways of subsurface water and thus the location of sliding surfaces. Another example of an alternating sequence of hard and soft rocks related to deep-seated failures occurs in the folded mountains near Nagano, Japan [*Kamai*, 1989, 1998]. Highly weathered and permeable conglomerates overlie a thin (<1 m) layer of weak, altered tuff that has a concave shape following a syncline. Underlying the thin tuff layer is a deep, compacted conglomerate deposit. This relatively impermeable lower conglomerate promotes the accretion of a water table that extends into the upper, more permeable conglomerate and weakens the interbedded tuff deposit. The concave shape of the weak tuff layer facilitates the initiation of rotational slumps along the wing of the syncline. These deep-seated slumps activate a much larger and gently sloping, deep-seated slide-flow that moves approximately perpendicular to the direction of the slumps [*Kamai*, 1989; 1998].

In the Fukushima area of Japan, some landslides occur in highly permeable beds of pyroclastic deposits (e.g., ash, scoria, and pumice) that overlie tuff, andesite, and ignimbrites. The underlying, relatively impermeable mudflow deposits form a barrier to water that is percolating downward through pipes and preferential flow channels in the overlying pyroclastics, thus triggering landslides via the development of excess pore pressures at the hydrologic discontinuity [*Chigira*, 2002b]. Shirasu (pyroclastic) deposits along the margins of steep hillsides in Kyushu, Japan, exhibit two types of unstable slopes where altered and permeable materials overlie competent and impermeable pyroclastic deposits. In the first case, a highly weathered mantle of pyroclastic deposits develops on the surface; as the weathering front progresses to a critical depth, these sites fail during rainfall events, sometimes leading to widespread stripping of soil mantles [*Shimokawa*, 1984; *Shimokawa et al.*, 1989]. Recent research suggests that halloysite formed as a weathering product of the volcanic glass is transported through the weathered deposits, forming clay bands at depth which impede the infiltration of water and contribute to slope failure [*Chigira and Yokoyama*, 2005]. Once the weathered Shirasu is stripped, the newly exposed deposits begin to rapidly weather, forming enough soil material in as little 12 to 30 yr to set in motion a new cycle of mass erosion [*Shimokawa*, 1984]. The second common example is where an altered, permeable pumice deposit overlies an unweathered, low-permeability pyroclastic deposit, leading to the accretion of pore water pressure at the ash–Shirasu interface and subsequent slope failure [*Iwamatsu et al.*, 1989].

Shallow debris slides, avalanches, and flows are common occurrences in coastal Alaska and British Columbia in terrain sculpted by recent glacial retreat, where hillslopes have been oversteepened and relatively impermeable glacial till has been deposited over bedrock, forming a barrier to water penetration and an effective sliding surface [*Bishop and Stevens*, 1964; *Swanston*, 1969; *Gomi et al.*, 2004; *Roberts et al.*, 2004]. Soil mantles formed on such hillslopes are typically shallow (<1 m) and have low cohesion [*Swanston*, 1969; *Sidle and Swanston*, 1982; *Fannin and Rollerson*, 1993]. During autumn rainstorms, high pore water pressures develop within these shallow soils; shallow landslides occur at the soil–glacial till interface [*Bishop and Stevens*, 1964; *Swanston*, 1969; *Sidle*, 1984a]. In steep terrain in western South Island, New Zealand, thin soil mantles overlying poorly permeable and thick-bedded to massive sandstones experience high rates of landslide erosion [*O'Loughlin and Pearce*, 1976]. During heavy rain in this area, high pore water pressures in geomorphic hollows initiate shallow landslides [*O'Loughlin and Gage*, 1975; *O'Loughlin and Pearce*, 1976]. The smooth upper surface of the poorly jointed sandstone forms a barrier to the penetration of both water and tree roots, thus dictating the sliding plane [*Sidle et al.*, 1985].

Another type of bedding sequence that is very susceptible to deep-seated mass movements is a relatively hard caprock underlain by weathered volcanoclastic rocks. In the western Cascades of Oregon such altered and weathered volcanoclastic rocks, capped by competent andesitic and basaltic flows or welded tuff, form clay-rich regoliths that are conducive to slump-earthflows and soil creep [*Peck et al.*, 1964; *Swanson and Swanston*, 1977]. Water penetrates through fissures in this caprock, accelerating the chemical weathering processes in the underlying volcanoclastics [*Swanson and Swanston*, 1977]. Similar basaltic flows overlie weathered volcanoclastics in the Columbia Gourge [*Palmer*, 1977] and in the nearby eastern portion of the Oregon Coast Ranges [*Beaulieu*, 1974; *Swanston*, 1978] creating settings for many deep-seated slumps, earthflows and active soil creep. In northern Kyushu, Japan, relatively strong surface lava flow deposits (basalts) punctuated by fissures and fractures allow water to percolate into the underlying Tertiary sedimentary rocks promoting weathering of this weaker material. During heavy periods of rainfall, deep-seated slumps and earthflows initiate or reactivate; examples include the large, relatively slow-moving Washiodake and Ishikura earthflows in Nagasaki Prefecture [*Nakamura*, 1996].

Tectonics

The relative strength of the regolith is strongly influenced by the past tectonic setting as well as contemporary weathering [e.g., *Julian and Anthony*, 1996; *El Khattabi and Carlier*, 2004]. As already discussed, neotectonics contributes to slope instability by fracturing, faulting, jointing, and deforming foliation structures [e.g., *Ibetsberger*, 1996; *Pachauri et al.*, 1998]. Weak rock types frequently occur together with active tectonics and mountain building processes to produce high rates of landslide erosion. Neotectonic compression may affect the alignment of valleys, hillslopes, and crests by imposing stress at the base of the slope; this stress may then propagate upwards inducing slope failure [*Julian and Anthony*, 1996].

In active tectonic regions, much of the landscape may be formed by mass movement processes [e.g., *Khazai and Sitar*, 2003; *Duman et al.*, 2005]. Strong feedbacks exist amongst the processes of mountain uplift, climate, erosion (often dominated by landslides), sediment transport, and fluvial incision. Although earlier studies proposed that mountain denudation rates were primarily a function of either elevation or slope gradient [e.g., *Ahnert*, 1970; *Koons*, 1989], more recent research indicates that a strong interplay exists amongst gravitational mass movement, tectonic stress, regolith strength, and climate in tectonically active terrain [*Beaumont et al.*, 1992; *Schmidt and Montgomery*, 1995; *Roering et al.*, 2001; *Montgomery and Brandon*, 2002]. Such combinations of factors seem to control the sculpting of landforms in regions of active uplift (e.g., the

Himalaya, Taiwan, Olympic Range, New Zealand, Japan Alps) where high rates of mass wasting occur. Recent findings suggest that in high relief-terrain, changes in tectonic uplift rates affect the frequency of landsliding; this non-linear coupling of slope gradient and mass movement in regions of rapid uplift produces threshold slopes that increase very little in gradient with increasing uplift rates [*Montgomery and Brandon*, 2002].

The Darjeeling Himalaya represent an area where rapid tectonic uplift, frequent high-intensity storms, unstable regoliths, and extensive forest clearance may have precluded a steady-state adjustment in hillslope evolution [*Selby*, 1974; *Froehlich et al.*, 1990]. The Himalaya, in general, are mountains with arguably the highest levels of landslide erosion [*Starkel*, 1972a; *Shroder and Bishop*, 1998]. In and around the Taroko Gourge near Hualien, eastern Taiwan, where recent tectonic uplift is >6 mm yr^{-1} with sudden vertical tectonic displacements of 0.6 to 1.2 m during the October 1951 earthquake [*Hsu*, 1954; *Bonilla*, 1975], extensive landsliding and rockfall has occurred.

In the Raukumara Peninsula of North Island, New Zealand, the interactions amongst contemporary tectonic uplift, mechanical and chemical weathering processes, geologic structure, and bedding sequences have produced extensive regions of relatively low-gradient, deep-seated mass movements. Weak regoliths composed of severely crushed and deformed clay-rich rocks are uplifted at rates of 1 cm yr^{-1}. Rivers rapidly incise these soft-rock regoliths, destabilizing adjacent slopes (as gentle as 12–15°) and perpetuate the continual regrading of large-scale landforms by earthflows and deep-seated slide-flows down to the present riverbed level (Figure 3.1) [*Gage and Black*, 1979; *Pearce et*

Figure 3.1. Slopes on severely crushed, clay-rich rocks that have experienced continual regrading by earthflows and deep-seated slide-flows down to the present riverbed level, Raukumara Peninsula, North Island, New Zealand (New Zealand Forest Service photo by H. Hemming).

al., 1981]. Where these same rivers cut through stronger rocks, valley incision in the last 10,000 yr has created gorges up to 100 m deep with nearly vertical sides. Thus, there is a striking contrast between the areas with weak rocks, where slopes have been continually coupled to the stream system, and areas with stronger rocks, where slopes above gorges have been largely decoupled from the fluvial system for nearly 10,000 yr.

SOIL ENGINEERING, CHEMICAL, AND MINERALOGICAL FACTORS

Engineering Properties of Soils

Slope stability analyses by methods of limiting equilibrium (described in Chapter 4) require a quantitative determination of soil shear strength. This section provides insights into the important engineering properties that contribute to slope stability and the relationship of these properties to other characteristics of natural soils. Detailed analysis of mechanical behavior of soils is outside the scope of this monograph; readers are referred to *Terzaghi and Peck* [1967], *Wu and Sangrey* [1978], and *Fredlund and Rahardjo* [1993] for comprehensive treatments of the subject.

Soil shear strength (s) is a fundamental property that governs the stability of natural and constructed hillslopes; however, it is not a unique value but is strongly influenced by loading, unloading, and especially, water content. Shear strength is basically described as a function of normal stress on the slip surface (σ), cohesion (c), and internal angle of friction (ϕ) (see Chapter 4 for quantitative description) (Figure 3.2). The total normal stress represents the normal force acting on the soil element divided by the element area. True cohesion in soils arises from the bonding of fine-grained particles (clays and fine silts) primarily attributed to the following: (1) van der Waals forces; (2) electrostatic attraction between clay surfaces and edges; (3) linking of particles through cationic bridges; (4) cementation primarily by organic matter and iron and aluminum oxides; and (5) surface tension at air–water surfaces in unsaturated soil particles [*Baver et al.*, 1972]. *Yong and Warkentin* [1975] emphasize that the analytical parameter c shown in Figure 3.2 should not be confused with the property of cohesion in soils. The parameter c is actually a combination of "true soil cohesion" and other physical properties of soils such as soil moisture, grain size distribution, and relative density. Internal angle of

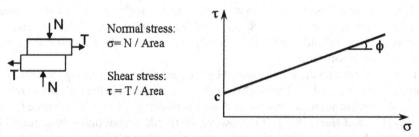

Figure 3.2. Mohr–Coulomb failure envelope describing the shear strength of soils, with the cohesion value (c) shown as the τ-axis intercept. N and T are the normal and tangential forces on the slip surface within a slice, respectively.

friction represents the degree of interlocking of individual grains or aggregates and is influenced by the shape, size, and packing arrangement of these particles. Angular particles have a larger ϕ compared to rounded particles because of their greater interlocking capabilities.

In practice, it is necessary to differentiate between total normal stress (σ) and effective normal stress (σ') when analyzing soils. Total normal stress is determined by the weight of the soil column at the depth of the failure plane and slope gradient. Infiltrating water influences the normal stress by increasing the unit weight of soil; for high-porosity soils, this weight increase may significantly contribute to failure of steep slopes [e.g., *Derbyshire et al.*, 1990; *Chigira and Yokoyama*, 2005]. Two stress-state parameters control the shear strength of unsaturated soil: net normal stress ($\sigma - u_a$) and matric suction (ψ_M), where u_a is the soil air pressure. Under atmospheric conditions, $u_a = 0$ (gauge pressure) and net normal stress equals total normal stress (σ). For unsaturated conditions, s can be expressed in terms of independent stress-state variables, ($\sigma - u_a$) and ψ_M, where u is the pore water pressure [*Fredlund et al.*, 1978]. The suction parameter (ψ_M) contributes to shear strength by providing an additional increase in the intercept of the Mohr–Columb failure line (Figure 3.2). Higher suction values (lower moisture) are associated with larger shear strengths. For saturated soils, the state parameter controlling shear strength is the effective normal stress (σ') given by the difference between total normal stress (σ) and positive pore water pressure (u). The effective normal stress may be viewed as the resultant of stresses acting along the contacts of soil solids.

Increasing soil moisture reduces soil cohesion (c) [*Baver et al.*, 1972]; thus, soil strength declines with decreasing suction. Cohesion typically increases proportionately to clay content because of the greater surface contact area associated with clay particles; thus, clay-rich soils are designated here as cohesive soils. Cohesive soils can be defined in context of Atterberg limits: (1) plastic limit, the minimum water content at which a soil can be deformed without rupture and below which soils no longer behave as plastic materials; (2) liquid limit, the water content above which the soil behaves as a viscous liquid (no measurable strength); and (3) plasticity index, the range in water content over which a soil is in a plastic state (i.e., liquid limit minus plastic limit). Cohesion reaches a maximum slightly above the plastic limit [*Baver et al.*, 1972]. Cohesionless soils essentially have zero plasticity indices. A thickening of water films around individual particles of cohesive soils with moisture contents within the plastic range results in a reduction in shear resistance. *Rogers and Selby* [1980] showed that the cohesion component of two landslide soils (clay and silty clay textures) decreased 18% and 70%, respectively, after saturation.

In field soils, cohesion is supplemented by the contribution of rooting strength (ΔC); the sum of these is called total cohesion (c + ΔC). Aggregation in the soil matrix reflects cohesive forces, but aggregation may increase interlocking, which increases the friction angle (ϕ) [*Yee and Harr*, 1977]. It is, however, difficult to quantitatively partition the contribution of aggregation into true cohesion and ϕ. The influence of clay mineralogy on cohesion and shear strength is discussed in the next section and root cohesion is discussed in a later section of this chapter.

Internal angle of friction is most important in low cohesion soils. True cohesionless soils, such as sands, form single-grained structure with their shear strength mainly dependent on intergranular friction and grain interlocking. Aggregated soils may exhibit cohesionless behavior when aggregates act as individual grains. *Yee and Harr* [1977] found that aggregation in two essentially cohesionless soils was largely responsible for their relatively high ϕ values (40–41°). Compaction of soil materials increases ϕ by rearranging particles into a tighter packing configuration.

Shrinkage is an important property of cohesive soils, and it affects slope stability. Expansive clays (e.g., montmorillonite and vermiculite), have a greater shrink-swell potential than do nonexpanding clays (e.g., kaolinite). Shrinking that accompanies drying increases with smaller particle size and greater plasticity index. *Rogers and Selby* [1980] calculated shrinkage in a cohesive soil that could effectively reduce slope stability by reducing the length of the failure surface over which shearing resistance is mobilized. Tension cracks in these soils allow positive pore water pressures to develop within the soil mantle during high intensity summer storms. Shrinkage potential of a soil can be related to the coefficient of linear extensibility (COLE):

$$COLE = \frac{\gamma_d}{\gamma_{1/3}} - 1 \qquad \text{(Equation 3.1)}$$

where γ_d is the dry unit soil weight and $\gamma_{1/3}$ is the unit soil weight at 1/3 bar moisture tension. *Ciolkosz et al.* [1979] correlated high COLE values with landslide-prone soils of southwestern Pennsylvania, particular those with high contents of expanding clays.

Sensitivity of clay-rich materials is the ratio of undisturbed to remolded strength and thus represents the potential for rapid loss of strength or collapse when the material is disturbed by ground shaking, rainfall, and other factors [e.g., *Smalley*, 1976; *Selby*, 1982]. While different criteria exist to denote sensitivity of clays, sensitivity values >16 generally indicate quick clay behavior—i.e., clay deposits that lose strength very rapidly upon disturbance [*Kerr*, 1979]. This rapid conversion from a relatively stable, solid phase to a semi-fluid substance upon disturbance is called thixotropy, and is a reversible process common in quick clay landslides.

Chemical and Mineralogical Properties

The mineralogy and chemistry of clays affect certain physical and engineering properties of soils and regoliths, thereby influencing the stability of cohesive materials [e.g., *Yatsu*, 1966; *Torrance*, 1999; *Duzgoren-Aydin et al.*, 2002]. While accumulation of clays in relict joints is associated with landslides [e.g., *Prior and Ho*, 1972; *Parry et al.*, 2000], clay mineralogy and chemistry can also provide indicators of potential sliding plane conditions [*Matsuura*, 1985; *Shuzui*, 2001; *Zheng et al.*, 2002; *Wen et al.*, 2004]. Clay mineralogy influences the stability of cohesive soils, especially in (1) slump–earthflow movements, (2) deep-seated soil creep, (3) shallow quick clay slides and flows, and (4) shallow slides in non-sensitive clay soils. The importance of clay mineralogy is minimal in shallow, rapid failures of cohesionless soils unless there is an underlying failure plane rich in clay. Failure mechanics in these soils are mainly controlled by pore water pressures and reductions in cohesion caused by decay of rooting systems.

Clay minerals are the most important chemical weathering products of the soil and regolith. These minerals form by the alteration of existing minerals or by synthesis from elements when minerals weather to their elemental form. Because of their small size (<2 μm), clay particles possess a net negative charge derived from isomorphous substitution within the crystal lattices of the clay minerals and broken bonds at the edges and surfaces. The two basic building units of clay minerals are connected silica tetrahedra and aluminum octahedral, forming tetrahedral and octahedral sheets, respectively. A 1:1 lattice clay mineral has one tetrahedral and one octahedral sheet, whereas a 2:1 lattice clay consists of two tetrahedral sheets and one octahedral sheet. Although a number of complex approaches have been suggested for grouping clay minerals [*Marshall*, 1977; *Brown and Brindley*, 1980; *Yatsu*, 1988], for the purpose of assessing slope stability, they can be broadly classified as follows: (1) 1:1 layer clays with major members that include kaolinite and halloysite; (2) 2:1 layer non-swelling clays, including illite and glauconite; and (3) 2:1 layer swelling clays, including vermiculite, montmorillonite, and beidellite. The name smectite is often used and refers to the group of clay minerals that includes, among others, montmorillonite and beidellite. The mineral group chlorite is not always considered a clay mineral but it often occurs together with illite in clay-rich soils. The 2:1 clays represent less advanced weathering stages than the 1:1 clays. In addition to these crystalline clay minerals in soil, amorphous clays such as allophane may be present. Allophane clays are structurally disordered aluminosilicates derived from weathered volcanic ash. Various physical and engineering properties of clay-rich soils depend on the interlinking of a specific number of silica tetrahedron and aluminum octahedron sheets present in the clay mineral together with the nature of isomorphous substitution and incorporation of water between the clay layers [*Mitchell*, 1993].

Many studies have attempted to link specific clay minerals to landslide susceptibility and slide type [e.g., *Prior and Ho*, 1972; *Ambers*, 2001]. For certain areas, such associations may be appropriate, but broad generalizations are difficult. In general, 1:1 layer and non-swelling 2:1 layer clay soils are much less susceptible to landslides than soils with similar contents of swelling 2:1 layer clays [*Ciolkosz et al.*, 1979; *Ambers*, 2001], although exceptions exist (e.g., quick clays).

Kaolinite is a component of more stable soils in the western Cascade Range of Oregon [*Paeth et al.*, 1971; *Taskey*, 1977; *Ambers*, 2001] and in California [*Borchardt*, 1976]. Structurally, the 1:1 clay units of kaolinite are firmly held together by hydrogen bonding, precluding the entrance of water between successive units. Cohesive soils high in kaolinite have relatively low shrink–swell potential, plasticity, and cation-exchange capacity. Halloysite, another member of the 1:1 layer clays, has somewhat random stacking of elemental sheets; however, these sheets are rolled into tubes, allowing water to penetrate between the clay layers. Unstable slopes have been associated with the hydrated form of halloysite in the Cascade Range in Oregon [*Taskey*, 1977] and in Shirasu deposits in southern Kyushu, Japan [*Chigira and Yokoyama*, 2005]. *Istok and Harward* [1982] found that hydrated halloysite and amorphous gels were abundant in earthflows in the Oregon Coast Ranges.

The 2:1 structural units of montmorillonite and vermiculite readily expand or contract, depending on the amount of water and cations present, and are referred to as swelling clays. The structure of the non-swelling illite is the same as for montmorillonite except that some of the silicons of the tetrahedral layers are replaced by aluminum, and the resulting charge imbalance is satisfied by the incorporation of potassium between adjacent units. The structure of vermiculite is similar to illite except that the interlayer charge imbalance is satisfied by hydrated magnesium ions. Chlorite is structurally similar to vermiculite except that interlayer hydrated magnesium is replaced by a crystalline octahedral sheet. The physical properties of chlorite are similar to illite.

The non-swelling 2:1 layer clays illite and chlorite have both been associated with quick clays of Scandinavia and Canada [Rosenquist, 1953; Kerr, 1963; Torrance, 1999]. Quick clay behavior involves a sudden structural alteration from a fairly stable, brittle solid to a liquid of negligible strength. Quick clays can suddenly turn to liquid when vibrated (e.g., by earthquakes or blasting). One explanation of this rapid loss of strength is the extensive leaching of saltwater cations which ultimately reduces interparticle bonding in the quick clay matrix [Rosenquist, 1953; Kazi and Moum, 1973]. Many quick clays were deposited in marine or brackish environments and thus developed a dispersed structure rich in sodium and other cations [Yatsu, 1967]. As these areas rose above sea level after glaciation, some adsorbed cations leached out of the clays, decreasing the ionic strength and changing the chemistry of the pore water [Torrance, 1979]. The lower cation concentrations within the soil matrix coupled with naturally high water contents (often exceeding the liquid limit) make quick clays highly unstable [Eden and Mitchell, 1970]. Shear resistance is much higher in sensitive marine clays that are saturated with calcium- compared to sodium-saturated clays [Torrance, 1999].

Smalley [1976] emphasizes that quick clay behavior can also occur in soils composed of the primary mineral particles (as opposed to true clays), in which case the rapid loss of strength may be more related to fracturing of interparticle cementing material (e.g., iron oxides, calcium carbonate, silica) during disruptions rather than to the long-term cation leaching phenomenon observed in marine clays. Disruption of cementation appears to be a greater cause of liquefaction in the Canadian quick clays than in the Scandinavian clays [Eden and Mitchell, 1970; Smalley, 1976]. Liquefaction has also been observed in soil deposits from shallow, rapid failures such as debris flows that have only small amounts of clay and in underconsolidated fine sands and silts.

The 2:1 swelling clays are strongly associated with landslides in many regions due to their affinity for water [Paeth et al., 1971; Ciolkosz et al., 1979; Matsuura, 1985; Ambers, 2001]. In Tertiary volcanics in Japan, the internal angle of friction (ϕ) has been shown to be inversely related to smectite content [Shuzui, 2001]. Soils prone to landslides in the Western Cascades of Oregon have high contents of smectite [Paeth et al., 1971; Taskey, 1977; Ambers, 2001]. Of several clay-rich soils in southwestern Pennsylvania, the Upshur soils with high percentages of both illite (25–50%) and vermiculite (15–35%) in the B and C subsoil horizons are more susceptible to landslides than the Cavode soils with higher illite (45–50%) and lower vermiculite (5–10%) [Ciolkosz et al., 1979]. Smectites are major constituents in deep-seated soil mass movements (e.g., earthflows) in New Zealand [Crozier, 1969] and Japan [Oyagi,

1977] and deep soil creep in California [*Fleming and Johnson*, 1975] and Japan [*Oyagi*, 1977]. Soils high in montmorillonite have been associated with bentonite debris flows in northern Alaska [*Anderson et al.*, 1969] and landslides in Montana [*Klages and Hsieh*, 1975].

The effect of adsorbed cations on the strength and swelling potential of 2:1 layer clays is complex. *Warkentin and Yong* [1962] noted higher shear strengths in sodium montmorillonite clays than in those saturated with calcium for similar void ratios. Sodium montmorillonite has weaker interparticle binding forces compared to calcium montmorillonite, thus it swells more. Shear strength increases with increasing net repulsive interparticle forces. These forces generate soil suction, which increases effective stress [*Yong and Warkentin*, 1975]. Activity values for sodium montmorillonite are much higher than for calcium montmorillonite [*Skempton*, 1953; *Egashira and Gibo*, 1988]. Although dispersed soils (sodium-saturated) may be inherently less erodible than flocculated soils (calcium-saturated) because of the mutual repulsion of individual clay particles [*Ariathurai and Kandiah*, 1978], the poorer drainage properties of dispersed systems could lead to unstable conditions.

Earlier work by *Yatsu* [1966] in Japan noted the difficulties of focusing only on the clays in the slip zone of landslides because these clays may be artifacts of reworking action along the sliding surface. Instead he proposed a categorization of clay minerals within landslip areas of Japan typical of different lithologies: (1) landslips in shales and tuffs are associated with montmorillonites and to a lesser extent halloysites; (2) landslips in metamorphosed crystalline rocks are associated with illites and chlorites and to a lesser extent swelling chlorites and montmorillonites; (3) landslips in serpentine rocks are associated with chrysotiles, chlorites, illites, montmorillonites, mixed layer minerals, and swelling chlorites; and (4) landslips in hydrothermally altered volcanic rocks are associated with montmorillonites and to a lesser extent zeolites, halloysites, cristobalites, and alunites [*Yatsu*, 1966].

Amorphous clays are of special interest because of their physical and engineering properties: (1) low bulk density; (2) high plastic and liquid limits; (3) low plasticity index; and (4) high 15-bar water content. Allophane, an amorphous mineral, has been associated with young volcanic ash soils of New Zealand, Japan, West Indies, Chile, and Oregon. *Fieldes* [1955] notes that allophane in ash soils will eventually weather to the kaolin group. Earthflows in the Cascade and Coast Ranges of Oregon have been related to the presence of amorphous clays [*Taskey et al.*, 1978; *Istok and Harward*, 1982]. The instability of these clays was partially attributed to their high water-holding capacity, which could presumably be released after disturbance and account for the fluid behavior of earthflows. Unlike in other clay-rich soils, there is a conspicuous lack of correlation between effective internal angle of friction (ϕ') and plasticity index in allophonic soils, presumably due to the unusual structure of the allophane particle [*Rao*, 1995].

Since deep-seated mass movement generally occurs in plastic soils, the activity of the soil [*Skempton*, 1953], defined as a ratio of the plasticity index to the percentage of clay in the soil, is useful to rate the relative stability of plastic soils. Examples of activity values for pure clays include kaolinite, 0.3–0.5; illite, 0.5–1.0; montmorillonite, 0.5–7.0; and allophane, 0–3.0. Generally, soils with high activity values are highly thixotropic and have low permeability and resistance to shear [*Grim*, 1962]. Higher activity values are associated with greater contributions of cohesion to shear strength.

Bonding forces attributed to flocculation are not strong enough to contribute directly to shear strength; however, they may help resist rearrangement of particles from a random pattern to a less stable parallel orientation [*Yong and Warkentin*, 1975]. Edge-to-edge and edge-to-surface bonds in clay soils can be greatly enhanced by cementing agents such as iron oxides. Such "bonded" clays have a structurally rigid mineral matrix that greatly enhances their stability [*Yong and Warkentin*, 1975]. Free iron oxide in acidic forest soils of western Oregon was noted to be an important factor in their resistance to shear failure [*Paeth et al.*, 1971]. *Pearce et al.* [1981] found that the degree of calcium carbonate cementation was directly related to stability of some calcareous sedimentary rocks in New Zealand. Moreover, they attributed the existence of several isolated landslides to acid leaching of the cement caused by weathering of pyrite. More fundamental research in mudstone and sandstone regoliths in Japan confirmed and linked these earlier studies in Oregon and New Zealand to specific weathering processes [*Chigira*, 1993; *Chigira and Oyama*, 1999]. In Japan, weathering processes in sedimentary deposits were characterized by an oxidized zone underlain by a dissolved zone where sulfuric acid (a product of pyrite oxidation) penetrates and dissolves calcite, zeolite, and volcanic glass, thus decreasing the regolith strength in the dissolved zone and creating a potential failure plane. However, the overlying oxidized zone becomes strengthened due to cementation of sand grains by iron oxide or hydroxide [*Chigira and Oyama*, 1999].

GEOMORPHIC FACTORS

Slope Gradient

Although slope gradient is important related to landslide initiation, other dynamic environmental factors also exert major influences. Numerous studies have related landslide risk to slope gradient alone in particular regions [e.g., *Lohnes and Handy*, 1968; *Swanston*, 1973; *Ballard and Willington*, 1975[1]]. While such relationships can be applied in some local situations, the wide range in the lower limit of slope gradients known to trigger shallow, rapid landslides and debris flows (20–67.5°; Figure 3.3) clearly shows that other geomorphic, hydrologic, geologic, and pedologic factors are important determinants of slope stability as well.

Tectonic uplift, volcanism, and glaciation all contribute significantly to oversteepening of hillslopes. Glaciers carve and oversteepen mountain slopes, subjecting them to mass wasting following glacial retreat. Additionally, past mountain building greatly influences the position and extent of glaciation. Throughout the formerly glaciated terrain of Austral–Asia, accelerated tectonic uplift following the early to mid-Pleistocene, greatly increased landsliding [*Page and Trustrum*, 1997; *Gupta and Virdi*, 2000; *Tamrakar et al*, 2002]. Glaciation is partly responsible for sculpting much of the steep, unstable terrain in coastal Alaska [*Bishop and Stevens*, 1964], northern Washington [*Waitt*, 1979],

---

[1] *Wolfe and Williams*, 1986; *Hylland and Lowe*, 1997.

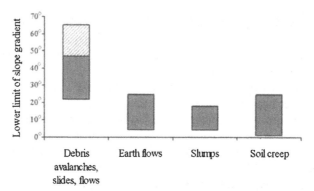

Figure 3.3. Ranges of lower limit of slope gradient from many sites worldwide for various types of landslides. The shaded area above the solid bar for debris avalanches, slides, and flows represents only one area in India [*Baruah and Mohapatra*, 1996].

New England [*Flaccus*, 1959], British Columbia [*Schwab*, 1983], northern Scandinavia [*Rapp*, 1960], Scotland [*McEwen and Werritty*, 1988], the European Alps [*Pellegrini*, 1979], the Carpathian Mountains [*Kotarba*, 1976], the Southern Alps of New Zealand [*Basher and Tonkin*, 1985], and the Himalayas [*Starkel*, 1972a]. Recently deglaciated hillslopes are typically very steep with thin, poorly developed soils and are often underlain with compact glacial till. The hydrologic discontinuity created by the presence of a till layer allows pore water pressures to develop in the soil mantle, thereby triggering shallow landslides during rainstorms or periods of rapid snowmelt [*Swanston*, 1970; *Sidle*, 1984a; *Gomi et al.*, 2004]. Following glacial retreat, valley sideslopes become susceptible to mass wasting because of the removal of the buttressing effect of the ice mass [*Haeberli and Burn*, 2002]. A notable exception has been reported in the northern Peruvian Andes, where steep (up to 37°) deglaciated hillslopes have remained stable for at least 8000 yr after initial gully development and subsequent deposition of eroded material as alluvial fans [*Miller et al.*, 1993].

There has been a resurgence of interest in slope gradient as an index of slope stability because of the availability of digital elevation models (DEM) that can electronically depict spatially distributed slope gradients [e.g., *O'Neill and Mark*, 1987; *Gao*, 1993]. Nevertheless, the inherent problems with the over-simplicity of such slope gradient analogues persist in spite of the high technology approach, not to mention potential problems associated with obtaining accurate estimates of localized gradient in the failure zone from coarse-scale DEM input data. While the appeal of such a simple slope gradient tool for predicting landslide hazard is apparent from a practitioner's point of view [e.g., *Wolfe and Williams*, 1986; *Hylland and Lowe*, 1997], the potential for error in such single-factor analysis is high. *Bhandari and Weerasinghe* [1996] clearly illustrate the problems associated with applying slope maps for assessment of landslide hazard in several districts of Sri Lanka.

Strong geomorphic arguments have been proposed supporting the concept of a threshold angle of slope steepness [e.g., *Carson and Petley*, 1970], which views slope angle as the main driving force of mass movement, especially for shallow landslides. However, this concept is largely based on observations of shallow mass movements; in particular, soil–rock assemblages over a suite of slope gradients combined with limit equilibrium analysis of shear forces and mobilized shear strength, thus, implying that slope steepness is controlled by soil strength. Similarly, *Schmidt and Montgomery* [1995] showed that large-scale bedrock strength could limit the development of local mountain gradients. A very constructive analogue of slope gradient related to mass wasting can be applied to cohesionless soils and weathered rock. In these materials, slope angle represents a reasonably accurate estimate of the internal angle of friction [e.g., *Carson and Kirkby*, 1972, p. 156]; e.g., dry talus deposits should equilibrate at a gradient close to the internal angle of friction of the particles.

A lower limit of slope gradient for various soil mass movements is difficult to ascertain, and wide variations in typical gradients for various processes have been reported (Figure 3.3). Furthermore, interpretations of these gradients are confounded by complex combination mass movements as well as inconsistent measurement and classification criteria. However, it is clear that debris slides, debris avalanches, and debris flows (shallow, rapid failure types) initiate on the steepest slopes, while earthflows, slumps, and soil creep (generally deep-seated mass movements) typically initiate on gentler slopes. Of the numerous data included in Figure 3.3, more than 60% of the sites with shallow, rapid failures occurred on slopes ≥34°. Although shallow landslides have occurred on slopes as steep as 70° [*Baruah and Mohapatra*, 1996], slopes steeper than 45° often have a lower frequency of landsliding [e.g., *O'Loughlin and Pearce*, 1976; *Varnum et al.*, 1991] because much to the susceptible soil mantle has already been removed by mass wasting leaving more stable bedrock exposed. However, notable exceptions have been reported where the lower limit of slope gradient that supports rapid landslides is much higher: hillslopes around Wellington, New Zealand (≥47°) [*Eyles et al.*, 1978]; Mylliem (≥65°) and Ryngngain (≥43°), India [*Baruah and Mohapatra*, 1996]; and Oahu, Hawaii (≥42°) [*Wentworth*, 1943]. While it is difficult to generalize, rapid mass movements are definitely more common on hillslopes >34°, assuming a soil mantle is present; if anthropogenic activities such as vegetation clearance, road construction, fire, and recreation occur, lower gradient slopes would be susceptible to shallow landslides. Most of the data included in Figure 3.3 are for more or less undisturbed sites, although a few exceptions exist.

Slow, deep-seated mass movements, such as slumps and earthflows, generally initiate on gentler slopes compared to shallow, rapid failures. Lower limits of slope gradient for the initiation of rotational slumps range from 4° to 18° [e.g., *Selby*, 1966; *Starkel*, 1976; *Mathewson and Clary*, 1977[1]]; those for earthflows typically range from 4° to 25° [e.g., *Dyrness*, 1967; *Zaruba and Mencl*, 1969; *Oyagi*, 1977[2]]. However, such ranges can be misleading because within the same landslide complex, slumps may have steeper gradients than the associated earthflow [*Bovis and Jones*, 1992]. Slumping or sliding around the perimeter of dormant earthflows can induce earthflow movement. Despite the predominance of deep-seated mass movements on moderate slopes (12° to 20°) in

---

[1] *Palmer*, 1977; *Hylland and Lowe*, 1997; *Massari and Atkinson*, 1999.
[2] *Palmer*, 1977; *Wasson and Hall*, 1982; *Bechini*, 1993; *Bisci et al.*, 1996; *Massari and Atkinson*, 1999.

western North America and New Zealand [*Palmer*, 1977; *Pearce*, 1982; *Wasson and Hall*, 1982; *Keefer and Johnson*, 1983], similar large failures have been reported in relatively steep terrain (20° to 30°) of central Honshu and Shikoku Islands, Japan [*Oyagi*, 1977; *Furuya et al.*, 1999].

Theoretically, soil creep occurs on any hillslope. Detectable measurements of soil creep have been made on slopes as gentle as 1.3° [*Finlayson*, 1981] (Figure 3.3). However, certain sites may require gradients as steep as 25° to initiate significant soil creep [*Burroughs et al.*, 1976]. From numerous investigations conducted worldwide, it is apparent that rates of soil creep are more related to soil engineering properties, clay mineralogy, local groundwater conditions, and natural perturbations (e.g., frost heave, shrink–swell potential) than to slope gradient [e.g., *Owens*, 1969; *Swanson and Swanston*, 1977; *Shimokawa*, 1980[1]]. Because soil creep is a gravity-driven process, steeper hillslopes will produce greater rates of creep in the same terrain [e.g., *Sonoda*, 1998; *Easterbrook*, 1999, pg. 89]; however, any hillslope gradient with the right combination of soil and environmental conditions can experience significant rates of soil creep. It is important to reemphasize that soil creep is often a precursor of deep-seated earthflows; this linkage was mentioned in *Sharpe and Dosch's* [1942] early work in the Appalachian Plateau, USA, and has later been noted in many other regions [e.g., *Swanson and Swanston*, 1977; *van Asch and van Steijn*, 1991; *Bisci et al.*, 1996; *Furuya et al.*, 1999].

Slope Shape

Slope shape exerts a strong influence on slope stability in steep terrain by concentrating or dispersing surface and primarily subsurface water in the landscape. On many vegetated hillslopes, rainfall or snowmelt infiltrates rapidly into the soil mantle; thus overland flow is rare except in saturated riparian corridors [e.g., *Hewlett and Hibbert*, 1963; *Tsukamoto and Ohta*, 1988; *Sidle et al.*, 2000a]. Thus, it is also important to consider substrate topography in areas where a relatively impermeable layer (e.g., bedrock, till) is present. Unfortunately, such information is rarely available without detailed geophysical surveys and it is generally assumed that substrate topography follows surface topography. Such assumptions are not always valid, as discovered in various hydrology studies [*McDonnell et al.*, 1996; *Thompson and Moore*, 1996; *Montgomery et al.*, 1997].

Three basic hydrogeomorphic slope units useful in assessing terrain stability are (1) divergent; (2) planar; and (3) convergent slope segments (Figure 3.4). With other site variables constant, divergent landforms are generally most stable in steep terrain, followed by planar hillslope segments and convergent or concave hillslopes (least stable). On slopes that are divergent or convex in plan form, subsurface (and surface) water is dispersed; thus, a perched water table is uncommon and pore pressures are typically much lower than in other slope plan forms. Hillslopes that are convergent or concave in plan form tend to concentrate subsurface water into small areas of the slope, thereby

---

[1] *Swanston*, 1981; *Matsuoka*, 1994; *Yamada*, 1997.

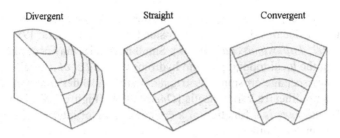

Figure 3.4. Plan form of divergent, planar (straight), and convergent hillslope segments.

generating rapid pore water pressure increases during storms or periods of snowmelt [*Sidle*, 1984a; *Fernandes et al.*, 1994; *Montgomery et al.*, 1997; *Tsuboyama et al.*, 2000]. The degree of topographic convergence for a slope segment can be calculated as the ratio of the upslope drainage area to outflow width [e.g., *Beven*, 1987; *Dietrich et al.*, 2001]. Thus, shallow, rapid landslides frequently occur in slope depressions and gullies and these geomorphic units (i.e., hollows) have been the topic of renewed interest. Planar slope segments are intermediate in susceptibility to landsliding between divergent and convergent slopes. Extensive rapid mass movement on planar as well as other hillslopes was reported in Hawkes Bay, New Zealand, following an extreme rainfall event [*Eyles*, 1971; *Sidle et al.*, 1985]. This apparent absence of topographic influence on landslide initiation may be related to hillslope areas in which shallow groundwater upwells through fractures into the overlying soil mantle.

One of the first studies that associated landslides with geomorphic hollows was in the Appalachian Mountains, USA [*Hack and Goodlett*, 1960]. This classic paper set the template for later research that developed linkages amongst hydrologic, geomorphic, and biologic processes in such settings worldwide [e.g., *Dietrich and Dunne*, 1978; *Tsukamoto and Minematsu*, 1987; *Trustrum and DeRose*, 1988; *Dunne*, 1991[1]].

Geomorphic hollows in steep terrain often exhibit evidence of repeated episodic evacuation by landslides [e.g., *Dietrich et al.*, 1982; *Marron*, 1985; *Reneau et al.*, 1989; *Shimokawa et al.*, 1989] (Figure 3.5). Once these concave sites fail, they are subject to infilling by a variety of erosion and biogenic processes as well as soil development via weathering over long periods of time (Figure 3.5). Thus, soil thickness in hillslope depressions and gullies may be significantly greater than on side slopes or intervening ridges [e.g., *Dietrich and Dunne*, 1978; *Dengler and Montgomery*, 1989]. The additional thickness of the soil mantle in hollows increases the tangential shear stress on steep slopes and, along with concentration of shallow groundwater, renders these geomorphic features more susceptible to mass movement than other slope positions. As soil thickening within hollows increases with time, eventually a landslide will occur [*Sidle*, 1987, 1988]. Even though soil accumulates in hollow troughs after failure [*Okunishi and Iida*, 1981; *Dengler and Montgomery*, 1989], it is difficult to generalize about the spatial

---

[1] *Reneau and Dietrich*, 1991; *Onda et al.*, 1992; *Fernandes et al.*, 1994; *Montgomery et al.*, 1997; *Torres et al.*, 1998; *Okunishi et al.*, 1999; *Yamada*, 1999; *Sidle et al.*, 2000a; *Gomi and Sidle*, 2003.

distribution of soil depth in these depressions due to the influences of the time since
last evacuation and localized failures and erosion [*Okunishi and Iida*, 1981; *Sidle*, 1988;
Onda et al., 1992; *Tsuboyama et al.*, 1994a]. Small landslides may occur in these valley
heads affecting the local topography and hydrology [*Onda et al.*, 1992]. Additionally,
basal subsurface seepage erosion may trigger small landslides that contribute to the
formation of amphitheatre valley heads [*Onda*, 1994] (Figure 3.6). Thus, soil depth
in geomorphic hollows is a function of a complex array of inter-linked geomorphic,
hydrologic, and biologic processes.

Because of the concave nature of geomorphic hollows, shallow groundwater tends
to accumulate along the longitudinal axis of these depressions when a hydrologic
discontinuity exists at the soil–bedrock contact [*Sidle*, 1984a; *Wilson and Dietrich*,
1987; *Onda et al.*, 1992[1]]; however, in some deeper colluvial hollows, expansion of the
saturated zone may dissipate pore pressures [*Fernandes et al.*, 1994]. This accretion of
shallow groundwater is both seasonal and event-driven. Evidence of a gradual accretion

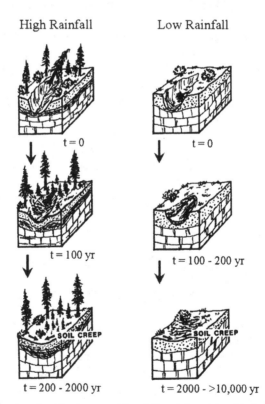

Figure 3.5. Processes involved in the repetitive failure of geomorphic hollows for conditions
of active infilling, weathering, or both (high rainfall) and slow infilling and weathering (low
rainfall) [after *Sidle*, 1988].

---

[1] *Montgomery et al.*, 1997; *Yamada*, 1999; *Tsuboyama et al.*, 2000.

Figure 3.6. Geomorphic hollow in Hitachi Ohta Experimental Watershed, Japan; site of an old landslide.

of shallow groundwater in hollows throughout rainy seasons has been shown in several areas [*Sidle*, 1984a; *Sidle and Tsuboyama*, 1992; *Fernandes et al.*, 1994; *Yamada*, 1999]. Such groundwater accretion is a precursor for shallow landslide initiation. Typically, as the rainy season progresses, hollows recharge with groundwater and the pore water pressure response during rainstorms is rapid and much greater than during drier periods [*Sidle*, 1984a; *Onda et al.*, 1992; *Tsuboyama et al.*, 2000]. This dynamic hydrologic response may be facilitated by preferential flow networks in the soil that expand during wet conditions [*Sidle et al.*, 2000a, 2001] and by convergent flow in the hollow [*Onda et al.*, 1992; *Yamada*, 1999; *Tsuboyama et al.*, 2000]. The extent of the pore pressure response appears to be related to soil depth, topographic incision, the interconnectivity of the preferential flow network, and the "leakiness" of the bedrock or other substrate [*Sidle*, 1986; *Tsukamoto and Ohta*, 1988; *Onda*, 1992; *Onda et al.*, 1992[1]]. If the pore pressures developed in the hollow reduce the soil shear strength to a critical level, a landslide will occur. Thus, hollows are susceptible sites for the initiation of debris slides, debris avalanches, and debris flows [*Hack and Goodlett*, 1960; *Dietrich and Dunne*, 1978; *Benda*, 1990[2]].

Hollows may or may not be directly coupled to headwater streams, but when a landslide occurs, a channel can be created on the hillslope linking the hollow to an existing channel [*Benda*, 1990; *Fannin and Rollerson*, 1993; *Benda and Dunne*, 1997b; *Gomi et al.*, 2002]. This linkage is more likely to occur if the landslide is large and mobilizes directly into a debris flow. Such channel extension or scouring can facilitate increased peak flows [*Thomas and Megahan*, 1998; *MacFarlane and Wohl*, 2003]. In the case of smaller landslides in hollows, the depositional material may remain on the hillslope, at

---

[1] *Sidle et al.* 2000a; *Tsuboyama et al.*, 2000.
[2] *Dunne*, 1991; *Yamada*, 1999.

least temporarily [*Sidle and Swanston*, 1982; *Fannin and Rollerson*, 1993; *Sidle*, 2005]. Some small landslides in and around hollows serve to recharge the scour zone.

Once a landslide occurs in a geomorphic hollow, a recovery process is set into motion [*Dietrich et al.*, 1982; *Shimokawa*, 1984; *Sidle*, 1987, 1988[1]]. The infilling or accretion of soil in hollows may span a period of several decades to several tens of millennia [*Stout*, 1969; *Okunishi and Iida*, 1981; *Reneau et al.*, 1986[2]] (Figure 3.5). This soil accretion process is a complex set of hydrologic, geomorphic, and biological feedbacks that involve surface wash, dry ravel, slumping and sloughing around the original headwall, regolith weathering, small landslides, frost heave, inputs of woody debris, bioturbations (from animals), and soil creep [*Dietrich and Dunne*, 1978; *Lehre*, 1981; *Okunishi and Iida*, 1981; *Swanson and Fredriksen*, 1982[3]]. Generally, the infilling process is rapid at first, dominated by inputs of sediment and wood, and becomes progressively slower as sloughing and erosion give way to chronic processes such as soil creep (Figure 3.5). The infilling process has been characterized by the following sigmoid function for the purpose of modeling soil depth recovery [*Sidle*, 1987, 1988]:

$$H_t = H_0 + a_1 e^{b_1 t^{-1}}$$ (Equation 3.2)

where H_t is the vertical soil depth at time t after the landslide, H_0 is the depth of soil remaining in the hollow (after the landslide), $a_1 = H_\infty - H_0$, H_∞ is the upper limit of soil accretion in the hollow, b_1 is $-2t_i$ (where $[t_i, H_i]$ is the inflection point of the recovery curve), and t is the time in years since initial failure in the hollow.

Only a few studies have characterized the dynamics of the infilling and most of these have only inferred the episodic sequence of loading and unloading of hollows based on dating colluvial deposits. Of the studies conducted, it appears that drier sites have much longer periods between sequential landslide occurrences in hollows, while humid environments and those with high rates of geologic weathering promote a rapid recurrence of failures (Figure 3.5). *Reneau et al.* [1986, 1990] found that the probable periodicity of loading and unloading of colluvial hollows in the northern San Francisco Bay area and southern Sierra Nevada Mountains of California was about 9000 to 15,000 yr based on radiocarbon dates. Using similar techniques *Kelsey* [1982] reported residence times of debris on hillslopes on the northern California coast to be ≈15,000 to 50,000 yr. Average soil residence times of 5000 to 6000 yr were estimated in the Oregon Coast Ranges [*Reneau and Dietrich*, 1991]. In Redwood Creek basin, northwestern California, *Marron* [1985] found that colluvium in hollows was at least 2500 to 7000 yr old. In contrast to these studies in western USA, rapid reoccurrence of landslides have been reported in Kyushu, Japan, based on more direct aerial photo interpretation and dendroecology measurements (tree age, tree scars) [*Shimokawa*, 1984]. Reoccurrence of landslides was estimated from very rapid (12 to 30 yr) in weathered ash and pumice (Shirasu deposits) to moderately rapid (200 to 1000 yr) in granodiorite lithology [*Shimokawa*, 1984]. The

---

[1] *Reneau et al.*, 1990; *Heimsath et al.*, 1997.

[2] *Shimokawa et al.*, 1989; *Miller et al.*, 1993; *DeRose*, 1996; *Heimsath et al.*, 2001.

[3] *Shimokawa*, 1984; *Heimsath et al.*, 1997; *Sasaki et al.*, 2000.

extensive shattering of granitic rock in the Rokko Mountains of Japan (near Kobe), due to crustal movement associated with active regional tectonics, has contributed to high rates of bedrock weathering and soil formation [K. Okunishi, 2003, personal communication]. As such, reoccurrence of landslides is also believed to be rapid (>50 yr) in this mountainous area that receives high and very seasonal rainfall; early studies [Watanabe and Seno, 1968] indicated that rain events with recurrence intervals of ≈30 yr initiated landslides in the Rokko Mountains. Similarly, in weathered granitic mountains in Asuka, Japan (near Nara), reoccurrence of landslides in unstable hollows is estimated to be several hundred years on average, but may be as rapid as ≈80 yr [Sonoda and Okunishi, 1994]. Somewhat intermediate in timeframe to the hollow evacuation studies in western USA and Japan are studies on steep slopes of Taranaki Hills, North Island, New Zealand, which documented average turnover rates of 450 yr based on dendroecology [Blaschke, 1988]. However, on slopes <32°, residence times ranged from 10,000 to 14,000 yr based on tephra dating [Trustrum et al., 1989].

The wide disparity of landslide reoccurrence estimates in hollows, particularly between Japan and western USA, deserves some discussion. Studies in western USA were mostly (but not exclusively) conducted in drier sites compared to Japanese sites. Japan experiences strong seasonal patterns of Baiu and typhoon storms, which trigger most landslides. Additionally, the regoliths studied in Japan are generally very thin compared to the colluvial hollows in western North America. The combination of easily weathered regoliths, seasonally high rainfall, and either coarse textured (i.e., non-cohesive) or weak soils in the Japanese sites would partly explain the rapid reoccurrence of very shallow landslides in hollows [Shimokawa, 1984; Sonoda and Okunishi, 1994]. Many of the deeper igbrinite deposits in Kyushu (Shirasu) rapidly weather to considerable depths once exposed by landslides [Shimokawa et al., 1989; Yokota and Iwamatsu, 1999]. Thus, a newly weathered regolith of sufficient depth to fail again may develop in 15 to 50 yr following an earlier landslide [Shimokawa, 1984]. In the deeper colluvial hollows of western USA, there is no way to ascertain the trigger mechanism of landslides that occurred many millennia ago; in fact, these failures could be related to highly irregular processes, such as large earthquakes, rather than more typical causal mechanisms like episodic storms. Another possible disparity between the two regions that has not been previously explored relates to methods of dating landslide reoccurrence. The older dates obtained in hollows from western USA were almost exclusively derived from radiocarbon dating of charcoal deposits. Therefore a history of fire is prerequisite to establishing a date. It is plausible that many small, shallow and possibly more frequent failures have been overlooked in such studies because no charcoal could be found due to a lack of fire during the shorter historical record. Most all of the studies in Japan used more direct dating methods such as dendroecology, archived records, aerial photographs, and actual observations to establish recurrence intervals for hollow evacuation.

Aspect and Altitude

Neither slope aspect nor altitude alone exerts a dominant control on slope stability, although each may interact with other factors to increase landslide susceptibility. Slope aspect strongly affects hydrologic processes via evapotranspiration and thus affects weathering processes and vegetation and root development, especially in drier environ-ments. In the Northern Hemisphere, north-facing slopes maintain higher and less variable moisture compared to south-facing slopes, which are generally drier and experience more frequent periods of wetting and drying [e.g., *Churchill*, 1982]. The higher moisture levels in north-facing slopes support greater weathering and thus deeper soil mantles. Therefore, it would be expected that north-facing slopes should have higher rates of rainfall-initiated landslides compared to south-facing slopes (the opposite true in the Southern Hemisphere). While such aspect characteristics have been attributed to increased landslide rates in some areas [*Churchill*, 1982; *Gao*, 1993; *Hylland and Lowe*, 1997; *Lan et al.*, 2004], other areas experienced either opposite trends [*Crozier et al.*, 1980; *Marston et al.*, 1998; *Lineback Gritzner et al.*, 2001; *Dai and Lee*, 2002; *Gokceoglu et al.*, 2005] or no significant trends [*Pachauri and Pant*, 1992; *Ayalew and Yamagishi*, 2005]. The commonly observed excep-tions to this anticipated aspect influence on landsliding are related to effects of prior mass wasting, direction of frontal precipitation, local lithology and structure, direction of seis-mic propagation, proximity to faults, soil drainage properties, and general physiographic trends. Aspect is closely related to bedrock structure, especially metamorphic foliation; landslides are common on aspects oriented in the same direction as the foliation planes [*Lee et al.*, 2002; *Fernandes et al.*, 2004]. Landslides also typically originate perpendicular to the general physiographic trend of mountainous areas, regardless of aspect [*Gokceoglu et al.*, 2005]. Cold regions in the Northern Hemisphere (e.g., permafrost zones) may experi-ence greater mass movement activity on south-facing slopes related to cycles of freezing and thawing [*Niu et al.*, 2005]. Aspects that cause more frequent freezing–thawing and wetting–drying cycles promote higher rates of dry ravel.

Altitude or elevation is usually associated with landslides by virtue of other factors, such as slope gradient, lithology, weathering, precipitation, ground motion, soil thick-ness, and land use. While many studies have drawn strong statistical relationships between elevation and landslide occurrence [e.g., *Pachauri and Pant*, 1992; *Lineback Gritzner et al.*, 2001; *Dai and Lee*, 2002], elevation alone provides no plausible physi-cal explanation for landslide occurrence. High elevation sites, especially mountain ridges, are susceptible to collapse during earthquakes [e.g., *Fukuda and Ochiai*, 1993], but this is related to amplification of ground motion. Thus, neither altitude nor aspect is considered to be a good general indicator of slope stability, unless other factors are evaluated concurrently.

Relative Importance of Slope Gradient Versus Soil Thickness

Although associations have been made between landslide type and slope gradient and between initiation mechanisms and slope gradient, the concept of a threshold angle of slope steepness for landslide initiation nevertheless underestimates the importance of variations in soil thickness on natural slopes over a time period when slope gradient remains relatively constant. Soil thickness can change rapidly within a century in hollows, especially just after evacuation of soil by mass failure [*Dietrich and Dunne*, 1978; *Lehre*, 1982; *Shimokawa*, 1984; *Sidle*, 1987, 1988; *Iida*, 1993]. Observations of repeated landslides in certain areas [e.g., *Mark et al.*, 1964; *Haruyama*, 1974] indicate that for some slopes and soil properties there may exist a threshold of soil thickness, beyond which failure must occur, provided the slope gradient is greater than the angle of internal friction of the failure surface [e.g., *Okunishi and Iida*, 1981; *Tsukamoto and Minematsu*, 1987; *Iida*, 1993]. Thus, in hillslopes with rapidly weathering or developing soil mantles, soil thickness as well as rock and soil properties and slope steepness must be considered in landslide hazard analysis.

Landslide–Debris Flow Linkages

Complex linkages exist between shallow landslides on hillslopes and debris flow initiation in headwater channels. Unfortunately the mechanics and processes controlling these interrelated mass movements have typically been investigated separately. While the distinction between landslides and debris flows can be described by the higher fluidity and pore water content of the latter [*Pierson*, 1980a; *Iverson et al.*, 1997], the linkages between landslides and debris flows in steep, vegetated terrain encompass a more complex set of site characteristics, including terrain roughness, woody debris dams in channels, and channel tributary junctions [*Nakamura*, 1986; *Benda and Dunne*, 1997b; *Gomi et al.*, 2001; *Lancaster et al.*, 2003] that cannot be attributed to rheology alone. Such geomorphic and biological heterogeneities in headwater ecosystems complicate the magnitude–frequency relationships of in-channel debris flows in steep terrain.

When small landslides occur on steep hillslopes or within geomorphic hollows, they may (1) initially mobilize into a debris flow and proceed downslope to a channel [e.g., *Rapp and Nyberg*, 1981; *Benda*, 1990; *Wang et al.*, 2002a; 2003b]; (2) enter headwater channels and immediately trigger a debris flow [e.g., *Wasson*, 1978; *Costa*, 1991; *Palacios et al.*, 2003; *Sidle and Chigira*, 2004; *Chen*, 2006]; (3) transport sediment and organic debris into the channel heads and channels, but not immediately initiate a debris flow [e.g., *Jonasson*, 1988; *Bovis and Jakob*, 1999; *Benda and Dunne*, 1997b; *Gomi et al.*, 2002]; or (4) deposit debris on the hillslope that can potentially be transported to the headwater channel during future hydrogeomorphic events [e.g., *Wieczorek et al.*, 1983; *Fannin and Rollerson*, 1993; *Peart et al.*, 2005] (Figure 3.7). From a geomorphic perspective, the first case is the most simple and corresponds to the total liquefaction of the mass failure where the hillslope soil is essentially saturated [*Iverson et al.*, 1997]. The role of sliding surface liquefaction in triggering liquefied slope failures that travel long distances has been already been noted [*Sassa*, 1996]; while the rapid mobilization of these failures may be related to shearing and grain crushing along the failure plane in some deep-seated landslides [*Wang and Sassa*, 2002], the generation of excess pore pressure during progressive shear displacement appears to be more related to soil contraction within the shear zone as a consequence of particle rearrangement during shearing [*Wafid et al.*, 2004]. Assumptions of rapid mobilization are commonly employed in physically and empirically based landslide models for simplicity [e.g., *Ellen and Fleming*, 1987; *Wu and Sidle*, 1995]. Practically, it is sometimes difficult to distinguish between landslides that almost immediately mobilize into debris flows (type 1) and those that mobilize into debris flows upon reaching channels (type 2), although this latter type should display no evidence of debris flow deposits in and around the failure zone. For these simpler combination landslide-debris flows (types

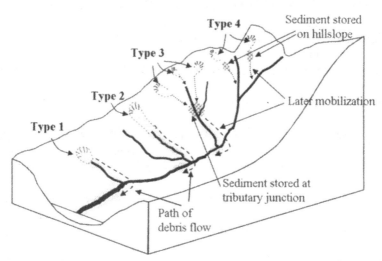

Figure 3.7. Relatively shallow landslides in steep terrain may (1) initially mobilize into a debris flow and proceed downslope to (and through) a channel; (2) enter headwater channels and immediately trigger a debris flow; (3) transport sediment and organic debris into the channel heads and channels but not immediately initiate a debris flow; or (4) deposit debris on the hillslope that can potentially be transported to headwater channels in future hydrogeomorphic events.

1 and 2), the geomorphic thresholds are related to the return interval of the triggering storm or snowmelt event [e.g., *Caine*, 1980; *Cannon and Ellen*, 1985; *Sidle et al.*, 1985; *Larson and Simon*, 1993] (Figure 3.8).

In contrast to type 1 and 2 debris flows that occur directly in response to landslides, many field observations attest to the importance of sporadic and chronic infilling of headwater channels and gullies in steep terrain as cumulative precursors for debris flows [*Dietrich et al.*, 1982; *Shimokawa*, 1984; *Benda*, 1990; *Miyabuchi*, 1993[1]]. It is apparent that critical levels of sediment and debris inputs are needed together with the occurrence of a large runoff event to trigger an in-channel debris flow [*VanDine*, 1985; *Benda and Dunne*, 1997b; *Palacios et al.*, 2003; *Imaizumi et al.*, 2005]. The complex nature of infilling may involve chronic processes such as surface wash, soil creep, freeze–thaw, bedrock weathering, and dry ravel (both sediment and organic debris), as well as episodic processes such as small debris slides, rockfall, blowdown of trees, earthflow inputs, and small slumps [*Okunishi and Iida*, 1981; *Shimokawa*, 1984; *Sidle*, 1988; *Trustrum and DeRose*, 1988[2]]. Because of the wide array of infilling processes, these debris flows in headwater channels (types 3 and 4) may be more frequent compared to hillslope landslides that mobilize directly into debris flows (types 1 and 2) (Figure 3.7). This more regular occurrence of in-channel debris flows is supported by studies in British Columbia [*Bovis and Jakob*, 1999], coastal Alaska [*Gomi et al.*, 2004], western Washington [*Buchanan et*

---

[1] *Benda and Dunne*, 1997b; *Bovis and Jakob*, 1999.
[2] *Reneau and Dietrich*, 1991; *Gomi et al.*, 2001; *Palacios et al.*, 2003.

Figure 3.8. A landslide that almost immediately mobilized into a debris flow near Juneau, Alaska: (a) photograph taken from near the initiation zone looking downslope; (b) beginning of the deposition zone of the debris flow.

al., 1990], Lapland [*Nyberg*, 1989], Japanese Alps [*Imaizumi et al.*, 2005], and the Buzău Subcarpathians of Romania [*Balteanu*, 1976], among other areas.

For cases where landslides do not immediately mobilize into debris flows, the timing of landslide initiation is separated, but not independent from, the timing of the debris flow initiation processes (Figure 3.9). Factors that contribute to landslide initiation on slopes (e.g., rain intensity, total precipitation, storm duration, antecedent soil moisture) may be more stochastic [e.g., *Iida*, 1993; *Sidle and Wu*, 1999; *Dhakal and Sidle*, 2004a] compared to the partly deterministic factors that control in-channel debris flows (e.g., sediment and woody debris accretion, channel storage sites, topographic and channel roughness) [e.g., *Benda and Cundy*, 1990; *Benda and Dunne*, 1997a; *Montgomery and Buffington*, 1997; *Lancaster et al.*, 2003]. Sediment deposited near channel heads by stochastically triggered hillslope landslides reaches a critical level of accumulation after which time the material fails as a debris flow during a large (but not uncommon) rainfall/flood event (Figure 3.9). Thus, it may be possible to predict such debris flows by a more deterministic approach. Similar phenomenon would likely apply to landslides and the resulting debris flows triggered by snowmelt or rain-on-snow events [e.g., *Megahan*, 1983; *Wieczorek et al.*, 1989; *Buchanan et al.*, 1990; *Toews*, 1991]. The role of small landslides that do not immediately enter channels (type 4) is poorly understood, but certainly contributes to the long-term sediment redistribution (including surface erosion) and recharge of these headwater systems [*Dietrich et al.*, 1982; *Gomi et al.*, 2002; *Peart et al.*, 2005] (Figure 3.7).

HYDROLOGIC FACTORS

Hillslope hydrology exerts a major control on landslide initiation. The most significant hydrologic processes include precipitation (spatial and temporal distribution of rain and snowmelt); water recharge into soils (and the potential for overland flow); lateral and

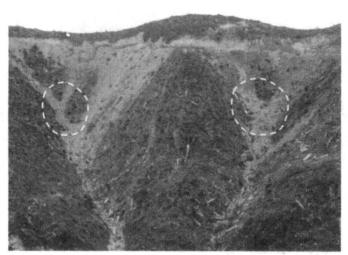

Figure 3.9. Example of a headwater channel that received sediment from prior landslides (in this case, road failures); after substantial accumulation of landslide material and woody debris near headwater junctions (see circled areas), debris flows occurred during a subsequent storm, Oregon Coast Ranges.

vertical movement of water within the regolith; and evapotranspiration and interception (discussed in Vegetation Influences section). The relative rates of these spatially and temporally distributed processes determine the transient level of groundwater in different portions of the hillslope and thus the potential for slope failure during rainstorms, snowmelt, or prolonged periods of water accretion. The rate of water movement into and through potentially unstable soil mantles is strongly controlled by physical properties of soil and bedrock, not only at the microscale as typically assessed, but also at the hillslope scale [e.g., *Anderson and Burt*, 1978; *Montgomery et al.*, 1997; *Sidle et al.*, 2000a; *Sulebak et al.*, 2000].

Precipitation

Spatial patterns of rainfall and snow accumulation (and melt) are closely associated with landslide initiation [e.g., *Campbell*, 1966, 1975; *So*, 1971; *Starkel*, 1976; *Li and Li*, 1985[1]]. Typically, higher mountain elevations experience larger volumes of precipitation, both rain and snowfall. In such areas, altitude may be used as an approximate surrogate for precipitation to help stratify landslide hazard because few remote regions have detailed distributed precipitation data; however, such approximations should be used cautiously. The spatial concentration of storm cells in dry regions may determine the occurrence of landslides [e.g., *Curry*, 1966; *Caine*, 1976; *Rapp and Nyberg*, 1981; *Wieczorek et al.*, 1983]; such distributions may need to be randomly specified for modeling purposes.

---

[1] *Keefer et al.*, 1987; *Wieczorek et al.*, 1989; *Buchanan et al.*, 1990; *Auer and Shakoor*, 1993; *Miyabuchi*, 1993; *Chigira*, 2001; *Hardenbicker and Grunert*, 2001.

Four rainfall attributes strongly affect landslide initiation: (1) total rainfall; (2) short-term intensity; (3) antecedent storm precipitation; and (4) storm duration. These attributes all influence the generation of pore water pressure in unstable hillslopes, including hollows [*Sidle and Swanston*, 1982; *Sidle*, 1984a; *Tsukamoto and Ohta*, 1988; *Iverson and Major*, 1987[1]]. Although many studies in mountainous regions have noted or implied a correlation of long-term precipitation with shallow landslide occurrence [e.g., *Endo*, 1969; *Eyles*, 1979; *Glade*, 1998; *Pasuto and Silvano*, 1998], other studies where more detailed precipitation data have been collected [e.g., *Sidle and Swanston*, 1982; *Keefer et al.*, 1987; *Larsen and Simon*, 1993; *Finlay et al.*, 1997; *Fuchu et al.*, 1999] have concluded that short-term rainfall intensity is a more important determinant. While detailed spatial and temporal precipitation inputs are most useful for predicting specific landslide location and timing, the more commonly available precipitation data (e.g., daily values) must suffice in many areas.

Recent modeling studies have permitted sophisticated testing of different rainfall scenarios on landslide probability at the catchment scale [*Baum et al.*, 2002; *Dhakal and Sidle*, 2004a]. Such simulations show that the number of landslides produced by different rainstorms and the relative importance of rainstorms in triggering landslides depend on the combined influence of mean and maximum hourly intensity, duration, and total rainfall amount (Figure 3.10). The temporal distribution of short-term intensity also affects landslide occurrence [*Dhakal and Sidle*, 2004b]. In all rainstorms, most failures occur after some threshold of cumulative rainfall and maximum hourly rainfall intensity. Such findings are consistent with field observations from around the world [*Campbell*, 1975; *Okuda et al.*, 1979; *Caine*, 1980; *Sidle and Swanston*, 1982[2]]. However, the influence of rainfall patterns on slope stability must be considered together with the hydrological characteristics and flowpaths of the soil and weathered bedrock.

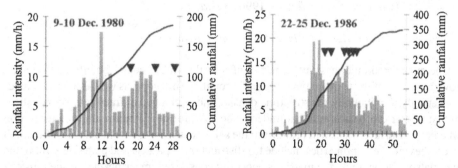

Figure 3.10. Influence of rainstorm characteristics on the timing and occurrence of landslides; simulations from Carnation Creek catchment, Vancouver Island, British Columbia. Solid line is cumulative rainfall; triangles denote the timing of simulated landslide occurrences in the catchment. Note that most landslides occurred after substantial rainfall and during or just following periods of relatively high intensity [after *Dhakal and Sidle*, 2003].

---

[1] *Larsen and Simon*, 1993; *Fernandes et al.*, 1994; *Montgomery et al.*, 1997; *Torres et al.*, 1998.
[2] *Schwab*, 1983; *Sidle et al.*, 1985; *Finlay et al.*, 1997; *Ekanayake and Phillips*, 1999a.

Short-term rainfall intensity can be included in stochastic models of landslide initiation [e.g., *Terlien*, 1998; *Sidle and Wu*, 1999; *Dhakal and Sidle*, 2004a], provided that sufficient precipitation data are available to either establish reliable estimates of return intervals or formulate simulation scenarios. *Keefer et al.* [1987] developed a real-time warning system for shallow landslides in the San Francisco Bay region of California based on a telemetered network of recording rain gages (see Chapter 5). Other similar real-time warning systems based on rainfall data (and antecedent moisture) have been proposed and tested in Hong Kong, New Zealand, Brazil, and Japan [*Hashino and Sasaki*, 1990; *Finlay et al.*, 1997; *Crozier*, 1999; *Fukuzono et al.*, 2003]. While such systems are very useful and may save lives and protect property in highly developed, densely populated urban settings, this level of data collection, transmission, and warning is not yet practical in most vulnerable regions of developing countries that need it the most. Additionally warning systems based strictly on rainfall cannot discriminate slope failure potential amongst different geologic materials or different land use–vegetation cover complexes. Advances in real time forecasting of spatially distributed rainfall that triggers landslides are possible with future developments in Doppler radar systems [e.g., *Klazura and Imy*, 1993; *Wei et al.*, 1998], although some problems exist related to orographic effects in mountainous terrain [*Gysi*, 1998]. As more of these radar networks come on line worldwide, the potential to use such data in landslide forecasts should increase.

For deep-seated mass movements, long-term precipitation records are more useful in estimating periods of activity as these failures or deformations generally initiate after a threshold of pore water pressure has developed within the regolith [*Wasson and Hall*, 1982; *Iverson and Major*, 1987; *Wasowski*, 1998; *Hardenbicker and Grunert*, 2001]. Thus, real-time and historical long-term precipitation data are useful to predict locations of such deep-seated movements as well as their overall probability of occurrence. Only a few studies have applied this approach to predict movement of deep-seated mass movements [*Collison and Anderson*, 1996; *Buma*, 2000].

Hydrological Properties of Soils and Weathered Bedrock

The most important physical properties of soils that affect slope stability are those that govern the rate of water movement into and through the hillslope, as well as the water holding capacity. Additionally, the structure, density, and orientation of fractures and interstices in bedrock or other substrate that underlie the soil profile are important in determining whether subsurface water will drain from the soil or enter the soil from below. At the small scale, these intrinsic properties include the distribution of particle and pore sizes within the soil matrix; for bedrock, the matrix is generally considered impermeable. At the hillslope scale where slope stability concerns arise, larger scale properties of the soil and regolith must be considered as well as these intrinsic properties. Many researchers have attempted to characterize hillslope hydrology with small-scale hydrologic attributes [e.g., *Freeze*, 1972; *Elsenbeer and Vertessy*, 2000; *Gautam et al.*, 2000; *Sherlock et al.*, 2000]. However, such assumptions may lead to discrepancies in spatially scaled hydrological behavior that necessitate heavy parameterization in models to achieve realistic predictions [e.g., *Stephenson and Freeze*, 1974; *Vertessy and Elsenbeer*, 1999; *Whitaker et al.*, 2003].

Of the larger-scale attributes that influence water intake and movement through the hillslope, the preferential flow network within the regolith is the most important and the most difficult to quantify [*Harp et al.*, 1990; *Luxmoore and Ferrand*, 1993; *Sidle et al.*, 2001; *van der Hoven et al.*, 2002]. Subsurface water may also enter the soil or weathered regolith from fractures or cracks in the bedrock [*Montgomery et al.*, 1997; *Noguchi et al.*, 1999; *Uchida et al.*, 2003], however, most studies of pore water pressure changes during storms or snowmelt have failed to monitor pressures within the bedrock. Of course, water must migrate into the fractured bedrock, but this is often facilitated in unstable slopes by the presence of tension cracks that act as conduits for surface water entry to bedrock [e.g., *Sidle et al.*, 1985; *Vandekerckhove et al.*, 2001; *Ocakoglu et al.*, 2002]. Thus, an upslope recharge zone appears necessary to develop sufficient pore pressures via "return" fracture flow to promote landslide initiation by this mechanism. Moderately deep (5–8 m) landslides in fractured bedrock provide field evidence of the importance of such preferential flow as a trigger mechanism for such large, rapid mass movements [*Sidle and Chigira*, 2004].

At the micro-scale, the rate of water movement in hillslope soils is best characterized by hydraulic conductivity (K), the subsurface flux of water per unit hydraulic gradient. The value of K varies non-linearly with volumetric moisture content for a wide range of soil textures, from near zero for dry conditions to a maximum at saturation (K_{sat}). Clayey soils and compact silty soils with very small interstitial pores have much lower values of K than coarser textured soils. While many studies focus on the role of dynamic pore pressures during rainstorms on landslide initiation, other investigators [*Haneberg*, 1991b; *Brooks and Anderson*, 1995; *Torres et al.*, 1998[1]] have noted the need to evaluate the effect of decreased soil suction above the wetting front on soil strength. *Torres et al.* [1998] showed that the dynamic response in the unsaturated zone contributes strongly to the timing and magnitude of pore pressure response in soils. *Iverson* [2000] describes a reduced form of *Richards'* [1931] equation for saturated–unsaturated flow in soils, with the z-direction of the three-dimensional coordinate system oriented normal to the slope, to assess the effects of rainfall infiltration on the vertical distribution of pore pressure development in soils. To characterize the permeability of soils in such wet, but not saturated conditions, it is necessary to assess unsaturated K values or the hydraulic diffusivity. Accordingly, the accurate specification of these unsaturated parameters may be important for slope stability but poses difficulties related to availability of such spatially variable data.

Values for K are typically measured in small core samples or via small-scale in situ tests such as the auger hole approach. While the latter of these two techniques is more representative of field conditions, both methods largely capture the flow properties in the soil matrix and emphasize unidirectional flow. Lumped values of K_{sat} can be derived from field studies where pits are excavated at the base of a hillslopes [e.g., *Whipkey and Kirkby*, 1978; *Tsuboyama et al.*, 1994b]. Such values may more accurately reflect the importance of preferential flow networks, but recent evidence has shown that the preferential flow may be spatially and temporally non-linear, thus complicating the use of such lumped parameters [*Harp et al.*, 1990; *Sidle et al.*, 1995, 2000a; *Woods and Rowe*, 1996[2]].

---

[1] *Iverson*, 2000; *Sasaki et al.*, 2000.

[2] *Noguchi et al.*, 2001; *Uchida et al.*, 2001.

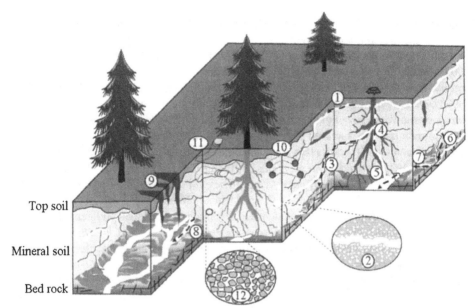

Figure 3.11. Schematic of three-dimensional connectivity of different types of preferential flow pathways in forest soils. Pathways include (1) preferential flow occurring in the organic horizon above the mineral soil; (2) macropores interacting with surrounding mesopores to enlarge these preferential flow paths during wet conditions; (3) connection of individual macropores via physical interaction; (4) preferential flow through buried organic matter and decayed roots; (5) flow along a lithic contact; (6) preferential flow into and through bedrock fractures or joints; (7) exfiltration of water from shallow bedrock fractures or joints; (8) flow over microchannels on the bedrock or other substrate (i.e., substrate topographic control); (9) flow through soil cracks; (10) flow in worm holes; (11) flow in animal (including insect) burrows or cavities; and (12) preferential flow through inter-aggregate spaces.

Preferential flow networks in regoliths are comprised of complicated combinations of vertical and slope-oriented pathways and features (Figure 3.11). Tension cracks that exist in and around potential failure sites are large, primarily vertical, preferential flow paths [*Sidle et al.*, 1985; *Chigira*, 2001; *Vandekerckhove et al.*, 2001; *Ocakoglu et al.*, 2002] (Figure 3.12). Soils high in clay content (particularly swelling clays) develop desiccation cracks that promote vertical transport of water [*Grisak and Cherry*, 1975; *Rooyani*, 1985; *Hallett et al.*, 1995; *Simoni et al.*, 2004]. Additionally, freeze–thaw action promotes vertical cracking in soils. In deeper tills, glaciotectonic processes can impart fracturing that is oriented in the direction of glacial recession [*Hicock and Dreimanis*, 1985; *McKay and Fredericia*, 1995; *Sidle et al.*, 1998]. Fracture flow also occurs in weathered, jointed, sheared, and foliated bedrock near the lithic contact of the soil [*Montgomery et al.*, 1997; *Noguchi et al.*, 1999; *Uchida et al.*, 2001; *Onda et al.*, 2004] (Figure 3.11). Preferential flow in these fractures may exchange into and out of the soil at the lithic contact, thus exerting positive pore pressures in areas of return flow [*Terajima and Moroto*, 1990; *Montgomery et al.*, 1997; *Noguchi et al.*, 1999]. In addition to shallow flowpaths, deeper water in bedrock fractures can exfiltrate into the failure zone; the area where the fractures intersect the failure plane as well as the hydraulic head imposed by the fracture flow are important factors affecting landslide initiation [*Montgomery et al.*, 1997; *Sidle and Chigira*, 2004]. Interaggregate spaces within the soil fabric from yet another type of preferential flow path, which can be oriented in various directions [*Mori et al.*, 1999]. Vegetation and fauna strongly contribute to preferential flow networks via live and decayed root channels [*Tsukamoto and Ohta*, 1988; *Kitahara*, 1989; *Noguchi et al.*, 1999], buried organic material [*Sidle et al.*, 2001], zones of loose soil conditioned by insects or other fauna [*Green and Askew*, 1965; *Tsukamoto and Ohta*, 1988; *Onda and Itakura*, 1997], earthworm passageways [*Edwards et al.*, 1988; *Sidle et al.*, 1998; *Botschek et al.*, 2002], and burrowing animals [*Aubertin*, 1971; *Wilson and Smart*, 1984]. These biological components of preferential flow networks all facilitate slope-parallel movement of water and, in some cases, vertical transport (Figure 3.11).

Large macropores (often called "soil pipes"), typically oriented downslope, can be created by subsurface erosion [e.g., *Terajima and Sakura*, 1993; *Onda*, 1994]. Natural large-scale interconnected pipe systems have been documented in only a few instances, largely in peat soils in the United Kingdom [*Jones*, 1981; *Smart and Wilson*, 1984; *Chappell et al.*, 1990], in deep sequences of silty material [*Billard et al.*, 1993; *Botschek et al.*, 2002], and in certain dry regions [e.g., *Fletcher et al.*, 1954]. While several forest hydrology studies have reported flow from large pipes [e.g., *Roberge and Plamondon*, 1987; *Kitahara et al.*, 1988; 1994; *Tsukamoto and Ohta*, 1988[1]], it is not clear whether these pipes are continuous over long slope distances. Even in sites where large soil pipes have either emerged from the ground or have been exposed by excavation at soil pits, there is almost no documentation of cases where individual pipes continued for distances greater than a few meters upslope [*Tsukamoto and Ohta*, 1988; *Kitahara et al.*, 1994; *Carey and Woo*, 2000; *Terajima et*

Figure 3.12. Tensions cracks around the head of a potential landslide site, Oregon Cascade Range.

[1] *Terajima and Sakura*, 1993; *Quinton and Marsh*, 1999; *Uchida et al.*, 1999; *Terajima et al.*, 2000.

al., 2000]. Thus, the commonly inferred notion [e.g., *McDonnell*, 1990; *Tani*, 1997] that long, continuous soil pipes drain hillslopes when uniform perched water tables develop appears inconsistent with findings from recent field studies where actual pathways have been measured [*Noguchi et al.*, 1999; *Carey and Woo*, 2000; *Sidle et al.*, 2001]. Such simplistic assumptions may grossly misrepresent predictions of pore pressure development in hillslope soils. Pipes can develop in unstable soils when high hydraulic pressures occur just prior to failure [e.g., *Harp et al.*, 1990]. Such spatial and temporal hydraulic responses during rainstorms pose serious problems for applying hydrologic models that assume isotropic, homogeneous, or constant permeability.

The water-holding capacity of the soil influences antecedent soil moisture and thus the amount of rainfall or snowmelt required to recharge the soil mantle. Finer textured soils hold more water in unsaturated conditions compared to coarse-textured soils. On steep slopes where debris avalanches and slides are the dominant mass movement types, the storage capacity of the soil may recharge quickly because of the shallow depth. Deeper mass movements in clay-rich regoliths (e.g., slump–earthflows) with higher water-holding capacities respond to long-term accumulations of water, giving rise to progressive movement in certain cases and to more rapid failure in extreme cases [*Oyagi*, 1977; *Bisci et al.*, 1996; *Hardenbicker and Grunert*, 2001]. The permeability of the confining layer underlying these unstable landforms regulates long-term drainage and thus controls moisture content in the unstable soil mantle [*Sidle et al.*, 1985]. When a permeable layer is confined within a clayey matrix, pore pressure can accumulate, leading to slope failure [*Hardenbicker and Grunert*, 2001]. Additionally, high porosity, moderately deep soils on very steep slopes may become unstable after extended periods of rainfall even if positive pore pressures do not develop.

To summarize, while the physical properties of the soil matrix are important as related to water movement and pore pressure accretion in unstable hillslope soils, it is often the larger-scale physical attributes of the soil catena and regolith that ultimately dictate where and when landslides will occur. Estimating the role of preferential flow systems over appropriate scales is indeed problematic; however, *Sidle et al.* [2000a, 2001] present a conceptual model for such an approach based on stochastic and complex behavior of such systems. Although water-holding capacity of the soil is an important consideration in slope stability, especially in deep-seated mass movements with clay-rich soils, the interaction of the soil matrix and preferential flow systems cannot be ignored. It is also very important to consider not only the accretion of pore pressure due to infiltrating rain water, but also the more rapid response sometimes observed in fractured bedrock (i.e., from below the soil). Such rapid inputs of subsurface water can lead to devastating landslides that may mobilize over long distances [*Montgomery et al.*, 1997; *Sidle and Chigira*, 2004].

Infiltration

The rate of water entry into the soil is highly influenced by soil physical properties (e.g., porosity, hydraulic conductivity, pore size distribution, preferential flow networks), vegetation cover, cultural practices, freezing phenomena, and macro- and micro-topography. The term infiltration rate refers to the actual flux of water into the soil and is dependent upon these physical, biological, topographic and cultural factors as well as the water delivery rate (i.e., rainfall intensity or snowmelt rate). Infiltration capacity refers to the maximum or potential water flux into the soil at any given time (infiltration capacity is always greater than or equal to infiltration rate). Many quantitative descriptions of the infiltration process exist; however, one of the original empirical descriptions remains widely used due to its simplicity and ease of application [*Horton*, 1933]:

$$f = f_c + (f_0 - f_c) e^{-kt}$$ (Equation 3.3)

where f is the infiltration capacity at some time t, f_c is the infiltration capacity at $t \rightarrow \infty$, f_0 is the infiltration capacity at $t = 0$, k is an empirical constant, and t is the time from the beginning of the rainstorm. The infiltration capacity thus decreases exponentially from the onset of the storm due to wetting, and the value of k is a lumped parameter for a particular soil that attempts to capture the combined influences of all the physical and site properties, including swelling, pore blocking, and changes in macropores as the soil becomes wetter. It is difficult to simulate the effects of initial soil crusting or hydrophobicity by equation 3.3 as these influences typically depress the initial infiltration capacity. Values of f_c and f_0 also depend on soil and vegetation characteristics of the site.

The infiltration models developed by *Green and Ampt* [1911] and later extended by *Philip* [1957] employ analytical solutions to the Richards' equation for one-dimensional vertical flow. The Green–Ampt infiltration model neglects the depth of surface ponding and assumes soils are homogeneous and infinitely deep. The commonly applied Philip's equation is based on only two parameters:

$$f = A_p + 0.5 \, S' \, t^{0.5}$$ (Equation 3.4)

where S' is sorptivity (a parameter relating to the penetration of the wetting front) and A_p is a constant related to K_{sat}. For early stages of the infiltration process, A_p is often assumed to be $1/3 \, K_{sat}$.

None of the infiltration models described herein explicitly account for effects of preferential flow pathways that give rise to spatial variability in infiltration capacity over rather small distances [e.g., *Jones*, 1990; *Buttle and McDonald*, 2000]. Given that such heterogeneity typifies unstable terrain, these models may have limited applicability in many cases for landslide prediction. Furthermore, the Green–Ampt and Philip equations fail to describe infiltration behavior at long times, while the Horton equation does not work well for short times [*Collis-George*, 1977].

In most natural and managed forests, the infiltration capacity and hydraulic conductivity of surface soils are relatively high due to continual inputs of organic matter from litter fall and detritus to the soil surface, together with decaying root and wood biomass in the subsurface. In temperate forest soils, highly permeable organic horizons are typically quite thick [e.g., *Bormann and Sidle*, 1990; *Noguchi et al.*, 1999]; however, in tropical and subtropical forest soils, because of the high decomposition rates, organic horizons are relatively thin [e.g., *Brown et al.*, 1994; *Hairiah*, 1999; *Sidle et al.*, 2006]. Thus, the protection of underlying mineral soils from raindrop impact during intense tropical storms depends on the delicate balance of maintaining this thin organic horizon.

The infiltration process is affected by the presence of a shallow water table in the hillslope. When the water table rises to the surface, saturated (non-Hortonian) overland flow occurs [e.g., *Dunne and Black*, 1970]. Basically if the recharge rate of water into the soil exceeds the rate of transmission within the mantle, a water table will develop and expand. Such perched water tables typically occur above a confining layer [*Whipkey*, 1965; *Palkovics and Petersen*, 1977; *Sidle*, 1984a; *Tsukamoto and Ohta*, 1988], but in some cases saturated zones attributed to preferential flow networks were noted above and disconnected from the lithic contact [*Sidle et al.*, 1995, 2000a; *Noguchi et al.*, 1999, 2001].

While important to hillslope hydrology, it is difficult to directly relate infiltration characteristics to slope stability. Theoretically, reducing recharge into the soil mantle should stabilize hillslopes by reducing pore pressures that develop during storms. While at the micro-scale this may be true, much unstable terrain is replete with tension cracks, especially around potential landslide initiation zones [*Rozier and Reeves*, 1979; *Ochiai et al.*, 1985; *Sidle et al.*, 1985; *Julian and Anthony*, 1996], thus providing preferential pathways for any overland flow that that does not infiltrate into the soil matrix (Figure 3.12). This rapid recharge route is especially significant if the infiltration capacity of the soil around the cracks is low, promoting flow into and through the cracks. Such water may be rapidly redirected to the failure planes of deep-seated or even shallow landslides, thereby actually increasing the probability of slope failure. Rapidly infiltrating water can also enter fractures in the deeper bedrock and propagate pore water pressures to downslope locations via interconnected fracture networks. Additionally, tension cracks can interact with the complicated existing network of preferential flow paths, enhancing pore water accretion in localized areas. Indeed, numerous examples have been cited where large soil pipes are present in failure headwalls, and linkages between preferential flow and landslide initiation have been assumed [*Blong and Dunkerley*, 1976; *Tsukamoto et al.*, 1982; *Pierson*, 1983; *Brand et al.*, 1986[1]]. In areas where the infiltration capacity of the soil is rarely exceeded by rainfall intensity, tension cracks provide "short-circuits" for intercepting rainwater and subsurface flow and diverting them into deeper horizons near the failure plane. Thus, maintaining a viable vegetative cover that promotes infiltration will generally benefit slope stability as peak inputs of water into the soil mantle will likely be dispersed.

Subsurface Flow Processes

Because subsurface flow processes control the movement of infiltrated water through hillslopes, they influence both the temporal and spatial characteristics of pore water pressure distribution. Numerous researchers have attempted to simulate subsurface flow at the hillslope or small catchment scale using either finite-element or infiltration-based models that rely on estimates of soil physical properties, particularly hydraulic conductivity or diffusivity. Such models are based on *Richards'* [1931] equation and Darcy's

---

[1] *Fannin et al.*, 2000; *Sidle et al.*, 2000b; *Uchida et al.*, 2001.

law for unsaturated and saturated flow, respectively. Darcy's law was originally applied to flow in saturated porous media, and for one dimension can be expressed as

$$v = K_{sat} (\partial \hat{H}/\partial l)$$ (Equation 3.5)

where v is the macroscopic or so-called Darcy velocity and $\partial \hat{H}/\partial l$ is the change in total hydraulic head per unit distance of travel (i.e., the hydraulic gradient). For saturated subsurface flow in most soils, the total head ($\hat{H}$) is the sum of the pressure head (h_p, always positive for saturated flow) and the elevation head (h_z). In steep terrain where shallow soils overlie a relatively low permeability substrate, the difference in elevation head in the downslope direction largely dictates the Darcy velocity or flux.

   Richards [1931] extended the use of equation 3.5 to unsaturated soils by allowing the hydraulic conductivity to vary with water content. In this case, pressure head (h_p) is negative for unsaturated flow and positive for saturated conditions. Then Richards combined Darcy's Law for saturated flow with the equation of continuity to derive a partial differential equation describing unsaturated flow in porous media. While the nonlinear form of Richards' equation is difficult to solve for layered (i.e., with different values of K) hillslope soils [*Freeze*, 1978], various analytical and numerical approximations have been applied to hillslope or catchment-scale subsurface flow problems [e.g., *Stephenson and Freeze*, 1974; *Nieber and Walter*, 1981; *Watanabe*, 1988] as well as redistribution of infiltrated rainfall [e.g., *Buchanan et al.*, 1990; *Sammori*, 1995; *Iverson*, 2000] on sloping ground. Most investigators [e.g., *Freeze*, 1974; *Zehe et al.*, 2001] have solved Richards' equation in two dimensions for computational efficiency:

$$\frac{\partial}{\partial x}\left[K(h_p)\frac{\partial h_p}{\partial x}\right] + \frac{\partial}{\partial z}\left[K(h_p)\left(\frac{\partial h_p}{\partial z}+1\right)\right] = C\left(h_p\right)\frac{\partial h_p}{\partial t}$$ (Equation 3.6)

where $K(h_p)$ is the hydraulic conductivity as a function of pressure head, C is the slope of the water retention curve, and z and x represent the vertical and lateral coordinates, respectively. *Sammori and Tsuboyama* [1990, 1992] developed a two-dimensional seepage model based on Richards' equation to estimate pore pressure and for different slope shapes and conditions. Later, *Sammori* [1995] modified this model to a quasi-three-dimensional model for more complex slope configurations. *Iverson* [2000] referenced the coordinate system in Richards' equation to a two-dimensional hillslope to assess transient pore water pressure response during storms. In such a longitudinal hillslope slice, only the vertical redistribution of infiltrated rainfall can be examined along with the downslope movement of subsurface flow along the slice; topographic complexity (both surface and subsurface) is not represented. None of the models described here simulate the effects of preferential flow pathways on pore pressure accretion. Preferential flow, both within the soil [e.g., *Tsukamoto et al.*, 1982; *Sidle et al.*, 2000a, 2001; *Uchida et al.*, 2001] and from underlying bedrock [e.g., *Montgomery et al.*, 1997; *Uchida et al.*, 2003; *Sidle and Chigira*, 2004] may exert a major control on pore pressure development in steep hillslopes, and thus affect landslide initiation.

Changes in soil physical properties within natural soil profiles influence both the vertical infiltration and the location of saturated subsurface flow in hillslope soils. In many field soils, K_{sat} declines with depth related to changes in bulk density, pore-size distribution, and preferential flow pathways [e.g., *Harr*, 1977; *Ahuja and El-Swaify*, 1979; *Tsukamoto and Ohta*, 1988]. Although the assumption of an exponential decrease in K_{sat} with depth has been incorporated into hydrologic models [e.g., *Beven*, 1984; *Vertessy and Elsenbeer*, 1999], such phenomena are by no means ubiquitous [e.g., *DeBoer*, 1979; *Gautam et al.*, 2000; *Basile et al.*, 2003; *Vieira and Fernandes*, 2004]. A hypothetical example of a hillslope forest soil is shown in Figure 3.13, with values of K_{sat} given for respective horizons. Lateral subsurface flow is typically generated when infiltrating or converging water encounters a zone of significantly lower K_{sat}. However, rates of input must exceed the rate of transmission of a particular subsurface layer for lateral flow to occur. Thus, for this simple two-dimensional example, if prolonged rainfall intensity exceeds 50 mm h^{-1}, then lateral flow may be expected at base of the B horizon; for more typical, lower rainfall intensities, lateral flow would occur at the base of the C horizon just above bedrock (Figure 3.13). For soils with lower K_{sat} values in the upper mineral horizons (e.g., ferric or podzolized layers) or where freezing occurs with depth, lateral flow has been observed primarily in the organic-rich horizon [e.g., *Quinton and Marsh*, 1999]. The extent of groundwater accretion above a layer of low K_{sat} depends on antecedent soil moisture, characteristics of the rain storm, and topographic influences, as well as the rate of lateral water movement in the more permeable soil layer. The failure plane for landslides will typically correspond to this hydrologic discontinuity when it is significant (i.e., more than an order of magnitude difference in K_{sat}) (Figure 3.13). Nonetheless, recent studies have demonstrated the influence of transient pore water pressures within deeper soils on the location of failure planes [*Iverson and Major*, 1987; *Haneberg*, 1991a; *Reid*, 1994; *Iverson*, 2000], while other studies have shown that saturated flow occurs in preferential flow pathways above and detached from perched water tables [*Tsuboyama et al.*, 1994b; *Sidle et al.*, 1995; 2000a; *Noguchi et al.*, 1999]. Thus, although the commonly accepted model of saturated–unsaturated flow and subsequent pore water changes may adequately explain the initiation of many landslides, other factors related to temporal and spatial pore water pressure development need to be considered in some cases.

Similar to vertical infiltration, slope-parallel movement of water is strongly affected by networks of preferential flow [e.g., *Mosley*, 1979; *Beven and Germann*, 1982; *Tsukamoto and Ohta*, 1988[1]]. As noted earlier, spatial and temporal variability in subsurface flow may be strongly linked to three-dimensional preferential flow networks at the hillslope scale [*Sidle et al.*, 1995, 2000a; *Woods and Rowe*, 1996; *Carey and Woo*, 2000[2]] (Figure 3.11); these networks in turn influence the spatial and temporal accretion of pore water pressure that is typically responsible for landslide initiation.

---

[1] *Kitahara et al.*, 1994; *Tsuboyama et al.*, 1994b; *Quinton and Marsh*, 1999; *Terajima et al.*, 2000; *Uchida et al.*, 2003.

[2] *Noguchi et al.*, 2001; *Uchida et al.*, 2001.

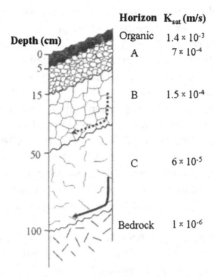

Figure 3.13. A typical forest soil profile on a hillslope showing decreasing hydraulic conductivity with depth and zones where lateral subsurface flow would be expected to occur (without the effects of preferential flow paths). Note that lateral flow would typically be generated at the C horizon–bedrock interface; only for storms with sustained rainfall in excess of 50 mm h^{-1} would any lateral flow occur at the B–C horizon boundary.

Hillslope or catchment-scale hydrology models based on Richards' equation or topographic approaches that attempt to incorporate effects of preferential flow systems by adjusting K_{sat} for the entire volume or a specified subvolume of soil [e.g., *Stephensen and Freeze*, 1974; *Yates et al.*, 1985; *Beven*, 1987] cannot capture the spatial and temporal dimensions or probabilities of pore water pressure response at such scales. In fact, such simplifications of preferential flow can lead to incorrect conclusions related to the dynamics of pore water pressure and thus landslide initiation. Unless such preferential flow pathways are explicitly or statistically generated within the hillslope or landscape that is being modeled [*Sidle et al.*, 2001], there is little chance to depict both the spatial and temporal response of pore water pressure that triggers landslides.

Some scientific debate has ensured as to whether saturated conditions are necessary for the occurrence of preferential flow. Such debates have relevance to whether or not laminar flow may occur in macropore systems. Many studies assumed that saturated conditions are necessary for preferential flow to occur and that turbulent flow in macropores and soil pipes is the norm [e.g., *Jones*, 1971; *Tsukamoto et al.*, 1982; *Smart and Wilson*, 1984; *Carey and Woo*, 2000]. However, it has been noted that preferential flow can occur within such networks as long as the point of entry and the perimeter of macropore systems are saturated [*Sidle et al.*, 1985, 2000a; *Uchida et al.*, 2001]. Recent theoretical research has shown that momentum dissipation may induce laminar preferential flow in the vertical direction in soils by not only the direct infiltration into macropores, but also the tearing off of water films from the bulk of the soil moisture reservoir [*Germann and Di Pietro*, 1999]. This phenomenon may relate to the high level of variability in vertical and slope-parallel preferential flow flux observed in both saturated and unsaturated soils [e.g., *Sidle et al.*, 1995, 2000a; *Woods and Rowe*, 1996; *Hollenbeck and Jensen*, 1998]. In large macropores (soil pipes), flow may be turbulent during major storm events and may not therefore conform to assumptions inherent in Darcy's Law or the Richards' equation. However, as previously mentioned, the spatial extent of these large pipe systems seems rather limited. In many cases, it appears that a rather diffuse, but spatially interconnected, network of many types of preferential flow pathways facilitate preferential flow fluxes within the hillslope [*Sidle et al.*, 2001] (Figure 3.11). Inputs of water from anthropogenic practices (e.g., irrigation, drainage diversions) can exacerbate this phenomenon. Such preferential flow networks may not connect over long slope distances, but instead become "dead end" or truncated, thus providing opportunities for pore water pressure accretion that can lead to landslide initiation [*Ohta et al.*, 1981; *Uchida et al.*, 1996]. Thus, the amount or flux of preferential flow within a hillslope is not necessarily a good indicator of the potential for pore water pressure accretion related to these pathways. Sites that are well-drained by preferential flow networks [e.g., *Tsukamoto et al.*, 1982; *Uchida et al.*, 2001] may not develop zones of concentrated pore water pressure during storms unless convergent flow exists. Convergent flow that overwhelms the transmission capacity of preferential flow paths likely cause the buildup of pockets of high pore pressure, thus leading to landslide initiation [*Tsuboyama et al.*, 2000; *Uchida et al.*, 2003].

Studies of preferential flow have included direct hydrometric measurements (typically at soil pits) [e.g., *Turton et al.*, 1992; *Sidle et al.*, 1995], stable isotope tracers [e.g., *Pearce et al.*, 1986; *Laudon and Slaymaker*, 1997], natural chemical tracers [e.g., *Elsenbeer et al.*, 1995; *Burns et al.*, 1998], applied chemical tracers [e.g., *Tsuboyama et al.*, 1994b; *Nyberg et al.*, 1999], staining tests [e.g., *Noguchi et al.*, 1999; *Zehe and Flühler*, 2001], and remote methods (soft x-rays, ground penetrating radar) [e.g., *Mori et al.*, 1999; *Holden et al.*, 2002], and even real-time mapping by listening to pipeflow (in large soil pipe systems) [e.g., *Atkinson*, 1978]. While each of these methods has certain advantages and disadvantages, often some combination of different methodologies yields the most accurate depiction of preferential flow behavior. To better understand how preferential flow (both vertical and slope-parallel) influences slope stability, it is necessary to evaluate or spatially simulate a likely array of pathways rather than simply preferential fluxes.

Pore Water Pressure Generation and Slope Stability

All of the hydrological factors discussed (precipitation, soil physical properties, infiltration, subsurface flow processes) together with other site attributes affect the generation of pore water pressure within hillslope soil mantles. Such pore pressures, generally depicted by the transient development of a perched water table within the regolith, are largely responsible for the initiation or acceleration of landslides. However, as noted earlier, some landslides can be triggered by the development of pore pressures within fractured bedrock underlying the more weathered regolith; for this failure mechanism, few data on pore pressure response exist.

In the case of slope stability affected by a perched water table at the base of the soil profile (or other hydrologic discontinuity), a number of field studies have monitored pore water pressure by nests of piezometers and tensiometers positioned above the hydrologic discontinuity [*Swanston*, 1967; *Gray and Megahan*, 1981; *Sidle and Swanston*, 1982; *Keefer and Johnson*, 1983[1]]. Of these investigations, only the study in coastal Alaska by *Sidle and Swanston* [1982] captured a natural, albeit small, landslide event, during which maximum piezometric response at the headwall was monitored. Additionally, the only field experiments that successfully measured dynamic pore pressures during slope failure were those of *Harp et al.* [1990] in Utah and California and *Ochiai et al.* [2004] in Japan, during which artificial rainfall or inputs of water were applied. A number of field studies have monitored pore pressures in deeper, slow-moving regoliths [e.g., *Oyagi*, 1977; *Ochiai et al.*, 1985; *Iverson and Major*, 1987; *Coe et al.*, 2003]. Only a few field investigations have attempted to measure pore pressures below the potential failure plane to assess the potential upwelling of shallow groundwater from underlying fractured bedrock [*Montgomery et al.*, 1997; *van Asch and Buma*, 1997]. In general, most flume experiments have ignored the notion of an upwelling of pore pressure and only dealt with the simple scenario of pore pressures generated by infiltrating rainwater [e.g., *Iverson and LaHusen*, 1989; *Reid et al.*, 1997; *Okura et al.*, 2002].

Geomorphic hollows are especially susceptible to the development of perched water tables due to convergent subsurface flow [e.g., *Anderson and Burt*, 1978; *Pierson*, 1980b; *Tsukamoto et al.*, 1982; *Tsuboyama et al.*, 2000]. In addition to convergent flow within the soil matrix, the timing and extent of pore water pressure accretion in hollows during storms is a function of antecedent moisture, substrate leakiness, extent and connectivity of preferential flow paths, and degree of topographic incision, as well as the timing and amount of rainfall [*Sidle*, 1984a, 1986; *Sidle et al.*, 2000a; *Tsuboyama et al.*, 2000]. Because of the history of mass wasting, root throw, and other bioturbations in these hollows, soils tend to be non-homogeneous and anisotropic, thus facilitating preferential flow. Rapid drainage of a hollow by preferential flow limits the development of a perched water table; however, if the transmission rates of preferential flow paths are less than water inputs, an increase in pore pressure will occur in the surrounding soil [e.g., *Sidle*, 1984a; *Uchida et al.*, 2001]. Several studies have shown artesian conditions occurring during major storms at various slope positions within geomorphic hollows [*Sidle*, 1984a; *Sidle and Tsuboyama*, 1992; *Tsuboyama et al.*, 2000].

Although numerous studies have assessed the relationship between rainfall characteristics and landslide initiation, only within the past few decades have detailed field studies been conducted to assess the dynamic response of pore water pressure within hillslope regoliths during storm events and snowmelt. Studies of piezometric response in the Idaho Batholith [*Megahan*, 1983] and in coastal British Columbia [*Kim et al.*, 2004] revealed that soils were nearly saturated during peak snowmelt. Likewise, investigations in various

---

[1] *Wieczorek*, 1987; *Harp et al.*, 1990; *Haneberg*, 1991a; *Fernandes et al.*, 1994; *Haneberg and Gökce*, 1994; *Onda*, 1994; *Montgomery et al.*, 1997; *van Asch and Buma*, 1997; *Torres et al.*, 1998; *Fannin et al.*, 2000; *Tsuboyama et al.*, 2000; *Kosugi et al.*, 2002; *Uchida et al.*, 2003; *Dhakal and Sidle*, 2004b.

parts of the world have shown that portions of hollows and concave hillslopes experience saturation during storms with return periods ranging from <1 to 4 yr [*O'Loughlin and Pearce*, 1976; *Pierson*, 1980b; *Sidle*, 1984a; *Tsuboyama et al.*, 2000]. Many of these storms triggered landslides within the respective areas. In most concave hillslopes, maximum piezometric head decreased in the upslope direction and temporal response was more dampened in the downslope direction. *Harr* [1977] observed completely saturated conditions and rapid piezometric response at the foot of slightly convex, unstable slopes during moderate winter storms (return periods <1.5 yr) in the Cascade Range of Oregon. Isolated areas of partial saturation at the surface soil-subsoil boundary were detected in midslope and upslope locations. Other studies have noted a high level of variability in dynamic piezometric response during water inputs to the soil [*Anderson and Burt*, 1977; *Tanaka*, 1982; *Sidle*, 1984a; *Harp et al.*, 1990[1]]. These findings refute the commonly held assumption of a wedge of groundwater accumulating from downslope to upslope during large inputs of water [*Freeze*, 1972; *Weyman*, 1973].

Examples of Groundwater Response

Few studies have monitored pore water pressures in the soil mantle at the time of slope failure, especially during natural rainfall events. Many steep hillslopes or hollows with shallow soils (<1 to 2 m deep) may require nearly saturated or even artesian conditions to induce slope failure [*O'Loughlin and Pearce*, 1976; *Sidle and Swanston*, 1982; *Sidle*, 1984a, 1986[2]]. In deeper regoliths prone to slow, progressive movement of the landslide mass, periods of accelerated movement typically correspond to the times when groundwater levels exceed some threshold [*Iverson and Major*, 1987; *Coe et al.*, 2003; *Simoni et al.*, 2004]. Where numerous piezometers have been installed within potential or actual landslide masses, interpretations of these data have been complicated by the temporal and spatial variability in pore pressure response [*Sidle*, 1984a; *Harp et al.*, 1990; *Johnson and Sitar*, 1990[3]]. In the less extensively monitored, deeper landslides [e.g., *Iverson and Major*, 1987], spatial variability of pore pressure response may be even greater than in shallow landslides. Nevertheless, useful generalizations related to groundwater response in both shallow and deep regoliths can be derived from the examples that follow.

Shallow soils. Numerous studies have monitored maximum pore water pressure in shallow, unstable soils. *O'Loughlin and Pearce* [1976] measured maximum pore pressures of 3 to 9 kPa in six hillslope hollows adjacent to failed hollows and concluded that pore pressures in excess of 10 kPa were needed to initiate failure. *Anderson and Kneale* [1980] observed pore water pressures ranging from 1.2 to 6.7 kPa in a site adjacent to a road embankment at the time of failure. In possibly the only study to capture the pore pressure during a natural landslide, *Sidle and Swanston* [1982] measured an increase in pore water pressure from 0.0 to 2.2 kPa (0 to 41.3 cm of piezometric rise) that triggered a small debris slide. *Pierson* [1980b]

[1] *Sidle and Tsuboyama*, 1992; *Montgomery et al.*, 1997; *Tsuboyama et al.*, 2000.

[2] *Harp et al.*, 1990; *Johnson and Sitar*, 1990; *Fernandes et al.*, 1994; *Tsuboyama et al.*, 2000.

[3] *Montgomery et al.*, 1997; *Dhakal and Sidle*, 2004b.

found that maximum piezometric response corresponded to rainfall inputs in geomorphic hollows and ranged from negligible to complete saturation of the soil mantle. A somewhat discontinuous time series of data from 15 piezometers in Carnation Creek, Vancouver Island, British Columbia, were analyzed over an 8-yr period; the ratio of maximum pressure head measured to the depth of soil (at a specific piezometer site) ranged from 0.50 to >1 (1 = total saturation; >1 = artesian) during the period [*Fannin et al.*, 2000; *Dhakal and Sidle*, 2004b]. Increasing topographic incision was found to increase maximum piezometric response from 11 to 37 cm in planar hillslopes to 63 to 92 cm in deeply incised hollows during a moderate storm (83 mm in 24 hr) in coastal Alaska [*Sidle*, 1986]. For five of the largest storms monitored during two autumn rainy seasons, at least one piezometer in the upper and lower portion of the most incised hollow indicated artesian conditions during maximum response [*Sidle*, 1984a]. For this coastal Alaska hollow, 99% of the variability in maximum pore pressure response could be explained by maximum 1-h rainfall intensity, 2-day antecedent rainfall and total storm precipitation [*Sidle*, 1992]. Similarly, *Pierson* [1980b] found that between 77% and 91% of the variability in maximum pore pressure within four hollows could be explained by 24-h rainfall alone; however, an improvement in prediction (R^2 = 0.92–0.93) was obtained when antecedent moisture was included in the regression model. While such relationships are useful for site-specific estimates of landslide assessment [e.g., *Sidle*, 1992], they must be extrapolated with great caution to other sites. All studies cited here indicate a strong influence of pore water pressure measured at the soil–bedrock interface on factor of safety for shallow landslides [e.g., *Pierson*, 1980b; *Sidle and Swanston*, 1982; *Harp et al.*, 1990; *Terlien*, 1997]. Another type of pore pressure propagation, derived from groundwater exfiltrating from underlying bedrock into the soil mantle, is also believed to be important in generating instabilities in shallow and deep regoliths [*Montgomery et al.*, 1997; *Uchida et al.*, 2003; *Sidle and Chigira*, 2004].

Recent research has documented dynamic responses of pore water pressure in unstable hillslopes during storms. In one of the few studies that monitored pore pressure dynamics during slope failures induced by irrigation, *Harp et al.* [1990] showed that pressures increased during the early stages of infiltration; however, just prior to failure, abrupt decreases in pressure were usually noted. These decreases may be due to the effects of removal of fine-grained material by piping as well as the dilatation of the landslide mass in the early stage of failure [*Harp et al.*, 1990]. At their Monroe Canyon, California, site *Harp et al.* [1990] applied water via a trench for 4 days (during daylight hours); on the fifth day, 55 min after irrigation commenced, a translational slide occurred at a depth of 1.3 m (total applied water ≈ 48,000 L). Most piezometers in the upper to middle portion of the failure zone responded within 25 min of the onset of irrigation; 5 min prior to failure, pore pressures dropped abruptly in most piezometers (Figure 3.14). The majority of the water discharging from the cut face was from fractures and macropores [*Harp et al.*, 1990].

Haneberg and Gökce [1994] measured piezometric fluctuations for several storms in a slower moving, yet shallow (≈2 m), landslide in southwestern Ohio. During five rather small storms in early spring, piezometric head increased by 23 to 54 cm compared to pre-storm conditions. A larger storm (40 mm precipitation) in mid-May, followed 6 h later by another smaller storm, produced cumulative responses in pore pressure peaks [*Haneberg*

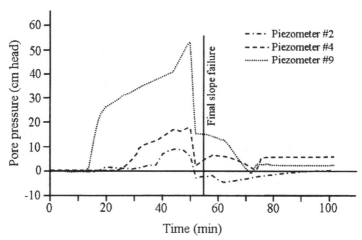

Figure 3.14. Piezometric response during a shallow slope failure in Monroe Canyon, California, induced after four discontinuous days of trench irrigation; failure occurred on the fifth day after only 55 min of irrigation (data shown in figure). Piezometer 2 is located within 0.5 m of the landslide scarp, piezometer 4 is about 1 m downslope, and piezometer 9 is between these two. All piezometers show a characteristic rise in pressure head that peaks about 5 min prior to failure; following this peak, pore pressures drop dramatically due to subsurface erosion (piping) and dilation of the landslide mass in the preliminary stages of failure [adapted from *Harp et al.*, 1990].

and Gökce, 1994]. The second peak was almost 20 cm higher (total increase above pre-storm level was ≈75 cm) compared to the first pore pressure peak. The lag time between peak rainfall intensity and peak pore pressure response ranged from 4 to 15 h.

Numerous studies examined pore pressure response in unstable soil mantles to varying rainfall inputs and antecedent moisture conditions. In the Briones Hills northeast of Berkeley, California, pore pressure responses in shallow (0.5–1.5 m) soils were compared for relatively dry and wet antecedent conditions [*Johnson and Sitar*, 1990]. During relatively dry conditions, the shallow tensiometers responded before the deeper instruments, indicating that the wetting front was advancing at a rate of about 3–4 cm h⁻¹. Lower on the hillslope, the deeper tensiometer responded earlier than the one installed at an intermediate depth, indicating the contribution of lateral flow (possibly preferential flow) [*Johnson and Sitar*, 1990]. For a series of storms preceded by wet antecedent conditions, the greatest increases in positive pore water pressure (≈50 cm) were observed during the two largest periods of rainfall near the middle to end of the sequence. These responses occurred in the tensiometer positioned at the soil-bedrock contact, and the advance of this pore water pulse propagated downslope at a rate of about 2–3 m h⁻¹.

In colluvial hollows in Brazil, *Fernandes et al.* [1994] found that pressure head along a hollow axis responded rapidly to storm inputs and was influenced by antecedent conditions; during a series of storms pore pressure increased to saturation after which no further pressure increases occurred. Such hydrologic response differs somewhat from those reported in other studies in hollows where excess pore water pressures can develop by means of exfiltration of groundwater from bedrock, "pressure-wave" propagation in the unsaturated zone, preferential flow pathways in the soil, convergent flow, or confining layers in the soil profile that may lead to artesian conditions [*Sidle*, 1984a; *Wilson and Dietrich*, 1987; *Montgomery et al.*, 1997; *Torres et al.*, 1998]. *Montgomery et al.* [1997] showed that local development of upward hydraulic head gradients from weathered bedrock into the overlying soil mantle in geomorphic hollows in the Oregon Coast Ranges was related to the patchy patterns of positive pore water pressure. These exfiltrating head gradients typically occurred in very wet soils during high intensity events and responded rapidly to peak rainfall inputs. *Torres et al.* [1998] attributed this rapid piezometric response to a "pressure-wave" type displacement of pre-existing pore water (nearly saturated) into the underlying saturated zone. This finding is similar to preferential flow passing through interconnected mesopores that facilitates hydrologic connectivity in the soil [e.g., *Luxmoore et al.*, 1990; *Wilson et al.*, 1990; *Luxmoore and Ferrand*, 1993; *Sidle et al.*, 2000a, 2001]. Other studies in various parts of the world have attributed rapid piezometric response in hollows more directly to the hydrologic connectivity induced by preferential flow networks, such as decayed root channels, animal burrows, subsurface erosion pipes, desiccation cracks, and other macropores [*Tsukamoto et al.*, 1982; *Sidle*, 1984a; *Brand et al.*, 1986; *Amen*, 1990[1]].

Detailed field investigations in coastal Alaska [*Sidle*, 1984a, 1986], British Columbia [*Fannin et al.*, 2000; *Dhakal and Sidle*, 2004b], northern California [*Wilson and Dietrich*, 1987; *Amen*, 1990], and Japan [*Sidle and Tsuboyama*, 1992; *Tsuboyama et al.*, 2000; *Uchida et al.*, 2003] also found variable pore water pressure response during storms in hollows and planar hillslopes, including the development of slightly artesian pressures under some conditions. A clear spatial pattern of maximum pressure head increase during storms was observed in response to increasing discharge at the outlet of a geomorphic hollow in eastern Japan [*Tsuboyama et al.*, 2000]. During periods of very low hollow discharge, pore pressures measured along the lower axis of the hollow increased, reaching a maximum level at modest discharges, similar to the saturated limit noted by *Fernandes et al.* [1994] in Brazil. However, during higher rainfall and wetter antecedent conditions pore pressures along the soil–bedrock contact in the upper portion of the hollow only developed after a relatively high discharge emerged and then increased abruptly [*Tsuboyama et al.*, 2000]. Similar spatial patterns of piezometric response have been noted in hollows of coastal Alaska [*Sidle* 1984a, 1986] and the Fudoji catchment in Japan [*Uchida et al.*, 2003]. At Carnation Creek, British Columbia, piezometric response generally corresponded to rainfall inputs for a large number of low to moderate intensity storms [*Fannin et al.*, 2000; *Dhakal and Sidle*, 2004b]. An example of the responses on steeper hillslopes is shown for a large (180 mm total rainfall), long duration (43 h), relatively low intensity (average and maximum rainfall intensities of 4.2 and 7.0 mm h^{-1}, respectively) event preceded by dry antecedent conditions (7-day antecedent rainfall = 5 mm) (Figure 3.15). Piezometric response did not commence until 6–7 h after the onset of light (2–3 mm h^{-1}) rainfall; higher intensities (6–7 mm h^{-1}) that occurred 10 h into the event sustained and enhanced pore pressures, but some declines were noted once intensity fell below about

---

[1] *McDonnell*, 1990; *Sidle and Tsuboyama*, 1992; *Terajima and Sakura*, 1993; *Uchida et al.*, 2001.

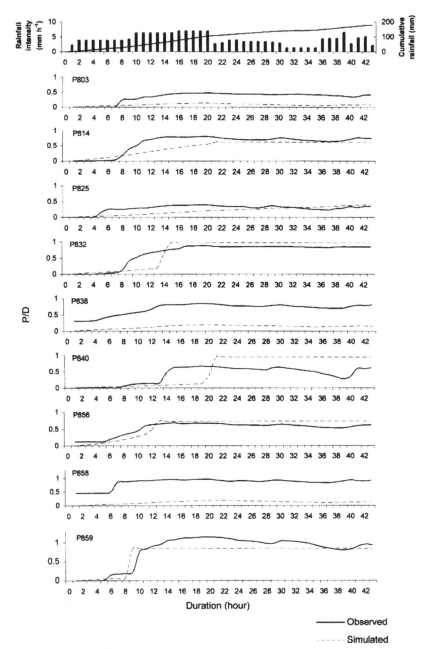

Figure 3.15. Example of measured piezometric response (solid line, plotted as a ratio of pore water pressure to soil depth) at nine sites during a long-duration, low-intensity storm on 6 December 1976, Carnation Creek, Vancouver Island, British Columbia. Dashed line shows simulated piezometric response using a stream-tube–based groundwater model (DSGMFW) [after *Dhakal and Sidle*, 2004].

3 mm h^{-1} [*Dhakal and Sidle*, 2004b]. At least a portion of the pore pressure response dynamics, including the site to site variability within the basin, was attributed to preferential flow [*Fannin et al.*, 2000].

Clearly, groundwater response in relatively shallow, unstable soils is affected by a complicated interaction of rainfall (or snowmelt) attributes, antecedent moisture, hydrologic properties of soils, and geomorphology. Nonetheless, shallow groundwater studies in unstable hillslopes share some common general conclusions: (1) timing and maximum level of piezometric response is influenced by storm characteristics (intensity, total rainfall) and antecedent precipitation as well as soil hydrology; (2) preferential flow pathways facilitate rapid pore pressure response to rainfall; these may consist of macropores or pipes within the soil or bedrock fractures that facilitate exfiltration of groundwater; (3) increasing topographic convergence facilitates greater pore pressure response; (4) in some unstable sites, artesian pore pressures are necessary to trigger landslides; and (5) the development of piping in soils prior to failure may be partly responsible for the coincident observed declines in pore pressure.

Deep soils. Although numerous investigations worldwide have documented the influence of cumulative rainfall on deep-seated soil mass movement, few studies have been conducted on the response of pore water pressure in deep landslides (e.g., earthflows) or deep unstable soil mantles. Several examples are presented here of the relatively sparse data on the interrelationships among pore water pressure response, precipitation inputs, and movement rates of deep-seated landslides.

Pore water pressures within the Davilla Hill earthflow complex in the Coast Ranges of central California were periodically measured from August 1975 to February 1978. No active earthflow movement occurred during the first 18 months of the study due to two consecutive very dry winter "rainy" seasons; positive pore pressures were found within the earthflow for only 1 day in 1975 (one piezometer) and two short periods in spring 1976: a 13-day period in March and a 20-day period in April [*Keefer and Johnson*, 1983]. During this dry period, water levels remained ≥0.8 m below the soil surface except for two isolated occurrences. In contrast, following more typical rainfall in the period from December 1977 to early January 1978, water levels in many piezometers rose to <0.6 m below the surface with a few instances of total saturation and artesian pressures [*Keefer and Johnson*, 1983]. These pore pressure increases reactivated two earthflows within the complex.

In Redwood Creek, northwestern California, where earthflows comprise about 12% of the basin area [*Pitlick*, 1995], the 4.5–7.7 m deep Minor Creek earthflow has been intensively monitored. Data from 1978 to 1980 clearly showed that the earthflow was activated in December following an accumulation of seasonal rainfall; however, highest movement rates (≈20–50 mm d^{-1}) did not occur until March and April after substantial winter rainfall had recharged the regolith [*M. Nolan*, unpublished data, U.S. Geological Survey, 1982]. Mean hydrographs for piezometers located at different depths showed that, in general, a downward hydraulic gradient persists [*Iverson and Major*, 1987]; however, for individual piezometers it was evident that on a few occasions hydrostatic or even upward hydraulic gradients may have briefly occurred. Generally pore pressures began to rise rapidly in October or November during the early rainy season; this rapid response persisted for 4–6 weeks, after which increases were limited to fluctuations induced by the temporal distribution of large storms. During the wet seasons, water levels in piezometers were <1 m below the soil surface. Periods of rapid earthflow movement (>50 mm month^{-1}) were noted when average water levels in piezometers were <2.4 m from the ground surface [*Iverson and Major*, 1987].

Movement of the deep (≈4080 m) Chausuyama earthflow–earthslide in central Japan was compared with groundwater levels in the landslide mass during a 5-yr period. Almost no movement occurred when water levels were >4.5 m below the ground surface [*Oyagi*, 1977]. Once groundwater level increased to about 4 m below the ground surface, movement occurred in most years; progressive increases in movement occurred as groundwater continued to rise (Figure 2.11). Maximum movement rate measured was about 1.2 cm d^{-1}, when the groundwater table was about 2.9 m below the soil surface [*Oyagi*, 1977]. Average movement rates since 1966 were about 0.8 m yr^{-1}; however, in the early to mid-1930s movement rate in the toe of the earthflow accelerated to 77 m yr^{-1} [*Oyagi*, 1977]. Such surges in deep-seated landslides have been noted elsewhere in response to seasonal rainy periods, heavy snow and rapid melt, and regional climate perturbations [*Kaliser and Fleming*, 1986; *Fleming et al.*, 1988; *Bovis and Jones*, 1992]. Numerous deep wells and drainage tunnels have been installed in the Chausuyama earthflow to remove groundwater and thus reduce the movement rate.

The moderately deep (7–10 m) Alani–Paty earthflow in Honolulu, Hawaii, is an example of a slow-moving mass movement that periodically activates during extended rainfall [*Baum and Reid*, 1995]. In a period of less than 2 yr, movement in the main body of the earthflow was recorded only twice: a storm in January 1990 (total displacement of 50–60 mm), and a rainy period in March 1991 (10–30 mm displacement) [*Reid*, 1994]. The water table in the lower portion of the earthflow was near the ground surface and did not respond to rainfall inputs; in the upper portion of the earthflow, the water table fluctuated between 1 to 2 m below the ground surface during drier periods. After periods of extended rainfall, the near-surface materials became saturated; thereafter, pore pressures at depth responded within several hours to large rainfall inputs [*Reid*, 1994]. The two highest pressure heads recorded during the 300-day monitoring period were during the March 1991 displacement and about one month prior to movement [*Reid*, 1994].

Several other investigations have contributed to the understanding of pore pressure response in deep-seated mass movements to precipitation inputs. Pore water pressures were continuously monitored near the crown of an earthflow complex in the Northern Apennines, Italy, during a dry year (August 2001 to June 2002) [*Simoni et al.*, 2004]. During the wet winter and spring periods, the soil mantle was saturated to within <1 m of the ground surface and pressure measurements indicated a downward gradient of flow. Almost totally saturated conditions were measured following the peak of several moderate to large storms. At shallower depths (≤2 m), pore pressure responded to individual rainfall events; initial response was typically within 10 h of the onset of precipitation [*Simoni et al.*, 2004]. The deep-seated Slumgullion earthflow in the San Juan Mountains of southwest Colorado moves at average rates ranging from 0.5 m yr^{-1} at the head to 6 m yr^{-1} in the narrowest section, predominantly in response to snowmelt [*Baum and Fleming*, 1996; *Savage and Fleming*, 1996; *Coe et al.*, 2003]. *Coe et al.* [2003] reported pore water pressure responses in the upper mantle of toe of the earthflow that coincided with snowmelt and heavy or prolonged inputs of rain; the three major pore pressure peaks recorded during the 3-yr monitoring period were all during maximum snowmelt. Velocity in the toe of the earthflow was similarly high during these snowmelt periods (0.85–1.15 cm d^{-1}) as well as in several rainfall events (0.75–1.05 cm d^{-1}) throughout the monitoring period. In general, pore water pressures and earthflow velocity respond quite rapidly (within one to several weeks) to rainfall, considering the depth of the failure near the toe (12–20 m) [*Baum and Reid*, 2000; *Coe et al.*, 2003]. *Lafleur and Lefebvre* [1980] indicate the importance of regional geology on pore pressure distribution within deep, sensitive clay deposits, where drainage characteristics of the underlying aquifer are particularly important. If the underlying aquifer is confined and located below the toe of the slope, artesian pressures will develop and will reduce shear strength. Conversely, if the aquifer is drained, the stability of the clay deposit may increase.

While it is agreed that the interactions amongst precipitation inputs, pore water pressures, and velocities of deep-seated mass movements are complex, some generalizations can be derived from the few detailed studies that have been conducted. Typically, deep-seated mass movement occurs only when groundwater is relatively near the soil surface and movement rates increase rapidly once a threshold of pore pressure is reached. While pore pressures within the landslide mass may respond to individual rainfall (or snowmelt) inputs, it generally takes some accumulation of precipitation to initiate movement; thus, these deep-seated mass movements often experience seasonal episodes of accelerated movement followed by periods of dormancy. The response times of pore pressure and mass movement velocity to individual rainfall events during a wet season are rapid if hydraulic conductivity is high and preferential flow pathways are active and the soil is relatively shallow. For deeper soils with low hydraulic conductivities, pore pressure responses and mass movement velocities are slow. Furthermore, in contrast to shallow, rapid failures, where a clear hydrologic discontinuity typically exists between the soil and substrate that promotes slope-parallel flow [e.g., *Harp et al.*, 1990], slower, deep-seated landslides generally have higher clay content, gentler topography, and predominantly vertical recharge [e.g., *Iverson and Major*, 1987]. Thus, it is evident that different modeling approaches may be needed to estimate pore pressure responses for these failure types.

VEGETATION INFLUENCES

Woody vegetation, particularly trees, augments the stability of hillslopes in primarily two ways: (1) removing soil moisture through evapotranspiration; and (2) providing root cohesion to the soil mantle [*Gray and Megahan*, 1981; *O'Loughlin and Ziemer*, 1982; *Riestenberg and Sovonick-Dunford*, 1983; *Greenway*, 1987[1]]. The first factor is not particularly important for shallow landslides that occur during an extended rainy season, except possibly in the tropics and subtropics, where evapotranspiration is high throughout the year [*Sidle et al.*, 2006]. In temperate regions, soils are near saturation and evapotranspiration is low during autumn and winter rainstorms when landslides

---

[1] *Tsukamoto*, 1987; *Phillips and Watson*, 1994; *Ekanayake and Phillips*, 2002.

TABLE 3.1. Relative influences of woody vegetation on slope stability (extensively modified from *Greenway*, 1987). 'A' denotes mechanisms adverse to stability, 'MA' denotes marginally adverse, 'MB' denotes marginally beneficial, and 'B' denotes beneficial mechanisms.

	Influences on types of landslides	
Mechanisms	Shallow, rapid	Deep-seated
Hydrological mechanisms		
1. Interception of rainfall and snow by canopies of vegetation, thus promoting evaporation and reducing water available for infiltration	B	B
2. Root systems extract water from the soil for physiological purposes (via transpiration), leading to lower soil moisture levels	B	B
3. Roots, stems, and organic litter increase ground surface roughness and soil's infiltration capacity	MA	MA
4. Depletion of soil moisture may cause desiccation cracks, resulting in higher infiltration capacity and short-circuiting of infiltrating water to a deeper failure pláne	MA	MA
Mechanical mechanisms		
5. Individual strong woody roots anchor the lower soil mantle into the more stable substrate	B	MB
6. Strong roots tie across planes of weakness along the flanks of potential landslides	B	B
7. Roots provide a membrane of reinforcement to the soil mantle, increasing soil shear strength	B	B
8. Roots of woody vegetation anchor into firm strata, providing support to the upslope soil mantle through buttressing and arching	B	MB
9. Weight of trees (surcharge) increases the normal and downhill force components	MA/MB	MA/MB
10. Wind transmits dynamic forces to the soil mantle via the tree bole	A	MA

usually occur; thus, soil water status may only be minimally affected by forest vegetation [*Megahan*, 1983; *Sidle et al.*, 1985; *Heatherington*, 1987]. Other effects of vegetation on slope stability, both positive and negative, are shown in Table 3.1 for shallow, rapid landslides as well as deep-seated landslides.

Evapotranspiration

Vegetation cover strongly influences evapotranspiration rates and thus the seasonal soil water balance, particularly when water is limiting or evapotranspiration demands are high. Effects of evapotranspiration on the soil water budget can be partitioned as follows: (1) canopy interception of rainfall or snow and subsequent evaporation loss to the atmosphere; (2) transpiration of infiltrated water to meet the physiological demands of vegetation; and (3) evaporation from the soil or litter surface. Different vegetation covers have different balances of these fundamental water loss processes. The first two of these processes are the most important related to the interaction of vegetation and slope stability. Evaporation from the ground surface is controlled by the depth to the water table, pore size distribution, and local heat budgets; in densely vegetated landscapes, variations are small compared with other evapotranspiration components [e.g., *Calder et al.*, 1986; *Jones*, 1997; *Scott et al.*, 2003]. Although difficult to assess in experimental studies, evaporation from soil surfaces under well-vegetated canopies may comprise about 10% or slightly more of total evapotranspiration; proportional rates in dryland areas with sparse vegetation could be much higher [*Campbell*, 1985].

Transpiration is the dominant process by which soil moisture in densely vegetated terrain is converted to water vapor. It involves the adsorption of soil water by plant roots, subsequent translocation of the water through the plant, and release of water vapor through stomatal openings in the foliage. Transpiration rates in vegetated landscapes depend on availability of solar energy and soil moisture as well as vegetation characteristics, including vegetation type (cf. conifer and deciduous), stand density, height and age, rooting depth, leaf area index, leaf conductance, albedo of the foliage, and canopy structure [*Reifsnyder and Lull*, 1965; *Dingman*, 1994]. Nevertheless, rates of transpiration are rather similar for different vegetation types if water is freely available [e.g., *Monteith*, 1976; *Whithead and Jarvis*, 1982; *McNaughton and Jarvis*, 1983]. Attempts to quantify transpiration have mostly focused on the use of Soil–Vegetation–Atmosphere Transfer (SVAT) models, where the movement of water from the soil through the plant to the atmosphere is represented by several resistances in series: (1) the integrated soil–root system; (2) the stem; (3) the branch; and (4) the effective stomatal resistance [e.g., *Campbell*, 1985; *Oltchev et al.*, 1996]. Eddy correlation techniques are commonly used to estimate transpiration fluxes, but extrapolation of such findings to the catchment scale has proven difficult. Recent studies indicate that vegetation under the forest canopy (i.e., understory) significantly controls transpiration variability where trees are deeply rooted and access groundwater [*Scott et al.*, 2003]. In spite of these conceptual understandings and advances, it remains difficult to separate transpiration fluxes from those of interception [*Jones*, 1997].

Interception by vegetation cover controls both the amount and timing of precipitation reaching the soil surface. As such, the interception storage capacity of vegetation complexes is important because intercepted water has a high surface area to volume ratio that promotes efficient evaporation by convection. Intercepted rainfall is mostly stored on the surface of foliage and stems, while intercepted snowfall often bridges across larger gaps in tree crowns facilitating an accumulation of snow over large surface areas of the canopy [*Satterlund*, 1972]. Interception and subsequent evaporation of water from vegetation cover is particularly significant in coniferous forests; losses (both snow and rain) from these dense canopies can account for up to 30–50% of gross annual precipitation [*Dingman*, 1994]. Less water is evaporated from understory or short ground cover (grasses and shrubs) compared to overstory canopies for two main reasons: (1) less precipitation (both snow and rain) reaches the understory in forests; and (2) understory or short ground cover has a low surface roughness and is often shielded from wind and sun by the overstory canopy (if present); thus, near-ground vegetation experiences lower turbulent exchange with overlying air compared to trees. Seasonal dormancy of deciduous trees greatly reduces canopy interception; dormancy typically coincides with winter rainy and snowmelt seasons in temperate climates, the period when landslides usually occur. Evaporation losses from snow intercepted by conifers are directly related to the period that snow persists in the canopy; conversely, deciduous canopies trap little snow [*Satterlund and Haupt*, 1970]. In contrast to temperate deciduous forests, multi-tiered tropical forests may intercept higher amounts of rainfall due to their structure and evergreen nature. A study in second-growth and old-growth conifer forests in the Pacific Northwest suggests that intense rainfall inputs to the soil may be buffered in the short-term by canopy interception; buffering effects in old-growth forests was more variable due their complex canopy structure [*Keim and Skaugset*, 2003]. For all environments, the proportion of rainfall intercepted by forest canopies is inversely related to both antecedent wetness and rainfall intensity. Gentle, short-duration rainfall may be almost totally intercepted, while interception may account for as little as 5% of precipitation during intense winter storms [*Satterlund*, 1972; *Jones*, 1997].

The potential for vegetation to affect slope stability through evapotranspiration depends primarily on vegetation cover and the biogeoclimatic setting. Since it is difficult to measure the separate fluxes of soil surface evaporation, transpiration, and interception, these losses are typically lumped together as evapotranspiration when assessing slope stability. In general, evapotranspiration rates in temperate regions are lowest from bare soil, several times higher from grassland, and 5 to 10 times higher (compared to bare soil) from forests [*Jones*, 1997]. The seasonal influence of evapotranspiration on soil water budgets, especially in temperate and dry regions, determines whether these losses influence landslide probability. During the winter rainy season in temperate climates, evapotranspiration likely has little effect on antecedent soil moisture because soils are already near saturation and transpiration is low [*Harr*, 1977; *Sidle et al.*, 1985; *Heatherington*, 1987]; thus, effects of evapotranspiration on shallow landslide initiation should be insignificant. Nevertheless, recent analyses of piezometric data in shallow soils in Vancouver Island, British Columbia, indicate that increases in pore pressure measured during moderate rainstorms after logging may be attributed to reductions in evapotranspiration [*Dhakal and Sidle*, 2004b]. However, for large rainstorms occurring during very wet conditions, evapotranspiration had little effect on pore pressure response. In temperate environments, where most shallow landslides occur during periods of extended winter rainfall, evapotranspiration (or the lack thereof) could alter the "window of susceptibility" for landsliding if a large storm occurred near the beginning or end of the rainy season [*Megahan*, 1983; *Sidle et al.*, 1985]. Additionally, when large landslide-producing storms occur during drier conditions, evapotranspiration can be a factor related to shallow landslide initiation. Rainforests in the tropics likely play a more significant role in altering soil moisture content since evapotranspiration is high throughout the year [*Greenway*, 1987; *Sidle et al.*, 2006].

In regions where winter snow accumulates, canopy structure largely controls interception of snow [e.g., *Toews and Gluns*, 1986; *Berris and Harr*, 1987]. In areas with deep snowfall, snow intercepted by and evaporated from canopies of dense conifer stands will reduce the recharge of meltwater into the soil mantle [*Harestad and Bunnell*, 1981; *Golding and Swanson*, 1986]. The longer that snow remains on canopies, the more evaporation can be expected. While this canopy-snow interaction has been shown to be important in controlling runoff from high-elevation forested catchments [e.g., *Troendle and King*, 1985], the effects of evaporation of snow from canopies on reducing the extent of landsliding has not been examined in detail. Following clearcutting and burning in the snow zone of Idaho, average and maximum piezometric levels relative to total soil depth increased by 68% and 41%, respectively [*Megahan*, 1984]. The probability of occurrence of this relative piezometric response increased by an order of magnitude after logging and burning. Generally, the timing of snowmelt occurs later with trees than without trees, but the specific timing and thus peak groundwater inputs are heavily dependent on the thermal conditioning of the snowpack.

Because deep-seated mass failures occur primarily during the wet season, water losses due to evapotranspiration will likely not play a major role in modifying movement rate in temperate regions. In tropical and subtropical environments, where evapotranspiration is high year-round, this influence has the potential to alter the timing and period of movement of deep-seated landslides. A simple simulation using daily, steady-state estimates of evapotranspiration rates in the tropics indicates that removal of forest cover could significantly extend the period of high soil moisture that may be associated with deep-seated mass movement (Figure 3.16). Additionally, threshold levels of soil water (that

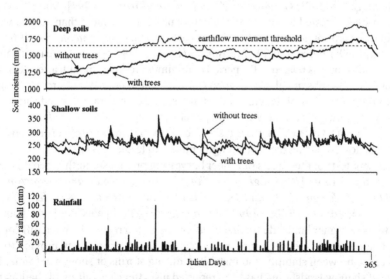

Figure 3.16. Estimated seasonal changes in soil water based on rainfall data in Peninsular Malaysia; simulations with and without trees (i.e., forested and clearcut scenarios) for deep soils (depth = 5 m; maximum water holding capacity = 2500 mm) subject to earthflows, and for shallow soils (depth = 1 m; maximum water holding capacity = 500 mm) subject to shallow, translational landslides.

activate mass movement) are exceeded earlier in the season for cleared areas compared to forest sites. Thus, the period of seasonal movement of deep-seated landslides appears to be decreased by evapotranspiration losses from tropical forest cover. In contrast, simulated moisture conditions in shallow soils in the same tropical environment were only slightly higher for cleared conditions compared to forested conditions during large storms preceded by wet conditions [*Sidle*, 2005]. Only during storms preceded by dry conditions were significant differences in soil moisture between forested and cleared sites detected [*Sidle*, 2005] (Figure 3.16). Such storms would be less likely to trigger landslides compared to similar storms that occur during wet periods.

Vegetation rooting depth in deep unstable regoliths is an important control on soil water depletion. Deeper-rooted vegetation species can sustain maximum transpiration rates for longer durations, thus drying the soil at greater depths compared to shallow-rooted vegetation [*McNaughton and Jarvis*, 1983]. Based on these limited findings, it appears that evapotranspiration (and consequently vegetation cover change) has only a very small influence on the movement of deep-seated landslides in temperate regions, but may exert more significant controls on slope stability in the tropics [*Greenway*, 1987; *Phillips et al.*, 1990; *Sidle et al.*, 2006].

Root Strength

The contribution of vegetation roots to soil shear strength is generally recognized as more important in stabilizing hillslopes compared to evapotranspiration losses [*Wu et al.*, 1979; *Gray and Megahan*, 1981; *Greenway*, 1987; *Phillips and Watson*, 1994]. Many field investigations in steep forested terrain worldwide have noted a 2- to more than 10-fold increase in rates of mass erosion 3 to 15 yr after timber harvesting [*Bishop and Stevens*, 1964; *Endo and Tsuruta*, 1969; *Fujiwara*, 1970; *Swanson and Dyrness*, 1975[1]]. This increase in landslide frequency and volume is related to the period of minimum rooting strength after clearcut harvesting and prior to substantial regeneration (Figure 3.17). Thus, the root strength minimum does not imply that a landslide will occur at any specific time; rather, it indicates that the probability of a landslide is higher given the likelihood of a triggering rainfall or snowmelt event. Described a bit differently, the threshold for a landslide triggering storm at a particular site would be lowest during the root strength minimum [*Sidle*, 1992].

Independent tests of the effects of timber harvesting on root strength based on mechanical straining of roots [*Burroughs and Thomas*, 1977; *Ziemer and Swanston*, 1977; *O'Loughlin and Watson*, 1979[2]] and shear tests [*Abe and Iwamoto*, 1986a; *Terwilliger and Waldron*, 1990; *Abe and Ziemer*, 1991; *Ekanayake et al.*, 1997] have confirmed empirical observations of higher landslide frequency after vegetation removal. When hillslope soils are in a tenuous state of equilibrium, reinforcement by tree roots may provide the critical difference between stability and instability during storms or snowmelt [*Sidle*, 1992]. Evidence of shallow landsliding has been reported just after clearcutting of shallow-rooted

---

[1] *O'Loughlin and Pearce*, 1976; *Megahan et al.*, 1978; *Amaranthus et al.*, 1985; *Marden and Rowan*, 1993; *Bergin et al.*, 1995; *Wu and Sidle*, 1995; *Jakob*, 2000; *Guthrie*, 2002.

[2] *Wu et al.*, 1979; *Watson et al.*, 1997.

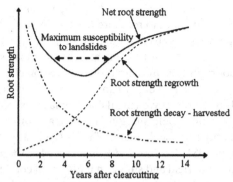

Figure 3.17. An example of typical changes in rooting strength of forest vegetation after timber harvesting (clearcutting). Root decay and recovery curves are based on numerous data worldwide. Net rooting strength is the sum of the decay and recovery curves.

conifers in southeast Alaska [*Sidle*, 1984a; *Johnson et al.*, 2000]. Such failures occurring prior to maximum root strength deterioration may be related to extensive disturbance of the root-reinforced organic horizon by highlead yarding that exposed weaker mineral soil [*Sidle*, 1984a] or the occurrence of an episodic storm [*Johnson et al.*, 2000] or both.

Measurement of root strength. The rooting strength of vegetated soils can be quantified or estimated in several ways: (1) tensile strength and shear strength measurements of individual roots with differing diameters [*Burroughs and Thomas*, 1977; *Ziemer*, 1978; *O'Loughlin and Watson*, 1979; *Wu et al.*, 1979[1]]; (2) direct shearing of individual roots imbedded in soils in the laboratory [*Abe and Ziemer*, 1991]; (3) in situ measurements using metal shear boxes in soils with various levels of root reinforcement [*Ziemer*, 1981; *O'Loughlin et al.*, 1982; *Abe and Iwamoto*, 1986a[2]]; (4) direct shear measurements of root-permeated columns in the laboratory [*Waldron and Dakessian*, 1981; *Waldron et al.*, 1983; *Terwilliger and Waldron*, 1990, 1991]; (5) uprooting tests of stumps or trees [*Somerville*, 1979; *Kitamura and Namba*, 1981; *Phillips and Watson*, 1994]; and (6) back-calculations of previously failed slopes [*Swanston*, 1970; *Sidle and Swanston*, 1982; *Preston and Crozier*, 1999]. The first two methods are similar and typically quantify tensile strength values used in models of rooting strength based on root density and size distributions (see next section). The third and fourth methods are similar and provide direct estimates of root reinforcement in laboratory or small-scale field tests. The last two methods represent rather large-scale aggregated (but indirect) estimates of rooting strength based on either external forces imparted on trees or actual slope failures.

Controlled measurements of breakage of individual roots of different diameters have been used to estimate rooting strength. Numerous researchers [*Burroughs and Thomas*,

---

[1] *Abe and Iwamoto*, 1986b; *Commandeur and Pyles*, 1991; *Lin*, 1995; *Watson et al.*, 1997; *Schmidt et al.*, 2001.

[2] *Ekanayake et al.*, 1997; *Zhou*, 1999.

1977; *O'Loughlin and Watson*, 1979; *Wu et al.*, 1979; *Waldron and Dakessian*, 1981[1]] have employed mechanical devices to measure the tensile strength of individual roots. Individual roots are clamped on both ends and then subjected to a tensile force, typically via a hydraulic cylinder or jack screw (usually at a constant load speed), and the tensile force at the time of root rupture is measured by a dynamometer or load cell. Roots of different sizes and states of relative decay can be tested to develop weighted estimates representative of field conditions (given available biomass distribution data). Problems with this methodology include deformation and slippage of roots in the mechanical device, as well as limitations on the diameter of roots (generally <15–20 mm) [*Abe and Iwamoto*, 1986b; *Ziemer*, 1981; *Phillips and Watson*, 1994]. Additionally, roots need to be carefully excavated in the field and stored in moist, cool conditions to preserve their strength characteristics, and only short, straight root segments can be tested by such methods [*Phillips and Watson*, 1994; *Watson et al.*, 1997]. *Ziemer* [1978] developed a simple apparatus to measure shear strength directly on roots up to 50 mm in diameter to alleviate some of these constraints. Shear strengths of individual roots were highly correlated ($R^2 = 0.97$) with root tensile strength by the following relationship [*Ziemer and Swanston*, 1977; *Ziemer*, 1978]:

$$T_R = -7.6 + 2.2\, S_R \qquad \text{(Equation 3.7)}$$

where T_R is the tensile strength of individual roots and S_R is the shear strength of individual roots. *Abe and Ziemer* [1991] later sheared several roots together imbedded in soil in a direct shear box to overcome some of the limitations of tensile strength testing. All of these indirect methods of estimating rooting strength, whether based on tensile strength or shear strength tests on individual roots (or small groups of roots), require field data on root biomass and density distribution of root size to calculate contributions to soil strength (see next section).

   In situ measurement of the shear strength of soil–root systems has revealed information about the mechanics of root reinforcement. Direct shear tests have been conducted both in the field and the laboratory on root-permeated soils. In the field, blocks of root-permeated soil have been isolated and enclosed within relatively large shear boxes [*Endo and Tsuruta*, 1969; *O'Loughlin*, 1974a,b; *O'Loughlin and Pearce*, 1976; *O'Loughlin et al.*, 1982[2]]. Typically, the sides of the shear box are open so as not to disturb the lateral root reinforcement. The encased soil is then weighted and shear stress is applied at a constant displacement rate by a hydraulic jack or other device and measured on a gauge. Contributions of roots to soil strength have been estimated by shearing in situ blocks of soil with and without roots and evaluating the maximum differences in the respective shear displacement curves [*Ziemer*, 1981; *O'Loughlin et al.*, 1982; *Ekanayake et al.*, 1997]. Direct shear tests on root-permeated soil columns are conducted similarly along pre-designated shear planes in columns [*Waldron*, 1977; *Waldron and Dakessian*, 1982;

---

[1] *Riestenberg and Sovonick-Dunford*, 1983; *Abe and Iwamoto*, 1986b; *Commandeur and Pyles*, 1991; *Schmidt et al.*, 2001.

[2] *Ziemer*, 1981; *Abe and Iwamoto*, 1986a; *Ekanayake et al.*, 1997.

Waldron et al., 1983; *Terwilliger and Waldron*, 1990, 1991]. Shear tests of root-permeated soil in both the laboratory and field offer the advantage of direct measurement of shear strength enhancement. This is particularly advantageous in the field, where the shearing is conducted in natural rooting systems. One disadvantage of this methodology, especially for tests in soil columns, is the small cross-section of soil sampled relative to complex root networks in the field, thus giving rise to the potential for high spatial variability in root reinforcement results [*Sakals and Sidle*, 2004].

It is difficult to quantify the effect of the soil–root bond on root pull-out resistance. *Riestenberg and Sovonick-Dunford* [1983] noted that about 40% of sugar maple roots had pulled out along landslide scarps; however, most of these roots were small and located in loose topsoil. In shear tests on root-permeated soil columns, *Waldron and Dakessian* [1981] concluded that root slippage (i.e., the soil–root bond) rather than breakage (due to tensile strength) was the limiting condition in root reinforcement in saturated, fine textured soils. On the other hand, based on tests in a glass-bottomed shear apparatus, *Abe and Ziemer* [1991] observed that roots imbedded in sand generally deformed rather than slipped with applied shear force and caused a widening of the shear zone. Other studies also reported that deformation and stiffness of root fibers may significantly reinforce soil shear strength [*O'Loughlin et al.*, 1982; *Gray and Ohashi*, 1983]. However, numbers and sizes of roots are inversely proportional to deformation in such controlled studies [*Abe and Ziemer*, 1991]. In spite of these detailed laboratory experiments and selected field observations, most investigations of landslide scars reveal that the majority of the roots break in tension, indicating that the pull-out resistance exceeds the tensile strength of roots in many cases [*O'Loughlin*, 1974b; *Gray and Megahan*, 1981; *Ziemer*, 1981[1]]. Even in the subsoil of the landslide study in Ohio [*Riestenberg and Sovonick-Dunford*, 1983], most roots broke in tension rather than slipped. Part of the reason for the apparent discrepancies in laboratory and field findings is due to the scale of the experimental or observational conditions. Root pull-out may occur more easily in laboratory tests than in the field, where complicated branching, interlocking, and bonding with stones occur over large distances [*Ziemer*, 1981; *O'Loughlin et al.*, 1982; *Schmidt et al.*, 2001]. In a subsequent larger (1.22 m) column study, *Waldron et al.* [1983] reported that shear resistance of pine-rooted soil increased steadily with displacement over the entire test displacement range, illustrating the effect of increasing scale on the diminished importance of root pull-out. At a shear displacement of about 75 mm, the shear resistance of pine-rooted soil (10.8–11.8 kPa) was approximately twice that of soil without roots.

Stumps and trees are widely used as anchors or tailholds in cable logging systems, thus, the resistance to uprooting has also been investigated in this context [e.g., *Peters and Biller*, 1986; *Pyles and Stoupa*, 1987]. Additionally, root anchoring of trees has been studied from the perspective of resistance to wind-induced forces that cause widespread windthrow in forests [e.g., *Somerville*, 1979; *Cannell and Coutts*, 1988; *Watson*, 2000]. The few studies that assessed the interactions of uprooting strength

---

[1] *O'Loughlin et al.*, 1982; *Philips and Watson*, 1994; *Schmidt et al.*, 2001.

and slope stability assume that root reinforcement in soil is proportional to the resistance to uprooting [*Kitamura and Namba*, 1981; *Tsukamoto*, 1987; *Philips and Watson*, 1994]. In such tests, a wire cable is secured around the base of the stump or tree and is pulled (usually downhill) with a constant force measured by a load cell. *Anderson et al.* [1989] utilized a torque winch bar apparatus with a block and tackle system to twist tree roots, similar to a ring shear test. Maximum tensile forces can give a relative index of root cohesion for different aged trees or different stages of root decay, providing that some quasi-steady-state data are available for root cohesion for particular species [*Sidle* 1991, 1992]. While this method provides a relative measure of individual tree rooting strength, it is difficult to estimate the overall soil-root reinforcement of a larger area from such data.

Estimates of vegetation root strength have been made from back-calculations of previously failed hillslopes where geotechnical, soil, and hydrological parameters are known or assumed [*Swanston*, 1970; *O'Loughlin*, 1974a; *Gray and Megahan*, 1981; *Sidle and Swanton*, 1982; *Preston and Crozier*, 1999]. Such estimates typically rely on infinite slope analysis, where data for slope gradient, soil cohesion, density, internal angle of friction, and pore water pressure are supplied and root cohesion (or combined root and soil cohesion if soil cohesion data are not available) is calculated as a residual assuming a factor of safety of 1.0 at failure. These estimates may represent the best spatially distributed data available for root cohesion in the vicinity of the landslide, assuming other input data are accurate. Unfortunately, investigations in previously failed sites with accurate soil data are rare.

Models of mechanical reinforcement of soils by vegetation roots. A root-permeated soil can be analyzed by techniques similar to those for reinforced earth materials, where roots act as fibers of high tensile strength within a matrix of lower tensile strength [e.g., *Wu*, 1976; *Waldron*, 1977; *Abe and Ziemer*, 1991]. In such a conceptualization of root reinforcement, root systems contribute to the shear strength (s) by providing an additional cohesion component (ΔC) in the Mohr–Coulomb equation [*Endo and Tsuruta*, 1969; *Waldron*, 1977; *Gray and Megahan*, 1981; *O'Loughlin et al.*, 1982[1]]:

$$s = (c' + \Delta C) + (\sigma - u)\tan \phi' \qquad \text{(Equation 3.8)}$$

where c' is effective cohesion of the soil and ϕ' is the effective internal angle of friction. With a few exceptions [*Luckman et al.*, 1982; *Terwilliger and Waldron*, 1991], most studies have concluded that roots have a negligible influence on the frictional component of soil strength.

The most common method of modeling root reinforcement of soils is based on root tensile strength and root density data. A relationship was developed that is useful for determining ΔC, based on the stress distributions shown in Figure 3.18, by resolving the tensile strength of roots (T_R) into components normal and parallel to the shear zone [*Wu*, 1976; *Waldron*, 1977; *Wu et al.*, 1979]. Based on this conceptual model, the increase in shear strength related to root cohesion is described as

---

[1] *Abe and Ziemer*, 1991; *Sidle*, 1991; *Schmidt et al.*, 2001.

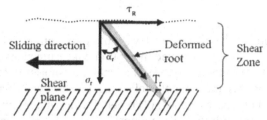

Figure 3.18. Stress distributions on individual roots.

$$\Delta C = \tau_R \left(\cos \alpha_r \tan \phi' + \sin \alpha_r \right) \qquad \text{(Equation 3.9)}$$

where τ_R is the average tensile strength of roots per unit area of soil and α_r is the angle of root deformation (see Figure 3.18). The root deformation angle appears to vary from about 48° to as large as 72° for conifers, but ΔC is relatively insensitive to changes in α_r [*Wu et al.*, 1979].

The average tensile strength of roots per unit area of soil (τ_R) can be estimated by multiplying the tensile strength of roots (T_R) by the proportion of the soil area occupied by roots [*Wu*, 1976; *Waldron*, 1977; *Wu et al.*, 1979]:

$$\tau_R = T_R \left(A_R / A \right) \qquad \text{(Equation 3.10)}$$

where A_R is the proportion of the soil cross section occupied by roots and A is the total soil cross-sectional area. Because numerous studies have shown that tensile resistance (as well as tensile strength) of roots varies with root size [*Wu*, 1976; *Burroughs and Thomas*, 1977; *Ziemer and Swanston*, 1977; *Gray and Megahan*, 1981[1]], it is desirable to quantify average tensile strength per unit area of soil (τ_R) based on available data for root size distribution [*Gray and Megahan*, 1981]. Thus, the average root tensile strength per unit area of soil (τ_R) can be determined as

$$\tau_R = \sum (T_i n_i a_i) / A \qquad \text{(Equation 3.11)}$$

where T_i represents the tensile strength of roots in size class i, n_i is the number of roots in size class i, and a_i is the average cross-sectional area of roots in size class i. When only the average tensile strength of roots (T_R) is available, equation 3.11 must be modified as

$$\tau_R = T_R \left(\frac{\sum n_i a_i}{A} \right) \qquad \text{(Equation 3.12)}$$

If an average value for α_r of 60° is assumed, along with a typical estimate for ϕ' (35°), root cohesion can be roughly estimated as [*Wu et al.*, 1979]

---

[1] *Riestenberg and Sovonick-Dunford*, 1983; *Abe and Ziemer*, 1991; *Schmidt et al.*, 2001.

$$\Delta C = 1.2\tau_R \qquad\qquad \text{(Equation 3.13)}$$

This simple, theoretically based model allows rooting strength to be estimated based on the proportion of the soil area occupied by roots and measurements of tensile strength or the resistance of the roots themselves. The model is subject to the assumptions of limit equilibrium analysis, including the following: (1) shear deformation along the slip surface is restricted to a narrow zone; (2) roots of different size classes (n_i) are flexible and linearly elastic with Young's modulus E; (3) roots are oriented perpendicular to the failure plane; (4) the full tensile strength of all roots is mobilized; (5) roots are well anchored and do not pull out when tensioned; and (6) the internal friction angle (ϕ) is unaffected by roots [*Waldron and Dakessian*, 1981; *Greenway*, 1987; *Phillips and Watson*, 1994]. While laboratory column experiments have documented root slippage as an important control on root reinforcement in saturated and homogeneous clay loam soils [*Waldron and Dakessian*, 1981], and some field shear tests have noted such phenomena [*Ekanayake and Phillips*, 2002], it is difficult to assess pull-out resistance of roots in the field due the complex architecture of the rooting system. *Schmidt et al.* [2001] suggest that tension may be transferred rapidly to the soil via shear prior to soil pull out in highly branched root systems.

Abe and Ziemer [1991] modified the preceding model (based on equation 3.9) that assumes roots fail in tension, by introducing a coefficient of deformation to account for the relative stiffness of roots subjected to shearing. In this model, ΔC is comprised of two factors: (1) the shear length applied to a root by earth pressure, and (2) the reinforced strength caused by tensile stress of a root. The coefficient of deformation is related empirically to A_R/A and to root diameter. As the coefficient of deformation increases, the stretching roots increase ΔC [*Abe and Ziemer*, 1991]. While these improvements to the model based on root density and tensile strength alone are appealing, it is difficult to obtain the necessary deformation data to incorporate them into field-scale models.

Several recent studies indicate that soil reinforcement by roots of woody vegetation may be greater than originally estimated from two-dimensional models based on limit equilibrium theory. Shear stress–displacement curves for root-permeated soils appear to have much flatter peak stress values than do soils without roots [*O'Loughlin et al.*, 1982; *Ekanayake et al.*, 1997; *Ekanayake and Phillips*, 1999b]. Based on a reexamination of in situ data for radiata pine and kānuka stands, an energy-based approach that incorporates the ability of roots to withstand strain during the shear displacement process has been included into a slope stability analysis [*Ekanayake and Phillips*, 1999b, 2002]. This method considers the characteristics of the shear stress–displacement curve of root-permeated soil only up to the beginning of the peak shear stress, and then approximates this relationship by an elliptical function [*Ekanayake and Phillips*, 1999b]. The model allows for the determination of the total energy capacity of the soil–root system as well as the amount of energy exchanged up to the peak displacement. In contrast to limit equilibrium analysis, where the contribution of roots is estimated as the maximum difference between stress–strain curves with and without roots at a point, the energy-based approach yielded greater strength contributions from tree roots (based on the two species examined) due to the increased peak resistance and the increased shear displacement [*Ekanayake and Phillips*, 1999b].

A simple empirical model of rooting strength based on root biomass was developed by *Ziemer* [1981]. Root reinforcement data were obtained by 18 independent in situ shear tests of *Pinus contorta* roots growing in sandy soils on the north coast of California. Of numerous parameters examined, the best regression relationship ($R^2 = 0.79$) to predict ΔC was

$$\text{Soil strength} = 3.13 + 3.31 \text{ biomass} \qquad \text{(Equation 3.14)}$$

where soil strength (kPa) is the combined strength of the soil plus root reinforcement (ΔC) and biomass (kg m^{-3}) is the dry weight of live roots <17 mm in diameter [*Ziemer*, 1981]. At this site, the mean biomass of roots <17 mm represented 64% of the total root biomass, thus equation 3.14 may be a good indicator of root reinforcement. Because many forestry studies measure root biomass, such a simple relation is attractive. Other studies lend support to this simple model. *Endo and Tsuruta* [1969] found that the root reinforcement by small (average stem diameter 16 mm) alder seedlings grown in a nursery increased proportionally to the fresh weight of roots per unit soil volume. Their data indicate that the presence of roots raised the soil shear strength from 5 to 10 kPa for root densities in the range of 4–12 kg m^{-3} (fresh weight). Limitations of such simple soil strength–root biomass models are obvious: they require testing and calibration for specific sites. However, the simplicity of these models allows the general relationship between these two parameters to be applied in many cases.

A plan-view, two-dimensional analysis of rooting strength based on root biomass distribution suggests that the interacting effects of individual tree root systems may enhance root cohesion for both undisturbed and selectively managed forest sites [*Sakals and Sidle*, 2004]. Such root cohesion benefits based on densely stocked indigenous stands of kānuka are also evident [*Ekanayake et al.*, 1997]. Furthermore, the importance of lateral root cohesion is emphasized in a recent study of shallow landslides in the Oregon Coast Ranges [*Schmidt et al.*, 2001]. In these sites, lateral root cohesion was estimated to be much greater than root cohesion along the basal shear plane, and lateral cohesion was much higher than reported values for similar species. Benefits of lateral root reinforcement have been suggested and described earlier [*Sidle et al.*, 1985; *Tsukamoto and Minematsu*, 1986; *Tsukamoto*, 1987]; however, it is generally believed that lateral roots impart much more protection against shallow landslides compared to deep-seated failures [*Swanston and Swanson*, 1976]. Tree roots may lend some stability to deeper soils by lateral reinforcement across planes of weakness [*Swanson and Swanston*, 1977; *Schroeder*, 1985]; however, this beneficial effect would diminish with increasing size (depth and area) of the potential failure site.

Soil arching has been noted in some non-cohesive hillslope soils where tree trunks are firmly anchored into the soil at appropriate intervals [*Gray and Megahan*, 1981; *Gray and Leiser*, 1982]. Individual trees may also act as buttresses to stabilize hillslopes [*Gonsior and Gardner*, 1971; *Thorne*, 1990]. Thus, the theory of describing lateral restraint in soils by means of piles embedded into hillslopes [*Wang and Yen*, 1974] has been used to quantify soil arching between trees [*Gray*, 1978; *Gray and Megahan*, 1981; *Greenway*, 1987]. *Gray* [1978] developed a nomograph to predict the critical spacing between trees as a function of soil cohesion for various values of residual friction and cohesion along the slip plane. Soil arching theory was applied to forested sandy slopes in the Idaho Batholith; a critical spacing between vertical root cylinders of ≈2 m was calculated as a prerequisite to soil arching [*Gray and Megahan*, 1981]. However, this calculation is highly sensitive to soil cohesion; thus, if soil cohesion is assumed to be only 2.4 kPa, with a residual cohesion along the sliding base equal to 0.3 kPa, then the critical spacing increases to 6.3 m [*Gray and Megahan*, 1981].

Examples of root strength for various vegetation types. Vegetation root strength has been estimated by several means in various regions. Because of their larger individual roots and deeper penetration, and thus greater contribution to slope stabilization, tree roots have been more extensively investigated than other vegetation types. Regions where substantial progress has been made in quantifying tree rooting strength include New Zealand, Japan, western USA (including coastal Alaska), British Columbia, and Taiwan. Comparisons of root strength measurements from site to site are complicated by the different measurements employed, how the effects of lateral roots are treated, and whether soil cohesion is measured in addition to rooting strength (as is the case for in situ tests). Table 3.2 summarizes some of the important published root strength data for different species together with the methodology used to quantify these data. It is apparent that roots of deciduous trees generally impart a greater reinforcement to the soil mantle compared to conifers for the same general age and size; however, much variability has been reported within individual species. Root strength data for grass species has not been included because although grass roots can contribute to soil strength and slope reinforcement at small-scales and shallow depths, they are not effective in stabilizing soil mantles deeper than about 10 cm in most cases [e.g., *Marden and Rowan*, 1993; *Bergin et al.*, 1995]. In contrast, woody understory shrubs can more effectively reinforce hillslope soils due to their deeper root penetration [e.g., *Marden and Rowan*, 1993]. A recent investigation in the Oregon Coast Ranges, however, has attributed locations of many landslides initiated by major storms in February 1996 to patches of alder and the associated weaker root mass of this species compared to mature Douglas-fir in managed forests [*Roering et al.*, 2003]. While these findings would appear to be in disagreement with the relatively high tensile strengths [7–90 MPa; *Greenway*, 1987] and root cohesion values [2–12 kPa; *Endo and Tsuruta*, 1969] previously reported for alder, the greater rooting depth and root radius of conifers could possibly explain this discrepancy. In partial contrast to these findings, a widespread landslide inventory on the East Coast region of North Island, New Zealand, following Cyclone Bola in March 1988, found that the largest numbers of landslides occurred in pasture, followed progressively by very young (<6 yr old) and young (6–8 yr old) exotic pine plantations, regenerating shrubs, older exotic pine plantations (>8 yr old), and indigenous (>80 yr old) closed-canopy secondary broadleaf forests [*Marden and Rowan*, 1993]. Landslide damage after Cyclone Bola was 65% and 90% less for 10- and 20-yr-old tea tree stands, respectively, compared to pasture [*Bergin et al.*, 1995]. *Hawley and Dymond* [1988] evaluated aerial photos following a major storm near Gisborne, North Island, New Zealand, and found that variably spaced popular trees within hillslope pastures

TABLE 3.2. Root strength data for different species or vegetation cover types from sites around the world, grouped by similar methodologies.

Tree or vegetation type	ΔC (kPa)	Study area	Method of estimation	Reference/notes
Young alder	2.0-12.0	Nursery in Japan	Root density and tensile strength data	*Endo & Tsuruta,* 1969
Rocky Mountain Douglas-fir Coastal Douglas-fir	2.0-13.0 5.0-22.0	Central Idaho West Oregon	Root density and tensile strength data	*Burroughs & Thomas,* 1977
Sitka spruce	5.9	Prince of Wales Island, coastal Alaska	Root density and tensile strength data	*Wu et al.,* 1979
Rocky Mountain Douglas-fir	10.3	Central Idaho	Root density and tensile strength data	*Gray & Megahan,* 1981
Lodgepole pine White fir-mixed conifer	3.0-21.0 6.0	Northern California coast	Root biomass and tensile strength data	*Ziemer,* 1981
Sugar maple	5.7	Cincinnati, Ohio	Root density and tensile strength data	*Riestenberg & Sovonick,* 1983
Acacia Candlenut Chinese Banyan	7.4 14.3 15.3	Hong Kong	Root density and tensile strength data	*Greenway,* 1987
Formosan alder Roxburgh sumac	78 24	Central Taiwan	Root density and tensile strength data	*Chang & Lin,* 1995
Yunnan pine	17.6	Hutiaoxia Gorge, China	Root density and tensile strength data	*Zhou,* 1999
Douglas-fir, hemlock, alder forest 1. Natural forest 2. Second-growth	Lateral root ΔC 11-94 6.8-23	Oregon Coast Ranges	Root density and tensile strength data	*Schmidt et al.,* 2001
2-yr-old radiata pine 16-yr-old kānuka	10.5-20.6 12.1-22.1	Northeast coast of North Island, New Zealand	Root density and energy-based analysis of in situ direct shear tests	*Ekanayake & Phillips,* 2002

TABLE 3.2. *Cont.*

Tree or vegetation type	ΔC (kPa)	Study area	Method of estimation	Reference/notes
Douglas-fir, hemlock, alder forest	Lateral root ΔC = 15.2	Oregon Coast Ranges	Root density at landslide sites and tensile strength data	*Roering et al.,* 2003
Spruce-hemlock forests on well-drained podzol soils	3.3-4.3	Prince of Wales Island, coastal Alaska	Back-calculation after failure	*Swanston*, 1970 (includes soil cohesion)
Spruce-hemlock-fir forest on podsolized hillslope soils	1.0-3.0	Southwestern British Columbia	Back-calculations	*O'Loughlin*, 1974 (includes soil cohesion)
Understory vegetation of spruce-hemlock forest	2.2	Chichagof Island, coastal Alaska	Back-calculation after failure	*Sidle & Swanston*, 1982 (includes soil cohesion)
Beech-podocarp-hardwood forest	6.6 (soil c included)	South Island, New Zealand	In situ direct shear tests	*O'Loughlin* et al., 1982 ΔC ≈ 3.3 kPa
6-yr old yellow pine seedlings grown in clay loam soil packed into columns	3.7-6.4	San Pablo, California	Direct shear test on 122-cm cores with roots; density and tensile strength data	*Waldron et al.,* 1983 (corrected for soil cohesion)
6-yr old Sugi grown in nursery soils (loamy sand)	2.9-4.4	Ibaraki Prefecture, Japan	Large-scale, direct shear test	*Abe & Iwamoto*, 1986 (corrected for soil cohesion)
Chaparral 1. Unburned 2. Burned 2-yr prior	0.1-0.8 0.3-1.0	Transverse Ranges, California	Direct shear tests on 25-cm root-permeated cores	*Terwilliger & Waldron*, 1990 (excludes soil c)

exerted a significant contribution against landslide erosion; the influence decreased with distance from individual trees. These findings from New Zealand compare well with other results that indicate hardwood and shrub species lend higher root reinforcement to unstable slopes compared to young planted conifers, and that grass roots exert a negligible reinforcement against landslides.

Root strength dynamics. Changes in vegetation rooting strength after cutting, during site regeneration, resulting from stand tending or on-going site management practices, and following natural and other human disturbances clearly affect the temporal stability of hillslopes. The fundamental knowledge of root decay and regeneration characteristics is necessary to assess the dynamic contributions of tree root systems on potential landslides. Several studies have evaluated the rates of root strength deterioration after vegetation removal. Data from a wide range of such studies have been conceptually summarized in Figure 3.17 to show the general form of root deterioration and regrowth common to most forest and rangeland sites. *Sidle* [1991, 1992] demonstrated that rooting strength of harvested trees follows an exponential decay function

$$D = e^{-kt^n} \qquad \text{(Equation 3.15)}$$

where D (dimensionless) is actual root cohesion divided by maximum root cohesion for site vegetation, t is time since vegetation removal (years), and k and n are empirical constants. Root strength recovery after tree removal (or death) is based on the rate of regrowth of planted or invading vegetation. Actual recovery rate is a function of species mix of regenerating vegetation, soil fertility, and other site conditions. *Sidle* [1991] described root strength regrowth by a sigmoid function

$$R = (a + b^{-kt})^{-1} + c \qquad \text{(Equation 3.16)}$$

where R (dimensionless) is the actual root strength divided by the maximum root strength and a, b, c, and k are empirical constants. The constants can be determined in a mathematical model [*Sidle*, 1991] by specifying the inflection point (t_i) of the sigmoidal recovery curve, the percentage of root strength recovery at 2 t_i, the maximum root cohesion, and the initial conditions. Net rooting strength is the sum of the decay and recovery curves (Figure 3.17) and is the critical root strength parameter that needs to be assessed related to temporal changes in vegetation reinforcement at unstable sites [*Kitamura and Namba*, 1981; *Sidle* 1991, 1992; *Wu and Sidle*, 1995; *Watson et al.*, 1999].

While investigations of rates of root strength deterioration following cutting or tree death have been limited, most show similar trends. *O'Loughlin and Ziemer* [1982] summarized earlier studies of root tensile strength declines after cutting for a variety of tree species in western North America and New Zealand (Figure 3.19). Although comparisons amongst these data may be complicated by slightly different experimental methods employed, it is clear that all of the conifer species lose more than half of their initial rooting strength within 3 yr after cutting, whereas the two hardwood species (rata and beech) have slower rates of root strength loss. *Sidle* [1991] analyzed root deterioration data from several sources and species and fitted these data to the decay function (equation 3.15) using a non-dimensional root cohesion. While the degree of fit varied from species to species, these standardized relations provide useful information on temporal changes in rooting strength following tree removal (Figure 3.20). Roots of radiata pine [*O'Loughlin and Watson*, 1979] and coastal Douglas-fir [*Burroughs and Thomas*, 1977] decay most rapidly, while those of sugi [*Kitamura and Namba*, 1981] and Sitka spruce–western hemlock [*Ziemer and Swanston*, 1977] forests decay more slowly. Intermediate decay rates were found for beech [*O'Loughlin and Watson*, 1981], white

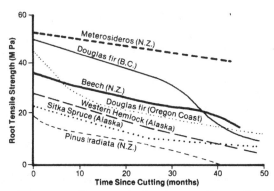

Figure 3.19. Changes in mean tensile strength of various conifer and hardwood species during a 4-yr period after cutting [after *O'Loughlin and Ziemer*, 1982].

fir–mixed conifer [*Ziemer*, 1981], and Rocky Mountain Douglas-fir forests [*Burroughs and Thomas*, 1977]. Decay coefficients (k and n) for the various tree species shown in Figure 3.20 are based on best fits obtained from equation 3.15. These values can be used for each tree species regardless of the maximum or equilibrium root cohesion for the site since D is a non-dimensional parameter. In contrast to the uniform exponential decline in tree root strength after cutting, mean tensile strengths of rata and kānuka roots, both indigenous hardwoods in New Zealand, actually increased in the 12–15-month period following cutting [*Watson et al.*, 1997]. This increase in tensile strength was attributed to a loss of root moisture to the surrounding soil and a possible decrease in root diameter. Not until about 4 and 5 yr after cutting did kānuka and rata roots, respectively, lose half of their initial mean tensile strength [*Watson et al.*, 1997]. In contrast, the widely planted softwood exotic, radiata pine, lost half of its tensile root strength within about 18 months of cutting [*O'Loughlin and Watson*, 1979].

Fewer studies have quantified the recovery of root strength following cutting or tree death. In temperate mixed conifer forests, *Ziemer* [1981] noted that it took 15 to 25 yr for a regenerating clearcut to restore half of its original rooting strength. An indirect example of root strength recovery can be derived from uprooting tests of stumps and live sugi trees in Japan [*Kitamura and Namba*, 1981]. If uprooting resistance data are converted to root cohesion and assessed for a span of tree ages, then a root strength recovery curve can be generated (Figure 3.21). This example (for sugi) indicates that about half of the root strength recovery occurs within 12 yr after cutting and 90% of the original rooting strength is restored within 20 yr. The data clearly fit the sigmoid relationship of equation 3.16.

A potential surrogate for the lack of data on root strength recovery is root biomass. Several investigators have established strong measured [*Ziemer*, 1981] and implied [*Watson et al.*, 1995; *Roering et al.*, 2003] relations between root biomass and root strength. The greater availability of root biomass data and relationships developed between tree diameter (dbh) and tree crown–root network radius [*Eis*, 1974; *Watson and O'Loughlin*, 1990; *Roering et al.*, 2003] may facilitate the use of root biomass in slope stability models. A study in New Zealand noted that root biomass of planted stands of the indigenous white tea tree kānuka (*Kunzea ericoides*) exceeds that of radiata pine for the first 9 yr of growth, after which time

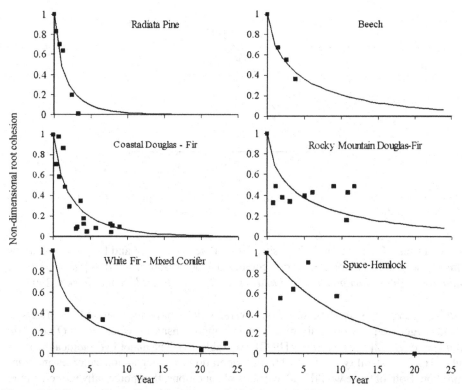

Figure 3.20. Decay of non-dimensional root cohesion for several forest stand types after clearcutting; coefficients (k and n) for the root decay curves are given [after *Sidle*, 1991].

radiata pine root biomass dominates due to its canopy development [*Watson and O'Loughlin*, 1990; *Watson et al.*, 1995]. *Roering et al.* [2003] suggests that small-scale mosaics of root strength caused by disturbances in temperate forests strongly influence the location of shallow landslides. Estimates of spatially distributed rooting strength derived from two-dimensional models based on root biomass data tend to support the idea that small-scale variability in root cohesion resulting from past forest management may affect landslide location [*Sakals and Sidle*, 2004]. Such timber harvesting legacy effects on slope stability are discussed in detail in Chapter 6.

Relationship Between Vegetation and Slope Stability

Although conclusive evidence has linked root systems of forests with enhanced soil strength, and less conclusive arguments have noted the potential benefits of enhanced evapotranspiration losses in forests, there remains a high level of variability in mass erosion from mountainous forested areas [*O'Loughlin and Ziemer*, 1982; *Sidle et al.*, 1985; *Guthrie*, 2002]. Clearly other natural factors such as lithology, geomorphology, hydrology (including precipitation inputs), and soil properties also strongly influence the stability of

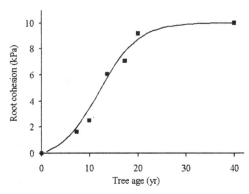

Figure 3.21. Recovery of root strength for sugi, based on uprooting tests [data from *Kitamura and Namba*, 1981].

forested sites. However, for a given site, the relative occurrence and landslide erosion is almost always lower in forested terrain compared to other land cover types [*DeGraff*, 1979; *Sidle et al.*, 1985; *Kuruppuarachchi and Wyrwoll*, 1992; *Froehlich and Starkel*, 1993[1]].

Influence of roots on slope stability. Several authors have summarized the effects of woody vegetation on slope stability, the most comprehensive of which are by *O'Loughlin and Ziemer* [1982] and *Greenway* [1987] (see Table 3.1). The major hydrological benefits related to slope stability are derived from evaporation from vegetation canopies and transpiration. Both of these would lead to drier soil conditions, thus potentially reducing pore water pressure accretion during storms or snow melt and have been discussed in detail. Conversely, vegetated land surfaces have higher infiltration capacities due to greater surface roughness, organic matter, and the presence of root systems and related preferential flow pathways. Additionally, depletion of soil moisture by deep-rooted vegetation may enhance the formation of desiccation cracks, especially in regoliths with high clay content, and thus may open preferential paths for rapid and deep infiltration. These cracks will typically close during the wet season but a legacy of higher hydraulic conductivity may remain [*Anderson and Kneale*, 1982]. Nonetheless, none of the mechanisms cited that would promote higher infiltration in vegetated sites is believed to greatly increase landsliding (Table 3.1). Lateral roots (dead and alive) and other slope-parallel preferential flow pathways in vegetated slopes could enhance drainage and thereby may preclude the development of excess pore pressures at critical locations within the slope [*Chandler and Bisogni*, 1999; *Sidle et al.*, 2001; *Uchida et al.*, 2001]. Thus, typically, the benefits of evapotranspiration and enhanced subsurface draining would outweigh disadvantages of greater infiltration, not to mention surface erosion benefits. A more likely impact of soil cracking due to water removal by deep-rooted trees may be the structural weakening of steep slopes or bluffs via fracture formation. It has been suggested that cracks caused by deep-rooted trees may weaken the regolith [*Natarajan and Gupta*, 1980].

---

[1] *Marden and Rowan*, 1993; *Bergin et al.*, 1995.

The major mechanical benefits of root systems for stabilizing hillslopes include (1) bonding or anchoring unstable soil mantles to more stable subsoil; (2) providing a membrane of reinforcement to the soil mantle; (3) binding across planes of weakness along the flanks of potential slope failures; and (4) providing localized centers of large reinforcement in proximity to individual tree trunks via soil arching and buttressing [*Gray and Megahan*, 1981; *O'Loughlin and Ziemer*, 1982; *Sidle et al.*, 1985; *Greenway*, 1987[1]] (Table 3.1). Numerous reports exist on the benefits of roots penetrating through relatively shallow soil mantles and anchoring into more stable substrate [*Wu et al.*, 1979; *O'Loughlin and Ziemer*, 1982; *Riestenberg and Sovonick-Dunford*, 1983[2]]. Fractures and discontinuities in bedrock and other substrate appear to control the ability of sinker roots to penetrate into this material [*Tsukamoto and Kusakabe*, 1984; *Greenway*, 1987]. While fractured substrates provide pathways for roots to anchor into more stable material, they may either dissipate or enhance pore water pressures in soils depending on orientation of fractures and dominant hydraulic gradients. Effect of root anchoring can be estimated by applying root density–tensile strength models to the potential failure plane.

Typically, forest soils have dense and interwoven root mats in the upper soil profile. These dense rooting systems are able to provide a laterally strong reinforcement system that affords protection against landslides [*O'Loughlin and Ziemer*, 1982; *Sidle et al.*, 1985; *Tsukamoto*, 1987; *Schmidt et al.*, 2001]. The membrane strength of these surface root systems appears to stabilize the underlying, and potentially weaker, soil. Landslides have been observed in uppermost soil mantle (just beneath the dense rooting zone) in New Zealand and Alaska, when surficial root reinforcement declined following clearcutting, indicating the influence of lateral rooting strength [*O'Loughlin and Ziemer*, 1982; *Sidle*, 1984a]. The lateral binding of unstable soil mantles across planes of weakness is basically a special case of lateral root reinforcement, where the lateral roots actually tie across the flanks of the potential landslide mass. Based on root density and diameter information collected around landslide scars and in stable soils in the Oregon Coast *Ranges, Schmidt et al.* [2001] concluded that lateral root cohesion was the dominant root reinforcement mechanism in these shallow soils. A study in the same area concluded that the larger roots contributed much less to this lateral support compared to roots <10 mm in diameter [*Roering et al.*, 2003]. Nevertheless, numerous studies worldwide have noted the stabilizing effects of large tree roots, both along the flanks and sliding surface of potential landslides [*Swanson and Swanston*, 1977; *Ziemer and Swanston*, 1977; *Wu et al.*, 1979; *Riestenberg and Sovonick-Dunford*, 1983; *Abe and Iwamoto*, 1990].

Tree surcharge and wind. In some cases, trees may impose minor mechanical instability on hillslopes due to surcharge and wind effects. For most forested slopes, the total soil weight above a potential failure plane greatly exceeds the weight of vegetation [*O'Loughlin and Ziemer*, 1982]; thus, the effect of tree surcharge on site stability is small. The weight (surcharge) of trees increases both the normal and slope-parallel force components on potential sliding surfaces. For cohesionless soils, if the slope angle (β) is less than the

---

[1] *Tsukamoto, 1987; Phillips and Watson, 1994; Schmidt et al., 2001.*

[2] *Phillips and Watson, 1994; Roering et al., 2003.*

internal angle of friction of the soil, tree surcharge should have a net stabilizing effect on hillslopes; when the slope angle exceeds ϕ, then surcharge would have a small desta- bilizing effect. Beneficial effects of greater surcharge on low cohesion soils are slightly greater when groundwater tables are higher [*Gray and Megahan*, 1981; *Sidle*, 1984b]. Likewise, tree removal may cause a small increase in stability when pore water pressure is low [*Sidle*, 1992]; however, in such cases hillslopes are typically stable. For most steep sites (especially when $\phi > \beta$) with some soil and root cohesion, decreasing surcharge by removing trees has only a minor stabilizing effect on hillslopes [*Bishop and Stevens*, 1964; *Gray and Megahan*, 1981; *Sidle*, 1992]. While some engineers have suggested that tree weight promotes failures along river banks [e.g., *Nolan*, 1984; *Thorne*, 1990], a study in Victoria, Australia, noted that the positive influence of root reinforcement is generally more important than any adverse tree surcharge effects related to bank stability [*Abernethy and Rutherford*, 2000]. For very steep, unstable banks, tree surcharge may cause bank collapse, but this process often occurs when banks are undercut.

The role of wind-generated shear stresses in forest hillslopes on landslide initiation has been discussed [*Brown and Sheu*, 1975; *Schwab*, 1983; *Greenway*, 1987; *Millard*, 2003], but theoretical findings are inconclusive due to difficulties in quantifying the dynamics of wind forces and their transmission into unstable regoliths. Isolated studies in steep sites with shallow soils have attributed landslide initiation to tree windthrow; however, findings are anecdotal with no clear indication of cause and effect [*Swanston*, 1967; *Schwab*, 1983; *Millard*, 2003]. Local wind drag coefficient, directly proportional to shear stress, is highest near forest edges, especially if wind direction is perpendicular to the edge [*Greenway*, 1987; *Chatwin*, 1994]. Thus, the most likely sites for wind-induced landslides are on shallow soils along the edges of forest openings. However, most instances of interactions of high winds and tree canopies result in blowdown or even "windsnap" (breakage of the trunk above the ground), and the association of such effects with landslide initiation is poorly documented [*Somerville*, 1979; *Greenway*, 1987; *Cannell and Coutts*, 1988; *Watson*, 2000].

SEISMICITY

Characteristics and Examples of Damages Caused by Earthquake-Triggered Landslides

Earthquake-induced landslides are less understood than rain-induced landslides because the contribution of rain to the latter phenomena is more easily determined than the contribution of ground motion by earthquakes. Although some recent advances have been made related to earthquake prediction, it is still not possible to forecast the details of locations and magnitudes of earthquake occurrences that may trigger landslides. Additionally, because earthquake-induced slope movements are much less frequent than landslides triggered by rainfall, there have been fewer opportunities to study the mechanisms of the former. At present, some of the best information related to earthquake-induced landslides is derived from retrospective analyses of landslide occurrence during periods with different magnitudes and conditions of ground motion [e.g., *Jibson and Keefer*, 1989; *Keefer*, 2002; *Duman et al.*, 2005]. In contrast to rainfall-initiated landslides, earthquake-induced landslides often occur in convex topography because convex landforms respond strongly to earthquake motion [e.g., *Harp et al.*, 1981; *Murphy*, 1995; *Okunishi et al.*, 1999] (Figure 3.22). Earthquakes also exert more complex stresses on slopes compared to rainfall, because seismic loads vary dynamically. The chronology of earthquake-induced landslides suggests that the occurrence modes, sites, and other characteristics differ from rainfall-induced landslides [e.g., *Jibson et al.*, 1994; *Harp and Jibson*, 1996; *Ochiai et al.*, 1996; *Khazai and Sitar*, 2003]. *Lin et al.* [2006] found that landslide erosion increased significantly during several years after the large Chi-Chi earthquake in west central Taiwan, primarily via retrogression of landslides triggered during the earthquake.

Virtually all types of landslides have been associated with different magnitudes of earthquakes in various settings [*Keefer*, 1984a; *Jibson et al.*, 1994]. In addition to triggering large numbers of new landslides, earthquakes can also reactivate dormant or slow-moving landslides [*Harp and Jibson*, 1996]. Many of the largest recorded rock slides, avalanches, and flows have been triggered by seismic shock, including those responsible for large losses of life [*Voight and Pariseau*, 1978; *Keefer*, 1984a,b, 2002]. While many of the catastrophic disasters have involved the failure of rock masses rather than soil materials, important exceptions exist. In 1920, a catastrophic M 8.5 earthquake triggered widespread landslides in dry loess deposits in central China, killing an estimated 180,000 people and causing devastation throughout the region [*Derbyshire et al.*, 2000]. The huge Alaska earthquake (M 9.2) of 1964 triggered major failures in soil materials in Anchorage, Valdez, and Seward, killing 52 people and damaging many homes and businesses [*Grantz et al.*, 1964; *Plafker and Kachadoorian*, 1966]. This large earthquake triggered all types of landslides and rock failures in the region, including very damaging, large lateral spreads in the highly susceptible Bootlegger Cove Clay in Anchorage, a formation consisting of silty clay, silt, and sand zones of low shear strength, high water content, and high sensitivity bounded above and below by relatively stiff clays [*Hansen*, 1966].

Figure 3.22. Landslides initiating on convex hillslopes during earthquakes: (a) a landslide that initiated at the point of slope inclination change near Okushiri Harbor, Japan, during the 1993 southwest Hokkaido earthquake (photo by the Regional Forest Office of Hokkaido); and (b) a landslide on a convex slope along the Shinano River, Niigata Prefecture, during the 2004 Chuetsu earthquake.

Earthquake-triggered landslides are most common in the circum-Pacific region and along the trans-Asiatic and Alpine Belt from Indonesia through the Himalaya Mountains to the Mediterranean Sea, where seismic activity occurs along the margins of lithosphere plates. Between 1904 and 1952, an estimated 75% of the energy of all shallow earthquakes was released in the circum-Pacific region and about 23% in the trans-Asiatic and Alpine Belt [*Gutenberg and Richter*, 1954]. An even greater proportion of energy release in intermediate (70–300 km) and deep-focus (300–700 km) earthquakes occurred in the circum-Pacific region [*Stacey*, 1969].

Due to its proximity to a triple-junction of tectonic plates, Japan experiences many earthquakes and resulting landslides. The 1974 off-Izu Peninsula earthquake and the 1978 Izu–Oshima Kinkai earthquake triggered extensive landsliding on the Izu Peninsula. After these earthquakes, much focus was placed on the problem of earthquake-induced landslides in Japan. In September 1984, the near-surface Western Nagano Prefecture earthquake (M 6.8) centered in Otaki village triggered various sizes of landslides, mainly on the slopes of Mount Ontake [*Ochiai*, 1997]. Large landslides occurred along the southeast ridge of Mount Ontake upstream of the Denjo River. One particularly large debris avalanche (3.1×10^7 m^3) transformed into a debris flow, killing 15 people and causing much damage to roads, houses, pipelines, and forests as it moved 11 km down the river [*Yanase et al.*, 1985]. Geologic formations consisting of volcanic materials were most susceptible to landslides, including failures on gentle slopes of Ontake Plateau [*Ochiai*, 1997]. Based on these experiences in Japan, it is assumed that many of the large-scale landslides, such as those that occurred at Unzen Mayu-yama in 1792 and at Mount Bandai in 1888, were caused by earthquakes.

More recent earthquakes in Japan have also caused much landslide damage. As a result of an M 7.8 earthquake off southwest Hokkaido in 1993, Okushiri Island was greatly damaged by tsunami, and many landslides and rockfalls occurred, including the collapse of a steep slope facing Okushiri Harbor [*Fukuda and Ochiai*, 1993; *Yamagishi et al.*, 1999] (Figure 3.22a). In 1995, the Hyogo-ken Nanbu earthquake impacted the greater Kobe metropolitan area, causing thousands of landslides, even when soils were relatively dry [*Ochiai et al.*, 1996; *Okunishi et al.*, 1999]. Many more landslides would have occurred if the soils were wetter. A series of earthquakes in Niigata Prefecture on 23 October 2004, the strongest of which was M 6.6, generated thousands of landslides due to the combined effects of three ground movements (all >M 6.2) within less than 40 min (the so-called "triple-punch"), the shallow focal depth of the earthquakes, soft regolith materials in tectonically active and steep terrain, the proximity of the three major epicenters, high antecedent rainfall, and land use activities [*Sidle et al.*, 2005].

Effect of Ground Motion on Landslide Occurrence, Size, and Type

The propagation of seismic waves causes a horizontal acceleration of the soil mantle. Recently improved monitoring networks of ground acceleration in Japan have measured values approaching or even exceeding gravitational acceleration during several earthquakes in the past 10 yr. The cyclic loading and unloading of soils during earthquakes depends on many factors—including earthquake magnitude, focal depth, and frequency—and exerts the major stress that causes landslides [e.g., *Updike et al.*, 1988; *Trandafir and Sassa*, 2005]. Cyclic loading of regoliths, especially in moist soft soils and unconsolidated deposits, may generate high pore water pressures that trigger landslides [*Seed and Lee*, 1966; *Wu and Sangrey*, 1978; *Ochiai et al.*, 1985].

Many factors related to earthquakes and the settings in which they propagate affect the number, sizes and types of landslides that result, as well as the distance from the epicenter. Such factors related to the intrinsic properties of earthquakes include (1) earthquake magnitude; (2) focal depth; (3) direction of seismic wave propagation; (4) seismic wave attenuation; and (5) aftershock or multiple mainshock distribution in space and time [*Voight and Pariseau*, 1978; *Khazai and Sitar*, 2003]. Factors related to the environments in which earthquakes occur include (1) inherent stability of the potential failure sites (including the geologic history); (2) existence of old or dormant landslides; (3) vegetation and land use; (4) orientation of potential failures in relation to the earthquake epicenter; (5) orientation of previously existing faults with respect to the direction of seismic wave propagation; (6) regolith wetness; and (7) slope gradient and other topographic factors [*Boore*, 1972; *Jibson and Keefer*, 1989; *Ochiai et al.*, 1996; *Tang and Grunert*, 1999; *Wang et al.*, 2003a]. While all of these factors appear to affect the frequency, size, and type of landslides that are triggered by earthquakes, at least to some degree, they are quite difficult to separate.

A useful simplification of the most important factors affecting landslide initiation by earthquakes was proposed by *Keefer* [1984a] (see Chapter 5). Data from 40 earthquakes worldwide indicated that the maximum area affected by landslides increases from about 0 at M 4.0 to 5.0×10^5 km^2 at M 9.2 [*Keefer*, 1984a]. Additionally the maximum distance from the epicenter where different types of landslides occurred generally increased exponentially with increasing earthquake magnitude, but varied slightly for different types of landslides. Rock falls, rock slides, soil falls, and disrupted soil slides are triggered by the weakest seismic activity; deep-seated slumps and earthflows are generally initiated by stronger (and probably of longer duration) seismic activity; and lateral spreads, debris flows, and subaqueous landslides require the greatest seismic activity [*Keefer*, 1984a, 2002].

Topographic Factors Affecting Landslide Occurrence During Earthquakes

During an earthquake, convex landforms exhibit stronger seismic amplification compared to other landform shapes, and earthquake-induced landslides tend to occur on and around convex topography [*Harp et al.*, 1981; *Fukuda and Ochiai*, 1993; *Murphy*, 1995; *Ochiai et al.*, 1996]. Particularly, mountain ridges shake strongly during earthquakes and shear failure may occur on these slopes, triggering a landslide [e.g., *Harp and Jibson*, 1996; *Nishida et al.*, 1996; *Tang and Grunert*, 1999; *Khazai and Sitar*, 2003]. The relationships between earthquake motion and topographic features have received attention because a maximum acceleration of 1.25 g was recorded at Pakoima Dam during the San Fernando earthquake in California in 1971. As a result, various investigations of relationships between topographic features or altitude and earthquake propagation directions have been conducted [*Boore*, 1972; *Bouchen*, 1973; *Davis and West*, 1973; *Griffith and West*, 1979]. The phenomenon by which earthquake motion is amplified in a convex landform is called a topographic effect. Studies have demonstrated that a significant topographic effect occurs when the wavelength of the seismic wave is the same as the length of the landform [*Boore* 1972, 1973; *Nishimura and Morii*, 1983]. Amplifications of ground acceleration by as much as 75% can occur, but in areas of complex topography the overall influence on ground motion cannot be easily predicted. *Wong and Jennings* [1975] found that the greatest amplifications of ground acceleration occurred when nearly horizontal waves traversed canyon topography where the distance between the convex portions of canyon walls was less than or equal to the wavelength.

Topographic effects on landslides were noted when the Great Kanto earthquake in 1923 caused extensive mountain collapse around Tanzawa, Japan [*Moroto*, 1925]. A survey of the earthquake disaster status found that more collapsed sites occurred at higher elevations, particularly on ridgelines and on convex landforms; lawns were detached as if they were talus slopes [*Moroto*, 1925]. A similar situation occurred during the 1993 earthquake-induced ridge collapse on Okushiri Island, Japan [*Fukuda and Ochiai*, 1993; *Yamagishi et al.*, 1999]. Here, rocks appeared on the surface, indicating very strong ground motion near the surface (Figure 3.23). In addition, a hotel overlooking Okushiri Harbor was struck by a landslide that initiated on a convex landform located near an abrupt gradient change on a tuff slope; this hillside reacted strongly to the ground motion during the earthquake (Figure 3.22a). Landslides near mountain ridges in Japan have also been reported as aftermaths of the 1949 Imaichi, 1948 off-Izu Peninsula, and 1978 Izu–Oshima Kinkai earthquakes. *Saito et al.* [1995] noted a similar trend of slope collapses on Okushiri Island after the 1993 earthquake; *Nishida et al.*

Figure 3.23. Segregation of the regolith causing rocks to appear on the ground surface of a narrow ridge on Okushiri Island, Japan, during the 1993 southwest Hokkaido earthquake.

[1996] further supported the phenomena via statistical analysis of slope collapses on Rokko Mountain following the 1995 southern Hyogo Prefecture earthquake. An assessment 50 yr after the M 7.7 Murchison earthquake in South Island, New Zealand, found that scarp slopes composed of dipping mudstones that faced away from the direction of oncoming seismic waves were most prone to landslides [Pearce et al., 1985]. A survey of earthquake-induced landslides in Guatemala reported that large ground accelerations associated with topographic effects often caused of landslides at sites where slope gradient changed and on convex landforms along streams [Harp et al., 1981].

In spite of the many empirical investigations that associate topographic effects with earthquake-induced landslides, the degree to which such effects amplify seismic motion and the influence of this amplified motion on the resultant stress have yet to be clarified due to the lack of detailed seismic observations in mountainous areas. In Japan, no such observations existed during the Southern Hyogo Prefecture earthquake in 1995. Hence, to determine the degree of the topographic effect, behavior of slopes during an earthquake was quantitatively modeled to estimate the acceleration response [Ochiai et al., 1995]. A large response to the acceleration waveform input from the bottom of the model in a direction orthogonal to the ridgeline was observed at a point on the ridge (Figure 3.24). The distribution of the large acceleration seen along the ridge represents the topographic effect from the earthquake motion that triggered the landslide (Figure 3.24).

Geologic Factors Influencing Landslide Occurrence During Earthquakes

During the 1949 Imaichi earthquake in Japan, landslides on granite slopes occurred near steep slopes and ridges [Koide, 1955], similar to the chain of slope failures on Rokko Mountain after the southern Hyogo Prefecture earthquake in 1995. However, landslides in pumice and ash layers on viscous soils tend to move as blocks on more gradual slopes of volcanic mountains [Koide, 1955]. A similar phenomenon appeared to occur in the 1914 Akita, the 1923 Kanto, and the 1930 Kita Izu earthquakes. During the 1984 Western Nagano Prefecture earthquake, weathered, low-cohesion pumice layers with small ϕ values acted as sliding surfaces [Tanaka, 1985]. Considering these findings, Ochiai et al. [1985] demonstrated the possibility of landslide initiation due to the liquefaction of pumice layers based on a stability analysis of a gradual slope at Ontake Plateau. An earthquake on Kozu Island south of Tokyo, Japan, in July 2000 caused extensive landslides in areas of jointed rhyolitic lava deposits, coarse pyroclastics, and crumble breccia [Miyazaki et al., 2005]. In southern Italy, numerous earthquake-related landslides occurred in weathered and foliated biotite gneiss [Murphy, 1995]. Many studies have noted the susceptibility of poorly consolidated sedimentary rocks and sediments to landslides during earthquakes. Most of the landslides on natural slopes during the 2004 Chuetsu earthquake, Niigata Prefecture, Japan, occurred in the regional geological structure consisting of silty sandstone and thin-bedded alterations of sandstone and siltstone [Sidle et al., 2005]. During the 1994 Northbridge, California, earthquake, most landslides occurred on weakly cemented Tertiary to Pleistocene clastic sediments [Harp and Jibson, 1996]. Likewise, most of the landslides initiated by the 1999 Chi-Chi earthquake in Taiwan (M 7.3) occurred in Tertiary sedimentary rocks [Khazai and Sitar, 2003]. During the 1996 Lijiang earthquake (M 7.0) in

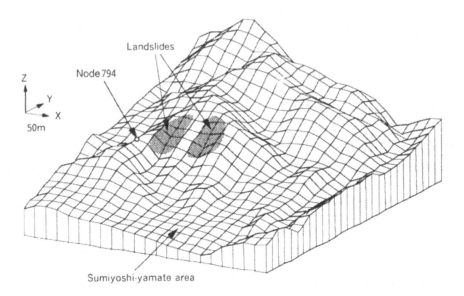

Figure 3.24. Model of the mountain landform at Sumiyoshi-yamate, Nishiomiya, Hyogo Prefecture, Japan, showing how accelerated ground motion along ridgelines can affect the location of landslides.

Yunnan Province, China, widespread debris slides occurred in weathered sandstone and mudstone [*Tang and Grunert*, 1999]. During a 35-yr period, landslides (mostly triggered by earthquakes) were most abundant in Cretaceous and Paleocene–Eocene flysch deposits in northwestern Turkey, an area bisected by the north Anatolian fault system [*Duman et al.*, 2005]. In all such geological settings, moist conditions at the time of the earthquake exacerbate landslide occurrence and damage.

Pore Water Pressure

Limited studies have been conducted related to the development of excess pore pressure during earthquakes in natural slopes. Pore pressure response depends strongly on long cycle elements of seismic acceleration [*Mavko and Harp*, 1984; *Ochiai et al.*, 1985; *Ochiai et al.*, 1987], but only a few records are available of such excess pore pressure response during earthquakes [*Fukuzono*, 1994]. As noted in the cases of debris flows into the Nebu River during the 1923 Kanto earthquake and landslides and debris flows during Western Nagano Prefecture earthquake (both in Japan), further investigation into the behavior of pore pressure during earthquakes appears necessary because pore pressure increases the risk of disaster when landslides mobilize.

Following the Nikawa landslide during the 1995 Hyogo-ken Nambu earthquake, soil samples were collected from the shear plane for ring shear testing. Based on these tests, volumetric shrinkage of pulverized particles in the shear zone was inferred as the cause of development of excess pore water pressure, thus facilitating the continuous movement of the landslide [*Sassa and Fukuoka*, 1996; *Wang et al.*, 2000]. *Sassa and Fukuoka* [1996] derived a model to explain the mechanism of mobile collapse on gradual slopes composed of volcanic products (e.g., pumice, but not sand) in terms of soil mechanics. Additionally, the landslide at Ontake Plateau was affected by the presence of groundwater from rainfall that occurred immediately before the failure. Furthermore, it was confirmed that groundwater existed in Nikawa area immediately after the landslide. Conversely, during the 1999 Chi-Chi earthquake in Taiwan, conditions were very dry, and earthquake-triggered landslides, as well as incipient failures, expanded greatly over the next few years during typhoons [*Lin et al.*, 2006]. These facts and observations indicate that previous groundwater conditions together with the geological setting play important roles related to the occurrence of earthquake-induced landslides.

Using the Seismic Coefficient in Stability Analysis

The seismic coefficient is very useful to analyze the factor of safety of earthquake-triggered landslides by limit equilibrium methods; this coefficient links the inertial seismic force on the sliding block to the static load on the block [*Koga*, 1989]. Originally the coefficient was defined as the average acceleration value and is determined by the spatial integral of the acceleration response. In addition, the coefficient is intended to convert the effect of earthquake motion to a static load acting continuously in one direction. Earthquake motion changes direction and magnitude repeatedly and irregularly over short time periods. Furthermore, because the rate of reduction in soil strength is not generally significant, it is unlikely that a slope will be totally destroyed by repeated varying loads at the peak of any particular earthquake. Hence, it is an overestimation to convert the instantaneous maximum value of response acceleration to a seismic coefficient for stability calculations; however, a reasonable value for an effective seismic coefficient may be less than the maximum value. *Ishihara* [1985] collected undisturbed samples from earthquake-induced landslides on a natural slope and compared their strengths as determined by static and dynamic loading tests. Assuming that there was a difference only in cohesion between the two loading methods and that the internal angle of friction was constant, *Ishihara* [1985] argued that the cohesion at dynamic loading could be obtained as the result of these tests. Considering the cohesion derived from the dynamic loading tests [*Ishihara*, 1985] may lead to the same results as an analysis based on seismic coefficient defined by *Koga* [1989].

VOLCANIC ACTIVITY

Causes and Types of Landslides

Volcanic eruptions are strongly associated with tectonic plate margins. During the creation of volcanic cones, not only does oversteepening occur, but also the discharge of magma loosens previously existing rocks and hydrothermal fluids promote rapid weathering, thus reducing the strength of the volcanic slope materials [e.g., *Major et al.*, 2001]. These steep, recently formed or forming slopes are subject to gravitational collapse; such flank collapses of stratovolcanoes threaten many populated regions, especially in developing nations, and many people have been killed in such disasters [*Siebert et al.*, 1987; *Reid et al.*, 2000, 2001]. In some cases, magma pressure may promote gravitational creep along the flanks of volcanoes, which can actually be mitigated by pressure relief following fracturing of the flank and local eruption [*Murray et al.*, 1994]. Planar, wedge, and arc-shaped rock failures bounded by faults, joints, bedding planes, and otherwise weak materials are common near the edifice of volcanoes [*Voight et al.*, 1983; *Iverson*, 1995; *Reid et al.*, 2000]. Processes associated with caldera collapse can trigger landslides or even lahars along the flanks of volcanoes [*Hürlimann et al.*, 1999]. Additionally, materials deposited along flanks of recently activated volcanoes are prone to extensive rockfalls and slides, along with slushflows [*Mills*, 1991].

Lahars (debris flows or hyperconcentrated flows) usually occur on the tall, steep cones of stratovolcanoes during or, more commonly, after explosive eruptions. Lahars typically occur in loose, recently deposited ash, volcanic ejecta, and pyroclastic debris on oversteepened volcanic slopes and are triggered by rainstorms, melting of snow and ice, or release of water from crater lakes, the latter two of which often reside near the summit of these volcanoes [*Pierson and Scott*, 1985; *Scott*, 1989; *Miyabuchi*, 1999; *Palacios et al.*, 2001]. Additionally, materials along the flanks of volcanoes are internally weakened by hot hydrothermal fluids [e.g., *Major et al.*, 2001], rendering them susceptible to lahar initiation. Lahars travel rapidly downslope and through channel systems, inflicting widespread damage because of their size and unexpected nature [*Suwa and Yamakoshi*, 1999; *Lavigne et al.*, 2000; *Lavigne and Thouret*, 2003; *Fagents and Baloga*, 2005] (Figure 2.3). Lahars concentrate in low-lying areas, but due to their high velocity (up to tens of kilometers per hour) and large volumes [*Major et al.*, 2001], they can also carve new channels into recently deposited ash and volcanic materials. As lahars proceed through lower gradient channels or valleys, several to tens of meters of sediments are deposited, causing most of the damage [*Miyabuchi*, 1999; *Major et al.*, 2001]. Rainstorms that trigger lahars are not necessarily very large because of the weak nature of the recently deposited ash on the steep flanks of the volcano [*Yamamoto et al.*, 1980]. Lahar occurrence is closely associated with the timing of volcanic eruptions; shortly after eruptions, lahar frequency is highest and declines exponentially as time passes after the eruption [*Suwa and Yamakoshi*, 1999]. Nevertheless, a number of cases exist where lahars have continued for a number of years or even decades after the initial eruption [e.g., *Kuenzi et al.*, 1979; *Manville et al.*, 2005]. Lahars often tend to reoccur in the same or similar locations in subsequent eruptions, although evidence of time intervals for such reoccurrence is difficult to obtain [*Scott*, 1989]. In Japan, such high-risk lahar tracks are incorporated into hazard warning systems where contemporary evidence has been collected and previous knowledge exists. Based on scaling analysis of 27 lahar paths at nine volcanoes (all but one in the USA), *Iverson et al.* [1998] developed semiempirical equations to predict inundated valley cross-section and planimetric areas affected as functions of lahar volume. These equations are useful to delineate potential hazard zones in areas where past records are not available; as expected, applications indicate that lahar hazards decrease as distance from the volcano and valley floor increase.

Recent Examples

During the 1980 eruption of Mount St. Helens, Washington, USA, one of the most intensively investigated volcanic events, a massive flank collapse occurred primarily due to gravitational collapse and resulted in a debris avalanche deposit of about 2.8 km^3 [*Voight et al.*, 1983; *Reid et al.*, 2000]. Also, during the eruption, rapidly melting snow and ice triggered two large lahars, which traveled 27 km from the crater and eroded the recently deposited pumice-flow deposits [*Pierson and Scott*, 1985]. Channel thalwegs were deeply incised (5 to 11 m) and remarkable sedimentation occurred as the lahars moved to the North Fork Toutle River. One of the most dramatic examples of a lahar disaster was during the 1985 eruption of the Nevada del Ruiz volcano, where rapidly melting snow and glacial ice triggered two large lahars that swept down different slopes of the volcano into valleys, killing about 20,000 to 24,000 people in Armero and about 1800 people in Chinchina, Colombia [*Pierson et al.*, 1990; *Mileti et al.*, 1991; *Voight*, 1996]. Such dramatic landform changes by mass wasting associated with volcanic eruptions are not rare and have occurred in Merapi volcano, central Java [*Lavigne et al.*, 2000]; Tenerife, Canary Islands [*Hürlimann et al.*, 2004]; San Vicente volcano, El Salvador [*Major et al.*, 2001]; Popocatepetl, Mexico [*Palacios et al.*, 2001]; Unzen volcano, Kyushu, Japan [*Miyabuchi*, 1999]; Pinatubo, Philippines [*Chorowicz et al.*, 1997]; and Hawaii [*Wentworth*, 1943], among other areas. The large volumes associated with volcanic landslides (particularly flank collapses) and lahars, together with their rapid velocity and temporal unpredictability, place these slope failures among the most hazardous in terms of loss of life and property.

4

Landslide Analysis

Landslides occur when the shear strength of a soil layer in a slope becomes smaller than the shear stress acting on the soil, resulting in shear failure of the layer and movement of the slope along the slip surface or at the boundaries of soil layers. Thus, factors that affect the shear strength of the soil and the shear stress acting on the soil are the causes of landslides. Various kinds of external factors that undergo changes within a short period of time are called triggers and include natural factors, such as changes in slope shape by fluvial cutting, changes in groundwater level due to rainfall or snowmelt, ground motion by earthquakes, and changes caused by human activities. Factors that form the basic characteristics of slopes that are prone to collapse and failure and need time to change are called primary causes, and include geomorphic factors (e.g., slope inclination, microtopography), geologic factors (e.g., presence of faults, bedding sequences), and soil properties (e.g., engineering and mineralogical characteristics) of the hillslope materials. Many of the external factors, as well as the primary causes of landslides and vegetation effects, are discussed in detail in Chapter 3. This chapter describes the initiation mechanisms of slope failures and landslides in terms of geotechnical engineering [e.g., *Oyagi*, 1992]. The methods briefly outlined herein employ limit equilibrium analysis to assess the stability of hillslopes [e.g., *Morgenstern and Sangrey*, 1978; *Duncan*, 1996].

ANALYSIS OF STRESSES WITHIN SLOPES AND INITIATION MECHANISMS OF LANDSLIDES

For simplicity, an infinite slope is assumed to evaluate the initiation mechanisms of slope failures and landslides (Figure 4.1). The slip surface is parallel to the slope and has an inclination of β. The length of the sliding surface is sufficiently greater than the depth of the moving soil mass so that the difference between the forces on the uphill and downhill sides of the sliding block are negligible (usually length is more than an order of magnitude larger than depth) [*Selby*, 1982]. To easily evaluate the forces acting on a slice, the length l along the sliding surface of a portion of the sliding mass is assigned a unit length (Figure 4.1). Here, the force between slices acting on the sides of the slice is balanced both horizontally and vertically. The water pressure acting on the sides of the slice is also balanced, and the water pressure acting on the bottom of the slice is u when the vertical water depth is h. The combination of loads acting on the soil mass give rise to the tangential (T) and

Landslides: Processes, Prediction, and Land Use
Water Resources Monograph 18
Copyright 2006 American Geophysical Union
10.1029/18WM05

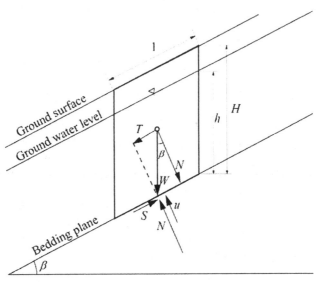

Figure 4.1. Forces acting on a slice of an infinite slope.

normal (N) forces along the sliding surface (Figure 4.1). The resistant force (S) along the sliding surface is given by the shear strength of the soil. Thus, S can be calculated by Coulomb's law, which for a soil column of unit length (Figure 4.1), is expressed in terms of forces as

$$S = c + (W \cos \beta - u) \tan \phi \qquad \text{(Equation 4.1)}$$

where c is the cohesion of the soil at the sliding surface, ϕ is the internal friction angle; for seepage parallel to the sliding surface, u is given as:

$$u = \gamma_w h \cos^2 \beta \qquad \text{(Equation 4.2)}$$

where γ_w is the unit weight of water, h is the vertical depth of the water table above the sliding surface, and W, the weight acting on the slice, is defined as

$$W = [\gamma_t (H - h) + \gamma_{sat} h] \cos \beta \qquad \text{(Equation 4.3)}$$

where γ_t is the unit moist (but not saturated) weight of the soil above the water table and γ_{sat} is the saturated unit weight of the soil.

The driving force that may cause slope failure is

$$T = W \sin \beta \qquad \text{(Equation 4.4)}$$

Thus, the safety factor of the slope (F_s) can be defined as the ratio of the resistant force (S) acting on the sliding surface to the driving force (T) acting on the potential slide mass:

$$F_s = S/T \qquad \text{(Equation 4.5)}$$

$$F_s = \frac{c}{W \sin \beta} + \frac{\tan \phi}{\tan \beta} - \frac{u \tan \phi}{W \sin \beta} \qquad \text{(Equation 4.6)}$$

Slope failure occurs when the resistant force (S) becomes smaller than the driving force (T), i.e., when $F_s < 1$. Equation 4.6 is known as the infinite slope model in soil mechanics and has been widely used to analyze the stability of natural and artificial slopes [e.g., *O'Loughlin and Pearce*, 1976; *Wu et al.*, 1979; *Schroeder and Swanston*, 1987; *Gabet and Dunne*, 2002], based on the underlying principles of limit equilibrium analysis [e.g., *Taylor*, 1948; *Lambe and Whitman*, 1969]. The condition $F_s = 1$ is not an absolute indicator of slope failure, but rather an index of likely failure because of the model assumptions and possible errors in the parameters.

For landslides that occur in shallow soils (e.g., debris slides and avalanches), the effects of cohesion are large. Because even sandy slopes contain cohesive components, the safety factor F_s of a slope decreases at greater depths in the soil profile and approaches $\tan\phi/\tan\beta$ in deep soils (i.e., the first and third terms of equation 4.6 become negligible). For shallow soils, the effects of pore water pressure (u) are large. When the soil mantle is deep, the effects of cohesion and pore water pressure become small, while the effects of slope inclination and internal friction angle become large.

For the case where no groundwater exists above the potential failure plane ($h = 0$), the stress state at the sliding surface in the slope can be expressed as

$$\sigma' = W \cos \beta = \gamma_t H \cos^2 \beta \qquad \text{(Equation 4.7)}$$

$$\tau = W \sin \beta = \gamma_t \sin \beta \cos \beta \qquad \text{(Equation 4.8)}$$

where σ' and τ are the effective stress and shear stress on the sliding surface, respectively. This stress state is expressed in Figure 4.2 as point A, which is located on a line passing through the origin with an inclination of β at a distance of $W = \gamma_t H \cos \beta$. When the failure envelope of the material is

$$\tau = c' + \sigma' \tan \phi' \qquad \text{(Equation 4.9)}$$

the safety factor F_s at point A is expressed as Y/X in Figure 4.2 [*Enoki*, 1993].

For circular slip surfaces, the Fellenius ordinary method of slices, which assumes that the resultant of the interslice forces is zero, provides the following approximate, but easily determined, factor of safety:

$$F_s = \frac{(c' + \Delta C) L_a + \tan \phi' \sum (W_s \cos \beta - u l)}{\sum W_s \sin \beta} \qquad \text{(Equation 4.10)}$$

where L_a is the total arc length of the slip surface, W_s is the total weight of the slice including vegetation, and l is the arc length of the slice. The Fellenius

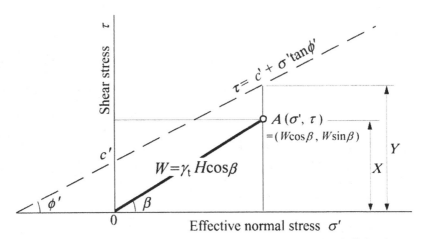

Figure 4.2. Stress conditions affecting the stability analysis of an infinite slope.

method uses only one equation that represents the moment equilibrium of the entire slide mass; it is only appropriate for circular slip surfaces and tends to yield low values of F_s [*Duncan*, 1996]. A more rigorous and accurate analysis of circular slip failures is Bishop's simplified method [*Bishop*, 1955], which assumes the resultant forces on the sides of the slices are horizontal. F_s appears on both sides of Bishop's equation, thus requiring an iterative solution:

$$F_s = \frac{1}{\sum W_s \sin\beta} \cdot \sum \left[\{(c' + \Delta C)b + (W_s - ub)\tan\phi'\} \cdot \frac{\sec\beta}{1 + \tan\beta \tan\phi'/F_s} \right]$$

(Equation 4.11)

where b is the width of the slice. Many advanced methods exist to analyze stresses and movements in slopes based on finite-element analyses [e.g., *Hungr et al.*, 1989; *Fredlund and Rahardjo*, 1993; *Duncan*, 1996]; however, such detailed geotechnical assessments are beyond the scope of this book.

QUANTIFYING LANDSLIDE TRIGGER MECHANISMS

The primary causes and triggers involved in the initiation of landslides can be understood as changes in stress conditions at a point in the slope and changes in failure conditions, such as the failure envelope. This section describes factors that act as triggers from the perspective of soil mechanics. It is usually a combination of factors rather than one factor alone that triggers a landslide.

Rise in Groundwater Level and Pore Water Pressure

To investigate shear failure in a slope consisting of materials with a shear strength τ_p it is necessary to define the cases where no groundwater exists above the potential failure plane ($h = 0$) and the condition when the water level is at the ground surface ($h = H$). When the water level is exactly at the potential failure plane ($A_{h=0}$ in Figure 4.3), stresses σ_0' and τ_0 are

$$\tau_0 = \gamma_t H \sin \beta \cos \beta \qquad \text{(Equation 4.12)}$$

and

$$\sigma_0' = \gamma_t H \cos^2 \beta \qquad \text{(Equation 4.13)}$$

Since $\tau_0/\sigma_0' = \tan \beta$, the stress state is $A_{h=0}$. When the water level is at the ground surface ($h = H$; $A_{h=H}$ in Figure 4.3),

$$\tau_0 = \gamma_{sat} H \sin \beta \cos \beta \qquad \text{(Equation 4.14)}$$

and

$$\sigma_0' = (\gamma_{sat} - \gamma_w) H \cos^2 \beta \qquad \text{(Equation 4.15)}$$

Thus, as the groundwater rises above the failure plane, the stress state changes toward point $A_{h=H}$ on a line with inclination $\tau_0/\sigma_0' = \tan \alpha$, and reaches the failure envelope at A_f, resulting in a landslide (Figure 4.3).

When the pore water pressure acting on a sliding surface increases by Δu due to various causes, the stress state shifts to the left and approaches the failure envelope (i.e., from point A_0 towards A_f in Figure 4.4). Possible causes of increases in pore water pressure (Δu) include infiltration from the surface, exfiltration from bedrock, preferential flow, and convergent flow, all of which cause local accretion of groundwater levels

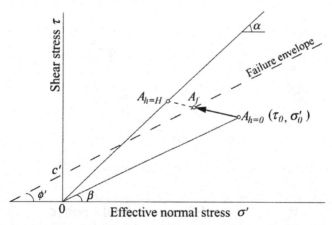

Figure 4.3. Stress changes and resultant slope failure for conditions of a rising groundwater table (i.e., from the failure plane to the ground surface).

(see Chapter 3). Pore water pressure increases can also occur due to dynamic loadings [*Mavko and Harp*, 1984; *Ochiai et al.*, 1985; *Trandafir et al.*, 2002].

Increase in Slope Inclination

Changes in slope inclination that act as possible trigger mechanisms for landslides are necessarily short-term. These changes can be caused by anthropogenic activities, such as cutting into hillslopes and overloading otherwise steep or potentially unstable slopes [e.g., *Thomson and Tiedemann*, 1982; *Smyth and Royle*, 2000]. Erosion of the toe of hillslopes by natural processes (e.g., fluvial incision, glacial retreat) can also create conditions that dictate slope failure [e.g., *Bailey*, 1971; *Edil and Vallejo*, 1976; *Haeberli and Burn*, 2002]. When slope inclination β increases by $\Delta\beta$ due to changes in landforms, the stress state approaches the failure envelope and the safety factor decreases (Figure 4.5). At $\beta + \Delta\beta$ in Figure 4.5, the stress state reaches the failure envelope and a landslide then occurs. In most situations such a slope failure is normally accompanied by some level of pore water pressure accretion; nonetheless, increases in slope inclination alone can trigger landslides.

Increase in Weight

The static weight of a slope can increase due to various reasons, the two most common of which are overloading the slope during earthworks and rainfall inputs. When the weight (W) of a soil mass increases by ΔW, the stress state changes as shown in Figure 4.6 and may approach the failure envelope, resulting in the decline of the safety factor. Similar to the case for slope inclination, other factors often accompany increases in static weight of soil to trigger a landslide. However, increases in soil weight during prolonged rainfall have been cited as a cause of landslides in Shirasu deposits with high pore volumes [e.g., *Chigira and Yokoyama*, 2005].

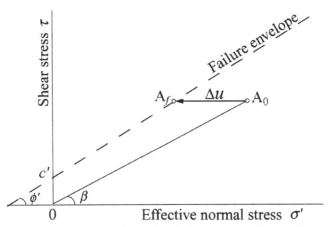

Figure 4.4. Stress changes and resultant slope failure for conditions of increasing pore water pressure.

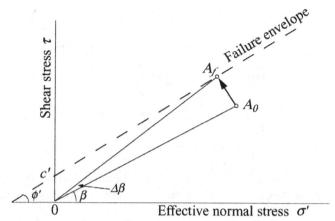

Figure 4.5. Stress changes and resultant slope failure due to an increase in slope gradient.

Earthquake Loading

The generation mechanisms of landslides and slope failures by earthquake loading are complex since earthquake loads change dynamically. An example of an earthquake load K_e, consisting of a horizontal component $k_h W$ and a vertical component $k_v W$, acting on an infinite slope, is shown in Figure 4.7 [*Ochiai*, 1997]. The sum force of the earthquake load K_e and weight (W) is W_e (see resultant vector "Sum W_e" in Figure 4.7) and can be represented as

$$W_e = W [k_h{}^2 + (1+k_v)^2]^{1/2} \qquad \text{(Equation 4.16)}$$

and, for a groundwater table below the sliding surface, the effective normal stress σ' and shear stress τ at the bottom of the block change to

$$\sigma_0{}' = W_e \cos(\beta + \alpha) \qquad \text{(Equation 4.17)}$$

and

$$\tau = W_e \sin(\beta + \alpha) \qquad \text{(Equation 4.18)}$$

where α is the angle between the weight vector (W) and W_e (sum vector).

The minimum earthquake load K_e necessary to trigger a landslide occurs when segment $A_0 - A_f$ is perpendicular to the failure envelope. This minimum load indicates the intensity and direction of the earthquake loading with the lowest safety factor in the infinite slope. For this condition, the angle between the weight (W) and earthquake load (K_e) vectors is θ, and can be expressed as

$$\theta = \frac{\pi}{2} + \phi' - \beta \qquad \text{(Equation 4.19)}$$

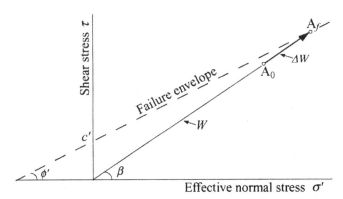

Figure 4.6. Stress changes and resultant slope failure due to an increase in total weight.

The direction θ of the earthquake load approaches $\pi/2$, or horizontal, thus supporting the use of horizontal earthquake loads to conduct slope stability analyses [*Ochiai, 1997*].

Effects of Vegetation

The effects of vegetation on the strength of soil are usually considered as reinforcement provided by root systems of trees and other woody species; these are calculated as an increase in cohesion (ΔC) (see Chapter 3). However, the lack of comprehensive data on rooting strength in soils and the high variability in reported values, together with the inherent problems in comparing values of ΔC derived from studies using different methodologies, have complicated the incorporation of rooting strength into landslide analyses [e.g., *Swanston, 1970; Abe and Iwamoto, 1986a; Terwilliger and Waldron, 1990; Schmidt et al., 2001*]. Nevertheless, it has been clearly shown that in some cases rooting strength provides the difference between stable and unstable slopes [e.g., *Riestenberg and Sovonick-Dunford, 1983; Abe and Iwamoto, 1990; Schmidt et al., 2001; Roering et al., 2003*].

To account for the effects of both root cohesion (ΔC) and weight of trees, the infinite slope model (equation 4.6) can be modified as

$$F_s = \frac{c + \Delta C}{W \sin \beta} + \frac{\tan \phi}{\tan \beta} - \frac{u \tan \phi}{W \sin \beta}$$

(Equation 4.20)

where W now includes the weight of vegetation. Such a modification of the infinite slope equation is useful for analyzing the dynamic effects of vegetation management (e.g., clearcutting and subsequent regrowth, partial cutting) on factor of safety [e.g., *Wu and Sidle, 1995; Dhakal and Sidle, 2003*]. Increases in the weight of trees generally have a minor effect on slope stability relative to other factors [*Gray and Megahan, 1981; Sidle, 1992*]. For shallow soil mantles ($H = 0.4$ m) on steep ($\beta = 42°$) slopes of coastal Alaska, the stabilizing effect of removal of tree weight due to harvesting was determined to be negligible in comparison with the destabilizing effect from the reduction in rooting strength [*Sidle, 1984b*]. Simulations of a series of alternate thinnings and clearcuts in a spruce–hemlock forest noted that a small but negligible increase in probability of failure

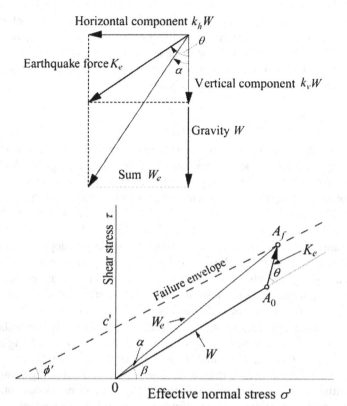

Figure 4.7. Stress changes and resultant slope failure due to earthquake loading.

(i.e., slightly lower F_s) occurred following thinning operations in the case where trees were not removed from the site compared to the case where trees were removed after thinning [*Sidle*, 1992].

Both live and decaying vegetation roots form macropores; additionally decaying wood on the forest floor contributes to macropore formation [e.g., *Kitahara et al.*, 1994; *Noguchi et al.*, 1999; *Sidle et al.*, 2001]. While the effects of preferential flow networks (derived from macropores) on saturated subsurface flow in forest soils has been investigated [e.g., *Tsuboyama et al.*, 1994b; *Sidle et al.*, 1995; *Uchida et al.*, 2001], the effect of preferential flow on landslide initiation is not fully understood. Recent research has focused on how simple pipeflow systems affect pore water accretion in soils [*Tsutsumi et al.*, 2005]; however, much work is still needed on this topic.

PRIMARY CAUSES OF LANDSLIDES

Characteristic factors observed in slopes where landslides occur or, in other words, factors that do not change within a short period of time but are based on the fundamental properties of such slopes, are referred to as primary or natural predisposing causes of landslides and slope failures. Major predisposing attributes include geomorphic factors, such as slope inclination, slope shape, and unevenness of the landform; geologic factors, such as presence of faults and tectonic activity; and soil properties, such as mineralogy and the ability of soil materials to resist slope movement. These factors have been discussed at length in Chapter 3; here they are briefly presented in terms of their specific effects related to quantitative landslide analysis.

Geomorphic Factors

Because slope movement is a phenomenon that occurs in a gravitational field, the steeper the slope, the larger the shear stress τ, and the more susceptible the slope is to failure. Since the effects of slope length, distribution of soil depth, and cross-section shape cannot be investigated for individual calculations based on the infinite slope model, statistical methods have mainly been used to study the conditions that may cause landslides and slope failures [*Yagi et al.*, 1988]. Numerical analyses of shallow landslides that incorporate a circular sliding surface indicate that for longer slopes, the lower end of the slope becomes wetter and the slope becomes less stable [*Sammori et al.*, 1993]. Also, simulations in isotropic sloping soils indicate that shallower soils are more susceptible to landslides than deeper soils because the infiltrating water reaches the impermeable layer more rapidly in shallow soils [*Sammori et al.*, 1993]. Simulations of the effect of infiltrating rainfall on the timing of slope failure revealed that deeper soils with low hydraulic conductivities had the longest lag time and shallow soils with high hydraulic conductivities had the shortest lag time [*Haneberg*, 1991a; *Iverson*, 2000].

The plan form of the hillslope is also an important factor affecting slope stability. Landslide initiation during rainfall is affected by the three-dimensional concentration and diffusion characteristics of the landform related to subsurface water. Statistical analyses and quasi-three-dimensional modeling [*Sammori*, 1995] have confirmed the results of most field investigations that report the higher susceptibility to failure of concave-shaped slopes (those that concentrate subsurface water) compared to convex slopes (those that dissipate subsurface water). On the other hand, convex-shaped slopes, especially those on ridges, are more prone to sliding and collapsing during earthquakes [*Nishida et al.*, 1996]. This is presumably because earthquake motion is amplified by ridge landforms and causes larger acceleration at ridges than in valleys [e.g., *Boore*, 1972; *Harp et al.*, 1981].

Geotechnical Properties of Soil Materials

Peak strength during shear, residual strength, and creep. Based on soil mechanics, it is possible to distinguish two types of mass wasting occurrences: those that occur at the residual strength of soil materials and those that occur at peak strength (Figure 4.8) [*Kobashi and Sassa*, 1990]. Since the internal friction angle (ϕ_r') of the soil at the residual strength is usually smaller than the internal friction angle at the peak strength (ϕ_p'), the inclination limit at which slope failure occurs is smaller at the residual strength than at the peak strength. Thus, landslides that initiate at the residual soil strength occur on gentler slopes; in Japan, these are called *jisuberi* and include deep-seated slides, some earthflows, and slumps. For failures that occur at peak strength, the shear resistance at peak strength drops after failure and a differential ΔF is generated between the shear stress and the shear resistance. In this case, the soil mass is accelerated to consume ΔF, and more rapid failure occurs—in Japan, these are called *hokai*, similar to shallow, rapid debris slides, avalanches, and flows on hillslopes. For landslides that occur at the residual soil strength, ΔF is almost 0, the soil mass is not accelerated, and the movement is slow.

Creep is caused by slow viscoplastic deformation of rocks and soil [e.g., *Owens*, 1969; *Fleming and Johnson*, 1975; *Furbish and Fagherazzi*, 2001]. Creep processes have been discussed in terms of geology and geomorphology (Chapter 3); however, only in a few cases has creep been thoroughly monitored. Because large changes in stress occur within a slope, the stress state does not reach the failure envelope, but the deformation of the slope progresses slowly by changes in groundwater levels and weathering processes [*Shimokawa*, 1980; *Jungerius et al.*, 1989; *Moeyersons*, 1989].

Fluidization processes in landslides. Slopes with sandy layers and volcanic soils that have large void ratios and low cohesion may experience dynamic shearing during undrained conditions when loads (e.g., ground settlement, earthquakes, sediment falling from upper slopes) are suddenly applied; following such loading, dynamic shearing fluidization occurs due to excess pore water pressure [*Sassa et al.*, 1996, 2004]. The mechanisms of fluidization are the same as those of liquefaction of sandy soils, but the slope movement phenomenon is called fluidization and is sometimes distinguished from liquefaction [*Iverson et al.*, 1997; *Okura et al.*, 2002; *Ochiai et al.*, 2004].

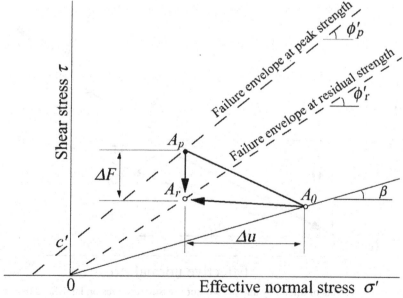

Figure 4.8. Slope failure occurring at peak strength and residual strength.

Prior to loading, the stress at some point on the slope is shown as A_0 in Figure 4.9. When loading is applied, a sudden increment in shear stress $\Delta\tau$ causes negative dilatancy and a resulting increase in pore water pressure (Δu, Figure 4.9). At first shear stress exhibits a peak, but then declines, finally reaching the fracture envelope (Figure 4.9). This final state is almost fluid, and the failed soil mass flows down the slope as a fluid body (i.e., debris flow) [*Sassa*, 1984; *Hutchinson*, 1986].

Unlike failures induced by sudden loading (e.g., earthquakes), landslides that occur during periods of rainfall do not encounter such rapid loading. Experiments conducted under controlled conditions have shown that shearing at the time of failure causes accumulation of excess pore water pressure due to negative dilatancy near the sliding surface, which in turn causes the soil layer near the sliding surface to fluidize and the soil mass to flow down the slope [*Iverson and LaHusen*, 1989].

Strength properties of unsaturated soil. In shallow slope failures, the effects of the strength of unsaturated soil on slope stability can be quite large. Slope failure has been found to occur when apparent cohesion by suction of unsaturated soil disappears during rain [*Kardos et al.*, 1944; *Abe and Kawakami*, 1987; *Crosta*, 1998; *Sasaki et al.*, 2000]. *Bishop et al.* [1960] expanded the principle of effective stress to unsaturated conditions and derived the following:

$$\sigma' = \sigma - u_a + \chi\,(u_a - u) \qquad\qquad \text{(Equation 4.21)}$$

where σ' is effective stress, σ is total stress, u_a is pore air pressure, u is pore water pressure, and χ is a parameter related to the degree of saturation of the soil. The magnitude

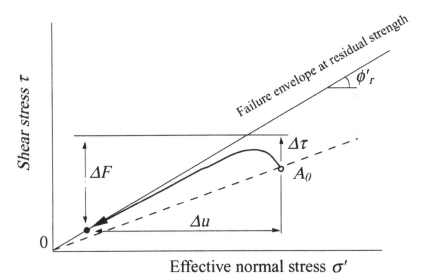

Figure 4.9. The effects of dilatancy and pore water pressure changes on landslides that involve fluidization.

of χ is unity for saturated soil and zero for a completely dry soil. If air pressure is ignored ($u_a = 0$) and the resulting simplified form of equation 4.21 is substituted into equation 4.9, we obtain

$$\tau = c' - \chi u \tan\phi' + \sigma \tan\phi' \qquad \text{(Equation 4.22)}$$

where the expression $-\chi u \tan\phi'$ denotes the apparent cohesion ($\Delta c'$); the general form of equation 4.22 can then be expressed as

$$\tau = (c' + \Delta c') + \sigma \tan\phi' \qquad \text{(Equation 4.23)}$$

Thus, higher values of $\Delta c'$ correspond to higher values of soil suction. Direct shear tests were conducted using sand and changing saturation conditions to investigate χ. Almost uniform $\tan\phi'$ values were obtained regardless of the degree of saturation; $\Delta c'$ values were always positive regardless of normal stress σ [Marui, 1981]. Since the effects of changes in cohesion c' are especially significant for shallow slope failures, rain causes the disappearance of suction, and the resulting decrease in apparent cohesion causes the slope to fail (Figure 4.10). Numerous studies have demonstrated the increase in soil cohesion for unsaturated conditions [e.g., Yatabe et al., 1986; Buchannan and Savigny, 1990; Sammori and Tsuboyama, 1990].

Hydrological Properties of Slope Materials

The hydrological properties of hillslope regoliths, especially the hydraulic conductivity of saturated and unsaturated soil and its distribution, affect the pore water pressure development within hillslopes and the initiation mechanisms of landslides [e.g., Reid, 1997]. For example, when a positive pore water pressure is generated in a slope above a semi-permeable layer by a sudden rise in groundwater during rainfall, shear failure of the basement layer occurs, resulting in a shallow landslide. Both the unsaturated hydraulic conductivity of the soil and the soil depth have considerable effects on the formation time and dynamics of the groundwater level [e.g., Iverson, 2000], which is the direct cause of slope failures. Slope irrigation experiments reveal that for shallower soil depths, the wetting surface reaches the impermeable layer more rapidly and thus groundwater accretes faster [Sammori et al., 1995]. A linear relationship was obtained between the soil depth and the time from the start of applied rainfall until the initiation of slope failure. The time until slope failure was also shown to be a function of volumetric water content of the soil, which determines its unsaturated hydraulic conductivity [Sammori et al., 1995].

The relationship between the slope movement mechanisms and groundwater has not been thoroughly clarified for deep-seated landslides with complicated structures because they are difficult to monitor. In the few studies where spatially distributed measurements of groundwater (or soil water) and earthflow velocity have been conducted, both have exhibited relatively high levels of variability [Iverson and Major, 1987; Reid, 1994; Coe et al., 2003; Simoni et al., 2004]. However, the overall principal initiation mechanisms

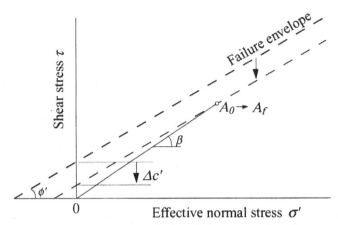

Figure 4.10. Disappearance of apparent cohesion ($\Delta c'$) upon wetting of unsaturated soils and subsequent slope failure.

should be similar to those of shallow slope failures. Thus, all landslides are affected by the interrelations between soil strength and pore water pressure. As such, monitoring unsaturated moisture and positive pore pressures in slopes is very important. *Reid* [1997] demonstrated that natural variability in hydraulic conductivity which impedes the downward movement of water can create regions of locally elevated pore pressure that may cause slope failure.

PROBLEMS IN APPLYING THEORETICAL STABILITY ANALYSIS TO NATURAL HILLSLOPES

The application of semi-quantitative stability analyses to natural hillslopes has difficulties and limitations which tend to reside in two general areas: (1) difficulties in characterizing and assessing the variability of factors that influence slope stability, and (2) inappropriateness of the inherent limit equilibrium methods for certain types of slope failures or strain conditions. Such limitations should be considered when applying the analytical methods discussed in this chapter to natural slopes.

All of the soil, topographic, geotechnical, and hydrological properties that affect stability analyses exhibit some level of anisotropy and heterogeneity in field sites. Additionally some parameters may vary over time or related to wetting conditions. Such natural variability can lead to errors in stability calculations. Most studies or summaries of the spatial variability of soil shear strength parameters indicate that c is more variable than ϕ. Coefficients of variability summarized for ϕ in gravels and sands were quite low: 7% and 12%, respectively [*Schultze*, 1972]. In contrast, *Fredlund and Dahlman* [1972] found much higher variability (coefficient of variability = 40%) in c for lakebed sediments in Canada. Values of ϕ measured in weathered granitic soils in Hong Kong varied from 33° to 41° for medium textured soils and from 21° to 40° for fine textured soils [*Lumb*, 1962]; cohesion measured in dry, decomposed volcanics varied more than 50-fold [*Lumb*, 1975]. Peak internal angle of friction values measured in sandy and gravelly sandy soils sampled from nine landslide scars in Washington ranged from 28.6° to 40.4° [*Buchannan and Savigny*, 1990]. Within a 20 km^2 study area in Jamaica, *Maharaj* [1995] examined the variability in physical and engineering properties of soils formed on several different geologic formations. Ranges of ϕ for five soil types were 24–33°, 28–37°, 30–31°, 28–32°, and 23–31°. Cohesion was not reported for all soils, but ranged from 1.00 to 5.27 kPa [*Maharaj*, 1995]. Mean cohesion and ϕ values and their respective standard deviations (in parentheses) were reported for several soil mapping units in coastal Alaska: Wadleigh, 8.6 kPa (±7.2 kPa) and 37° (±4.9°); Mitkof, 13.4 kPa (±10.5 kPa) and 33.3° (±5.2°); Kupreanof, 9.3 kPa (±6.7 kPa) and 38.2° (±4.1°); Tolstoi, 9.25 kPa (±6.7 kPa) and 36.2° (±3.8°); marine clay, 15.5 kPa (±5.17 kPa) and 30.3° (±3.0°); and Traitors, 9.73 kPa (±3.17) and 35.9° (±3.0°), respectively [*Schroeder*, 1983]. *Van Asch* [1984] back-calculated c and ϕ values from soil samples collected at rotational failures in southern Italy and compared these to values determined from direct shear tests. Back-calculated values in claystone regoliths (95% confidence intervals in parentheses) for c and ϕ were 0.13 kPa (±6.20 kPa) and 33° (±9.6°), respectively; in schist regoliths these values were 1.26 kPa (±8.90 kPa) and 52.2° (±20.3°), respectively. The back-calculated means for both soils were very similar to values obtained from direct shear tests [*van Asch*, 1984]. *Hawley* [1981] noted that the spatial variability of c and ϕ was higher in the upper 1 to 2 m of the soil profile. *Yee and Harr* [1977] found that ϕ in two cohesionless soils was strongly affected by wetting: for dry conditions, ϕ was much higher, ranging from 39.2° to 42.4° and from 39.3° to 42.1° for the two forest soils; for saturated conditions ϕ dropped to 30.2–33.6° and 27.9–30.4°, respectively. The higher ϕ values measured in dry soils were attributed to aggregation [*Yee and Harr*, 1977]. Based on many soil samples collected from a cutslope in New Zealand, both c and ϕ were normally distributed [*Pender*, 1976]. Similarly, *Burton et al.* [1998] found that values of soil shear strength (estimated by vane shear tests in the field) were normally distributed in Scotland. Soil shear strengths measured in the field in Italy depended strongly on the method used; strength decreased as wetness increased [*Zimbone et al.*, 1996]. Given that factor of safety is more sensitive to c than ϕ for typical ranges encountered in the field [*Gray and Megahan*, 1981; *Sidle*, 1984b], and given the generally higher natural variability in c measured in most studies, it is apparent that variations in c may strongly affect slope stability calculations. The high variability in c and the distribution of preferential flow pathways in deep, cohesive soils may partly explain their complex behavior upon wetting and loading [*Skempton*, 1970; *Wu and Sangrey*, 1978].

Variability of soil properties that influence water movement and the dynamics of pore water pressure can strongly affect the stability of materials with friction strength. In weathered granitic soils in Hong Kong, void ratios were highly variable with higher values in core samples compared to undisturbed soils [*Lumb*, 1962]. Hydraulic conductivity varied about three orders of magnitude in marine silts, about two orders of magnitude in weathered granite and volcanics, and one order of magnitude in red earth material in Hong Kong [*Lumb*, 1975]. Similarly, K values derived from pump tests in weathered granite in Idaho varied from 0 to 1.97 × 10^{-5} m s^{-1} [*Megahan and Clayton*,

1986]. Numerous investigations have reported that most soil hydrologic properties (e.g., hydraulic conductivity, infiltration capacity, water flux) are lognormally distributed [e.g., *Warrick et al.*, 1977; *Luxmoore et al.*, 1981; *Bronders*, 1994; *Mallants et al.*, 1997; *Regalado*, 2005]. Such variability is enhanced by the effects of macropores and interconnected preferential flow pathways [e.g., *Mallants et al.*, 1997; *Noguchi et al.*, 1999]. In unstable terrain in Brazil underlain by gneiss and granite, K varied by two orders of magnitude, sometimes at distances as small as 0.3 m. Smaller variability in K was reported in unstable residual tropical soils in Fiji [*Bronders*, 1994]. *Iverson and Major* [1987] reported that K varied by five orders of magnitude with little relation to soil depth in the Minor Creek earthflow. *Reid* [1997] showed that regolith layers with K values at least one order of magnitude lower than the overlying soil can induce failure due to loss of frictional strength. Recent measurements of pore water pressure in unstable hillslopes indicate a high degree of spatial variability that may be affected by local site conditions such as preferential flow paths, anisotropic K values, bedrock unconformities, soil heterogeneities, and topography [*Pierson*, 1980b; *Sidle* 1984a, 1986; *Iverson and Major*, 1987[1]]. Given the strong influence of pore water pressure on slope stability calculations, the temporal and spatial variability needs to be considered in such analysis. One possibility is to use a probability-based approach for estimating either the hydrological properties in the model or the pore pressure response to rainfall inputs [e.g., *Wu and Swanston*, 1980; *Reddi and Wu*, 1991; *Popescu et al.*, 1998; *Dai and Lee*, 2003]; however, such modeling approaches cannot hope to capture the spatial locations of actual landslides.

Many field investigations have shown that soil depth can vary up to an order of magnitude over several meters of slope distance [e.g., *Okunishi and Iida*, 1981; *Tsuboyama et al.*, 1994a; *D'Amato Avanzi et al.*, 2004]. At Monachyle Glen, Scotland, soil depth was highly variable and fit a Weibull distribution; depths ranged from 0 (exposed bedrock) to 1.2 m [*Burton et al.*, 1998]. Thus, because calculations of factor of safety are very sensitive to soil depth [e.g., *Gray and Megahan*, 1981], these variations need to be considered. Variations in soil bulk density (or unit weight) are generally not as high as other parameters [e.g., *Yee and Harr*, 1977; *de Ploey and Cruz*, 1979; *Schroeder*, 1983] and, given the relative insensitivity of factor of safety to typical ranges of unit weight [*Gray and Megahan*, 1981; *Sidle*, 1984b], realistic average values can often be applied.

Small-scale variations in rooting strength of vegetation have received little study but can contribute to large variations in factor of safety calculations over small slope distances [*Roering et al.*, 2003; *Sakals and Sidle*, 2004]. Based on back-calculated values of root cohesion from failed sites in Washington, *Buchannan and Savigny* [1990] reported values ranging from 1.65 to 2.87 kPa. The 12 values in this seemingly narrow range are based on composite estimates of failures; small-scale variability in ΔC could be much higher. *Sakals and Sidle* [2004] modeled small-

---

[1] *Amen*, 1990; *Harp et al.*, 1990; *Johnson and Sitar*, 1990; *Haneberg*, 1991a; *Sidle and Tsuboyama*, 1992; *Fernandes et al.*, 1994; *Thompson and Moore*, 1996; *Montgomery et al.*, 1997; *Fannin et al.*, 2000; *Dhakal and Sidle*, 2004b.

scale changes in root cohesion in Douglas-fir forests and found nearly an order of magnitude of variability in older stands and higher variability in selectively cut and strip-cut stands. Other studies suggest that past management practices as well as tree spacing and mixes of species can strongly influence the small-scale pattern of ΔC [*Schmidt et al.*, 2001; *Roering et al.*, 2003]. Because of the importance of rooting strength in stability calculations, variability due to tree species and location should be considered.

Because the estimated factor of safety of steep natural slopes, even under forested conditions, is often marginally stable (i.e., F_s in the range of 1.0 to 1.5), straining of soil materials via creep may create a condition that invalidates the use of limit equilibrium analysis. Additionally, although prediction of the initiation of failures such as earth-flows, debris flows, and debris avalanches using a factor of safety approach is possible, analysis of the deformation that takes place after failure is difficult. Developments in the fields of critical state soil mechanics and non-Newtonian fluid mechanics may lead to a better understanding and quantitative analyses of flow-type failures [e.g., *Painter*, 1981; *Coussot and Meunier*, 1996; *Iverson et al.*, 1997].

5

Hazard Assessment and Prediction Methods

OVERVIEW OF TECHNIQUES AND METHODOLOGIES

Methods of assessing landslide hazards can be roughly divided into four categories: (1) terrain stability mapping [e.g., *Ives and Messerli*, 1981; *Kienholz et al.*, 1984; *Howes and Kenk*, 1988[1]]; (2) simple rainfall–landslide and earthquake–landslide relationships [e.g., *Caine*, 1980; *Keefer et al.*, 1987; *Larsen and Simon*, 1993[2]]; (3) multi-factor, empirical landslide hazard assessments [e.g., *Aniya*, 1985; *Gupta and Joshi*, 1990; *Pachauri and Pant*, 1992[3]]; and (4) distributed, physically based models [e.g., *Montgomery and Dietrich*, 1994; *Miller*, 1995; *Wu and Sidle*, 1995[4]]. Some of these methods are more amenable to assessing relative landslide hazard at regional scales, others can be used as predictive tools for more specific sites, and yet others can be used to develop real-time warning systems. With each increment of specificity, the intensity of required data increases. Given the recent proliferation of landslide assessment techniques and models, this discussion focuses on the important attributes of each type with an emphasis on those that are useful in evaluating effects of land use.

The types of landslides emphasized herein, particularly those that are exacerbated by widespread land use practices, typically occur rather frequently and have small individual volumes [e.g., *Crozier et al.*, 1980; *Aniya*, 1985; *Froehlich and Starkel*, 1993; *Rosenfeld*, 1999; *Chang and Slaymaker*, 2002]. Such characteristics generally preclude high-cost geotechnical analysis and, particularly, structural control measures. Furthermore, many of the highest risk landslide regions are in developing countries where both technical expertise and financial resources are limited. Thus, there is a need to develop and implement landslide assessment methods for areas where certain critical data may be lacking and high-cost technology is unavailable.

TERRAIN HAZARD MAPPING

Terrain hazard mapping represents a somewhat general and qualitative level of landslide hazard assessment; topographic, geomorphic, and geologic information

---

[1] *Rupke et al.*, 1988; *McKean et al.*, 1991; *Wegmann*, 2003; *van Den Eeckhaut et al.*, 2005.
[2] *Jakob and Weatherly*, 2003; *Wieczorek and Glade*, 2005.
[3] *Dhakal et al.*, 2000; *Perotto-Baldiviezo et al.*, 2004.
[4] *Borga et al.*, 2002; *Brooks et al.*, 2002; *Wilkinson et al.*, 2002; *Dhakal and Sidle*, 2003; *Iida*, 2004.

Landslides: Processes, Prediction, and Land Use
Water Resources Monograph 18
Copyright 2006 American Geophysical Union
10.1029/18WM06

are utilized, as well as data on pre-existing landslides to generate maps with broad categories of landslide hazards [e.g., *Dobrovolny*, 1971; *Kienholz et al.*, 1984; *Howes and Kenk*, 1988; *van Den Eeckhaut et al.*, 2005]. Typically, such hazard mapping is developed or at least implemented by management agencies or regional governing bodies to evaluate the effect of various land uses on the occurrence of landslides. Thus, the main focus may not be to predict landslide occurrence, but rather to reduce the risk of landslide hazard related to a particular land use. If the assessment includes quantitative factors related to vulnerability of the elements at risk, then it can be considered a landslide risk assessment [e.g., *Alexander*, 1992; *Fell*, 1994; *Parise*, 2001]. Other types of terrain hazard maps may identify specific landscape or geologic features (e.g., geomorphic hollows; areas of windthrow; sensitive clays; vulnerable bedrock sequences) that are susceptible to slope failure. In this section, we discuss some examples of mapping procedures that have been used in various regions, but our treatment is not intended to be exhaustive.

Western Oregon Forest Practices Example

Following two devastating storms in western Oregon in 1996 and another event in early 1997, during which seven persons were killed by landslides and the estimated landslide damage exceeded $98 million [*Burns*, 1996; *Wang et al.*, 2002b], many questions were raised concerning timber harvesting and forest road-building practices in this region. In response to these concerns, the Oregon Department of Forestry conducted an inventory of landslide damage on State lands [*Robison et al.*, 1999] and drafted guidelines for mapping shallow landslide hazard sites and the potential for downslope propagation of such failures considering people and resources at risk [*Oregon Department of Forestry*, 2003a]. These evaluations are applied to areas being considered for timber harvesting and forest-road construction on State lands.

The evaluation procedure involves a multi-step assessment that includes an initial screening of the landslide/debris flow hazard at the operational site, mapping the locations of potential landslide/debris flow initiation and propagation, delineating the existence and exposure of structures and roads that may be impacted, the likely level of impact (related to human lives), assessment of public safety risk level based on hazard exposure and impact rating, and the determination of allowable timber harvesting and road construction for various safety risk levels [*Oregon Department of Forestry*, 2003a; Figure 5.1]. The identification of landslide hazard locations and downstream impact zones is somewhat subjective and is based on regional data related to hillslope and channel gradient, channel confinement, and existence of geomorphic hollows and debris fans. Other factors that are suggested for consideration (but not required in the final analysis) include tributary junction angles, likely entry angle of hillslope debris flows into channels, amount of large woody debris that may be entrained, and potential energy and its dissipation [*Oregon Department of Forestry*, 2003b]. While this assessment could certainly be improved by including a more detailed breakdown of different geological materials and zoning based on recurrence intervals of potential landslide-triggering storms, the procedure does address the element of human risk. However, although the Department of Forestry is obliged to inform opera-

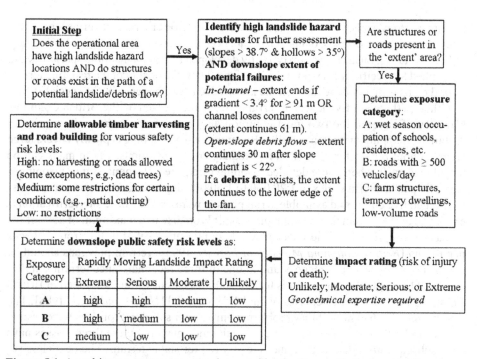

Figure 5.1. A multi-step assessment procedure used by the Oregon Department of Forestry to determine allowable timber harvesting and forest road construction based on landslide hazard exposure to people.

tors of known high landslide hazard locations [*Oregon Department of Forestry*, 2003b] within the operation areas and the State Forester can review impact ratings and geotechnical assessments, the ultimate responsibility for compliance with these guidelines rests with the operator.

British Columbia Terrain Hazard Assessment Example

Terrain stability maps that provide relative ranking of landslide hazard potential of an area large enough to delineate on aerial photographs or topographic maps have been developed as part of the Watershed Assessment Procedures in British Columbia to evaluate plans for timber harvest and road construction in areas susceptible to shallow, rapid landslides [e.g., *British Columbia Ministry of Forests*, 1995]. Mapping criteria are based on the origin of surface materials, the surface expression of the landform, and the geological processes that modify either the surficial materials or landforms [*Howes and Kenk*, 1988]. The surface materials can further be categorized by dominant textural conditions or organic components. Both geological processes and surface materials can also be qualified with regard to their current state of activity. Data required to develop these maps are typically obtained from aerial photographs, topographic maps, geologic maps, vegetation and ecosystems maps, and general field reconnaissance. Because of regional variations in climate, geology, soils, vegetation, and hydrology, specific criteria that define terrain stability classes across British Columbia have not been developed [*British Columbia Ministry of Forests*, 1995]. Such criteria are typically developed by terrain mapping specialists and are applied to specific areas within the Province. Thus, the quality of the resulting terrain hazard maps is highly dependent on the knowledge and experience of the specialist.

Although there are five levels of terrain survey recognized by the British Columbia Ministry of Forests, the two most common applications are "reconnaissance" and "detailed" mapping of terrain stability. Reconnaissance mapping typically covers large areas without regard for specific forest management operations. These maps can be useful to identify where more detailed mapping is required. Reconnaissance mapping is typically performed at scales of 1:20,000 to 1:50,000 depending on terrain characteristics, vegetative cover, and available aerial photo coverage. Very little site verification is employed at the reconnaissance mapping level: From 0% to a maximum of 25% of the delineated polygons are field-checked at a low intensity level of 0–0.1 ground checks per 100 ha [*British Columbia Ministry of Forests*, 1995]. Terrain polygons should be mapped in a single slope category and in one of three reconnaissance terrain stability classes. The original version of this terrain stability classification developed in British Columbia included five categories of landslide potential [*Howes and Kenk*, 1988]. The first two classes had no stability problems; only minor stability problems were associated with the third class. Hence, for reconnaissance mapping these three classes have been subsequently grouped into one category (S) for stable terrain. The next class (P) represents areas with a moderate probability of landsliding following road construction and logging. The most unstable category (U) denotes areas likely to experience landslides following road construction or logging [*British Columbia Ministry of Forests*, 1995]. Terrain categories P and U both require a field inspection by a qualified engineer or geoscientist prior to any activities that may adversely affect slope stability. However, before intensive field inspections, detailed mapping of terrain stability may be initiated on the more unstable categories, especially if they are directly impacted by proposed management practices. Because reconnaissance maps are generally quite conservative in terms of delineating unstable polygons, they usually show a higher percentage of unstable land area than would be mapped at a more detailed scale.

Detailed terrain stability maps identify areas that require on-site assessments of landslide potential prior to approval of specific forest management practices. Mapping is done on 1:15,000 to 1:20,000 aerial photographs, and the final terrain stability interpretations are typically presented on topographic base maps of similar scales. Slope gradient is presented as a "category", but field measurements or slope data from 1:5000-scale topographic maps increase the accuracy of these data compared to methodology used for estimating slope categories in reconnaissance mapping. General soil drainage classes as described in soil inventories are also included. Approximately 25–50% of the mapped polygons are field-checked at an intensity of 0.5 to ≥1.0 ground checks per 100 ha.

For either reconnaissance or detailed terrain stability mapping, this British Columbia system only broadly addresses forest management, basically equating timber harvesting impacts with those related to road construction. For a very general, conservative slope stability analysis, this procedure may be useful; however, it is apparent that such terrain analysis cannot delineate specific sites of high landslide risk nor can it evaluate even rough magnitudes of road- or harvest-related landslide potential. While the terrain stability maps are useful tools for initial assessment, they give no indication of the frequency, magnitude, or potential impact of the landslide hazard. Although such maps do not address the fine-scale spatial and temporal aspects of timber harvesting operations related to landslide initiation, they can be used as a first step in identifying susceptible terrain that needs further investigation prior to management actions.

Raukumara Peninsula, North Island, New Zealand, Example

The tectonically altered, soft, clay-rich rocks that underlie much of the Raukumara Peninsula predispose the landscape to progressive regrading by earthflows and slide-flow combinations [*Gage and Black*, 1979; *Pearce et al.*, 1981]. In some regions, continuous areas ≈50 to 100 km^2 have been influenced by mass movement over the whole landscape during a time span of 10,000 to 15,000 yr. Although most of this landscape has very high susceptibility to landslides, a more detailed assessment of stability may be required for certain land management activities. In >60 km^2 of this earthflow terrain, lithology and slope gradients (≈15° ± 2.5°) are similar [*Gage and Black*, 1979; *Pearce et al.*, 1981]. To map the relative landslide hazards, a detailed assessment of the age of the present land surface (using sequential aerial photographs, dendrochronology techniques, and dated tephra horizons), together with a very broad classification of geologic and geomorphic factors, was used. The presence or absence of the various tephra horizons (ages 1800–11,500 yr) provided reliable dates of various surfaces.

Based on drainage pattern, slope gradient, and erosion forms, *Gage and Black* [1979] divided their study site in Mangatu State Forest into areas dominated by mass movement and sites more strongly affected by fluvial erosion; this division closely corresponded to differences in lithology. Within the terrain dominated by mass movement, relative stability was assessed based on the recency of the land surface, as evidenced by tephra stratigraphy and other age criteria. Six levels of mass wasting were identified and mapped at a scale of 1:7820 [*Gage and Black*, 1979]. Minor modifications to this set of criteria have enabled application of this relative stability assessment to other parts of the Raukumara Peninsula, North Island, New Zealand.

Other Examples

Numerous attempts have been made to associate landslide susceptibility with geologic and geomorphic mapping units [*Swanston*, 1973; *Krauter and Steingötter*, 1983; *Hansen*, 1984; *Pomeroy*, 1982[1]]. While such maps may be very useful in specific regions, they are not applicable outside the immediate area for which there have been developed and tested. *Swanston* [1973] developed a very simple, but useful slope stability map for the area near Juneau, Alaska, using only slope gradient; highly unstable slopes were designated as >36°

---

[1] *Varnum et al.*, 1991; *Keaton and DeGraff*, 1996; *Hylland and Lowe*, 1997; *Parise*, 2001; *Lee et al.*, 2002.

based on prior experience of landslides in the area. In Anchorage, Alaska, four geological units were mapped related to landslide susceptibility [*Dobrovolny*, 1971]. The unit with the highest susceptibility consisted of bluffs (>24°) and flatter areas underlain by poorly consolidated clay, silt, and some discontinuous thin beds of fine sand (i.e., Bootlegger Cove Clay); these deposits may liquefy during strong ground motion and generate lateral spreads and rotational slumps. *Hylland and Lowe* [1997] developed landslide hazard mapping criteria for the east slope of the Wasatch Range in Utah using a simple combination of slope gradient and geological mapping units. Landslide susceptibility is strongly related to two types of incompetent bedrock in western Pennsylvania: (1) red mudstone, claystone and shale units in the Conemaugh Group, and (2) non-red mudstone and claystone of the Dunkard Group [*Pomeroy*, 1982]. In the Janghung area of Korea, landslide susceptibility mapping was conducted using remotely sensed landslide locations together with geological structure and topography [*Lee et al.*, 2002]. Mapping the characteristics of mass wasting deposits from aerial photographs with some supporting field work has been proposed as a hazard assessment tool for the early stages of land use planning in Norway [*Blikra*, 1990]. The Washington State Department of Natural Resources' Landslide Hazard Zonation Project is developing a set of landslide hazard maps based on comprehensive catchment analysis by a team of multidisciplinary specialists with landslide expertise [e.g., *Wegmann*, 2003]. The landslide susceptibility maps are intended as a guide for foresters and land managers to identify landslide-prone sites prior to operations. This developing program relies strongly on field investigations and represents one of the more advanced regional landslide susceptibility mapping efforts that incorporates uncertainty and estimates of sediment delivery to streams into the overall assessment.

In Sweden, where some 20,000 km^2 of sensitive marine clay deposits exist, about 25% of this area is susceptible to spontaneous landslides [*Stal and Viberg*, 1981]. Landslide hazard mapping based on aerial photographs, geologic and topographic maps, and site investigations in municipal areas has shown that sensitive clay areas with surface gradients <1.2° are stable, whereas slopes between 1.2° and 5.7° may have a minor risk of landslide and may require further inspection; slopes >5.7° are considered unsuitable for residential development [*Stal and Viberg*, 1981; *Berggren et al.*, 1991]. Sites in the intermediate slope range may become unstable if the toe of a hillslope is excavated or if changes in groundwater level occur as the result of development [*Berggren et al.*, 1991].

Recent Technology

Recent advances in remote sensing techniques and the development and application of contour-based digital elevation models (DEMs) can improve terrain hazard mapping, especially when processed within geographic information systems (GIS) [*Sakellariou and Ferentinou*, 2001; *Lee et al.*, 2002; *Dhakal et al.*, 2002; *Mizukoshi and Aniya*, 2002]. In particular, satellite imagery and light detection and ranging (Lidar or airborne laser scanning) are useful for assessing landslide locations as well as developing detailed DEMs and hillshade maps [e.g., *Lee et al.*, 2002; *Wills and McCrink*, 2002; *van Den Eeckhaut et al.*, 2005]. High-resolution DEMs produced from airborne laser altimetry can be used to assess surface characteristics of active and dormant landslides as well as delineating detailed topographic information useful in predicting areas of future landslides [*Dietrich et al.*, 2001; *Haugerud et al.*, 2003; *McKean and Roering*, 2004]. However, identification of landslide locations is subject to interpretation, and considerable differences are apparent when comparing results from hillshade map interpretations with those based on field surveys [*Wills and McCrink*, 2002; *van Den Eeckhaut et al.*, 2005]. Additionally, topographic conditions and landslides under dense forest cover present continuing challenges for remote sensing methodologies and costs of Lidar are high.

SIMPLE RAINFALL–LANDSLIDE AND EARTHQUAKE–LANDSLIDE RELATIONSHIPS

Landslide hazard assessment based on simple relationships with rainfall characteristics has been applied at both the global [*Caine*, 1980] and regional [*Cannon and Ellen*, 1985; *Canuti et al.*, 1985; *Larsen and Simon*, 1993[1]] scales. When coupled with real-time rainfall data, such analyses can provide the basis for early warning systems for shallow landslides [*Keefer et al.*, 1987; *Iiritano et al.*, 1998]. Thus, these analyses are quite different than terrain hazard assessments, which are more focused on developing general stability hazard maps. For earthquake-triggered landslides, simple relations between earthquake magnitude and distance to the epicenter have proved to be useful general indicators for landslide hazard assessment [*Keefer*, 1984a]. The major problem with earthquake analysis is uncertainties associated with future earthquake location, magnitude, and timing.

Rainfall Characteristics

At the most general level of prediction, 73 shallow landslide/debris flow occurrences from around the world were assessed on the basis of mean rainfall intensity and storm duration [*Caine*, 1980]. The lower bound of this log-linear relationship is given by

$$I = 14.82 \, D_s^{-0.39} \qquad \text{(Equation 5.1)}$$

where I is the mean intensity of the rainstorm (mm h^{-1}) and D_s is the storm duration (h). In principle, equation 5.1 could be used to assess the likelihood of shallow landslides and debris flows on hillslopes for different magnitudes and frequencies of rainfall. If at any time during a storm the average rainfall intensity exceeds the threshold value (I), shallow, rapid landslides may occur. Close inspection of *Caine's* [1980] analysis reveals that several storms near the short-duration end of the lower threshold were misinterpreted or based on data for in-channel debris flows. Additionally several of the data are plotted for "events" longer than 10 days; these are not likely individual events but rather artifacts of the rainfall records available.

---

[1] *Finlay et al.*, 1997; *Wieczorek et al.*, 2000; *Jakob and Weatherly*, 2003; *Fiorillo and Wilson*, 2004.

It appears that some of these data were excluded in Caine's original calculations because with these long- and short-duration events removed, the threshold calculation is only slightly modified:

$$I = 13.58 \, D_s^{-0.38}$$ (Equation 5.2)

In areas where rainfall intensity–duration–frequency relationships are well known, they may be overlain on these thresholds to identify return periods for various combinations of intensity and duration of precipitation that may trigger shallow landslides [*Sidle et al.*, 1985]. Such an application is complicated because such threshold relationships are based on landslides triggered by a minimum input of rainfall, i.e., presumably under nearly saturated antecedent conditions. Because such difficulties may occur in many areas, it may be beneficial to include an index of antecedent soil moisture to better utilize such simple intensity–duration relationships for shallow landslide prediction.

The rainfall intensity–duration concept has been adapted to landslides in specific regions of the world and compared with *Caine's* [1980] global threshold. Relationships developed for a set of 256 tropical storms in Puerto Rico over a 32-yr period show that for storms with durations <10 h, almost three times the rain intensity was needed to trigger landslides compared to temperate regions; after 4 days duration, the rainfall necessary to trigger landslides converges with the global threshold [*Larson and Simon*, 1993]. For pyroclastic deposits in Campania, Italy, intensity–duration thresholds all plot well above the global threshold, partly because selected data were excluded where only single debris flows occurred during low rainfall [*Fiorillo and Wilson*, 2004]. In the Piedmont Region of northwest Italy, a rainfall intensity–duration threshold that captured 90% of the landslides from four events plotted just below Caine's threshold for storm durations >10 h; for durations <10 h, the thresholds were similar [*Aleotti*, 2004]. An intensity–duration threshold based on 7 yr of rainfall data and individual failures in a small area of the Santa Cruz Mountains, California [*Wieczorek*, 1987], falls considerably below Caine's threshold. Shallow landslides modeled during the 86 most significant storms from 1972 to 1990 in a subcatchment of Carnation Creek, Vancouver Island, were compared to Caine's threshold; 65 storms produced only one landslide and 36 of these storms were below the threshold: all 21 storms that produced two or more landslides fell above the threshold [*Dhakal and Sidle*, 2004a]. *Jakob and Weatherly* [2003] plotted intensity values for different durations of 18 individual landslide-triggering storms in the North Shore Mountains of Vancouver, British Columbia. As expected, the low bound of these data fell well below the global threshold for landslide initiation; however, since the data represent incomplete segments of storms, it is difficult to compare them with studies based on average storm intensity or to interpret their significance related to landslide initiation.

Some interesting applications of the rainfall-duration concept for landslide initiation have been developed in the San Francisco Bay area of California. After dividing the region into areas of low (≤660 mm) and high (>660 mm) mean annual precipitation, thresholds were delineated on the mean intensity–storm duration plots that identified "abundant" landslide occurrence when thresholds were exceeded and once specified antecedent moisture conditions were met [*Cannon and Ellen*, 1985]. Because these thresholds were derived for abundant landslides, the high precipitation threshold is above Caine's threshold; however, the dry precipitation threshold is similar. *Keefer et al.* [1987] utilized these relationships between rainfall and landslide initiation together with antecedent moisture, geological indicators of instability, and real-time rainfall data to develop a warning system for landslides during major storms in the region. While such an advanced warning system depends on spatially distributed, accurate, and timely disseminated rainfall data, it is possible that similar applications could be successful in densely populated regions where local governments support regional telemetered networks of rain gages. Further advances in such warning systems can be considered in the future with advanced forecasting using improved Doppler radar systems [e.g., *Klazura and Imy*, 1993; *Wei et al.*, 1998].

Rainfall intensity–duration curves have also been used to assess the effects of various types of land use. Moderate-sized storms in December 1979 at Maimai, New Zealand ($I = 4.4$ mm h^{-1}; $D = 33$ h), and October 1980 at Freshwater Bay, Chichagof Island, Alaska ($I = 5.6$ mm h^{-1}; $D = 8$ h), generated landslides in recently clearcut areas, while nearby unharvested sites experienced no landslides [*O'Loughlin et al.*, 1982; *Sidle*, 1984a]. When plotted with Caine's data, these storms from New Zealand and coastal Alaska fell just above and below the global threshold, respectively, indicating that timber harvesting may have reduced the storm intensity necessary to trigger landslides for a given storm duration [*Sidle et al.*, 1985]. Rainfall intensity–duration relationships have been used in landslide risk analysis in sites affected by timber harvesting [*Rice et al.*, 1985], as well as for evaluating the effects of road construction on landslide hazards [*Shakoor and Smithmyer*, 2005].

To assess the effects of antecedent rainfall on the mean intensity–duration relationships with respect to shallow landslides, all data complied by *Caine* [1980] that included 2-day antecedent rainfall (only 12 events) were plotted along with new data from other studies (Figure 5.2). Two-day antecedent rainfall was used as an indicator of antecedent soil moisture due to data availability and because *Sidle* [1986, 1992] found this parameter strongly correlated with maximum piezometric response in unstable hillslope hollows. The total data set with 2-day antecedent rainfall (67 events) was divided into storms with "wet" antecedent conditions (>20 mm of rainfall in the 2 days prior to the event) and those with "dry" antecedent conditions ($\leq$20 mm). The lower thresholds of these data for wet and dry antecedent conditions plot below and above *Caine's* [1980] original threshold, respectively (Figure 5.2). The threshold for wet antecedent conditions is

$$I = 12.64 \, D_s^{-0.49} \qquad \text{(Equation 5.3)}$$

and the threshold for dry antecedent conditions is

$$I = 19.99 \, D_s^{-0.38} \qquad \text{(Equation 5.4)}$$

To account for the effects of antecedent soil moisture on landslide initiation during rainfall events, an Antecedent Water Status Model (AWSM) was developed that uses daily rainfall and estimates soil water status based on a climatic water balance [*Crozier*

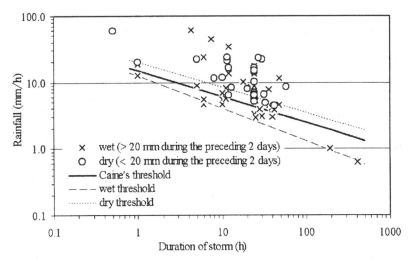

Figure 5.2. Average rainfall intensity–duration relationships for 67 worldwide events which are separated with respect to antecedent moisture conditions: dry, ≤20 mm of precipitation in the preceding 2 days (O); and wet, >20 mm in the preceding 2 days (×). Lower thresholds are presented for both of these relationships (different dashed lines) along with *Caine's* [1980] global threshold (solid line).

and Eyles, 1980; *Crozier*, 1999]. Antecedent soil moisture status is calculated based on pan evaporation rates, daily rainfall, and an assumed or measured soil moisture storage capacity (a value of 120 mm was initially assumed in New Zealand) [*Crozier and Eyles*, 1980]. Rainfall in excess of soil storage requirements is drained exponentially over a 10-day period to provide continual estimates of antecedent soil water status on a daily basis. When landslides and their associated daily rainfalls are plotted in relation to the soil water indices, linear thresholds can separate days with and without landslides (Figure 5.3) [*Crozier and Eyles*, 1980; *Crozier*, 1999]. The limitations of this empirical approach relate to the specific timing of storms and landslides, difficulties in representing site conditions with regional parameter estimates, and regional applicability of point scale climate values [*Crozier*, 1999]. Nonetheless, the AWSM effectively discriminated between landslide and stable conditions during a 1974 rainstorm in Wellington, New Zealand (Figure 5.3). Modifications to this model have been proposed, but at the expense of heavier parameterization and thus greater data requirements [*Glade*, 2000; *Glade et al.*, 2000]; results confirm that the model needs to be locally calibrated. Similarly, *Terlien* [1998] uses combinations of daily and antecedent rainfall along with intensity–duration relationships as a basis for modeling shallow, rapid landslides.

Earthquake Characteristics

Based on data from 40 major earthquakes in various settings around the world that ranged in local magnitude from 5.2 to 9.5, *Keefer* [1984a] found that landslide distribution and type were strongly correlated with earthquake magnitude and distance to the epicenter. To assess the lower limit of earthquake shaking levels associated with landslides, intensity data from several hundred earthquakes in the USA were examined. The lowest magnitude earthquake needed to trigger different types of landslides was ≈4.0 for rockfalls, rock slides, soil falls, and disrupted soil slides; ≈4.5 for soil slumps and soil block slides; ≈5.0 for lateral spreads, debris flows, subaqueous landslides, rock slumps, rock block glides, and slow earthflows; ≈6.0 for large rock avalanches; and 6.5 for soil avalanches [*Keefer*, 1984a, 2002]. The maximum area likely to experience some degree of landsliding during earthquakes ranged from 0 at M ≈ 4.0, to about 250 km² at M = 5.4, to a maximum of 500,000 km² at M = 9.2 [*Keefer*, 1984a]. Using the global data of *Keefer* [1984a] plus seven additional earthquakes, *Keefer and Wilson* [1989] calculated the following regression relationship:

$$\log_{10} A' = M - 3.46 (\pm 0.47) \qquad \text{(Equation 5.5)}$$

where A' is the potential area affected by landslides (km²) and M is earthquake magnitude in the range of 5.5–9.2. A similar relationship was developed for earthquake-triggered landslides in New Zealand [*Hancox et al.*, 2002]

$$\log_{10} A' = 0.96 (\pm 0.16) M - 3.7 (\pm 1.1) \qquad \text{(Equation 5.6)}$$

although only 22 events were included. The areas affected by landslides and the maximum epicentral distances for landslide impacts were generally smaller compared to the worldwide data of *Keefer* [1984a].

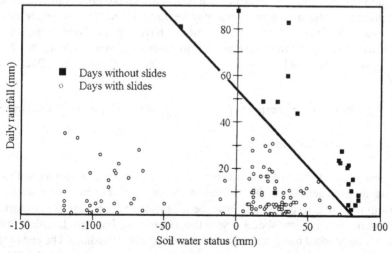

Figure 5.3. Landslides and their associated daily rainfalls plotted in relation to the soil water indices (either deficit or excess) for Wellington City, New Zealand [after *Crozier and Eyles*, 1980; *Crozier*, 1999].

Data on the maximum epicentral distance for landslides were analyzed for 47 earthquakes in Greece that occurred between 1650 and 1995 [*Papadopoulos and Plessa*,
2000]. These data fell is the same range as Keefer's data; the upper bound for the Greek
data was given as follows [*Papadopoulos and Plessa*, 2000]:

$$\log_{10}(R_e) = -2.98 + 0.75 M \qquad \text{(Equation 5.7)}$$

where R_e is the maximum epicentral distance (km) where landslides occurred for a
given magnitude earthquake (M 4.5 to 7.5). Although large errors were associated with
estimates of epicentral distance and magnitude prior to 1911 [*Papadopoulos and Plessa*,
2000], most of the maximum estimates (equation 5.7) were in the same range as the
worldwide data [*Keefer*, 2002].

While these general relationships between earthquake characteristics and landslide
occurrence (equations 5.5–5.7) are useful for designating the susceptibility of seismically active regions to different types of landslide disasters and for assessing the likely
damages from such disasters, the major problem of predicting the location and timing
of earthquakes still remains. Recent advances in analyzing changes in seismic activity
patterns (particularly related to non-linear dynamics), as well as ionospheric precursors of earthquakes, lend promise for advancing the state-of-the-art of spatial-temporal
earthquake predictions [e.g., *Bodri*, 2001; *Wyss and Matsumura*, 2002; *Peresan et al.*,
2005; *Pulinets*, 2006]. Nevertheless, because the stress level within the earth cannot be
directly measured and many precursors used in earthquake predictions lack quantitative definitions and continuous observations upon which statistical analyses can be
based, the science of real-time and spatial earthquake prediction is progressing slowly
[*Console*, 2001; *Wyss*, 2001; *Peresan et al.*, 2005]. *Keilis-Borok et al.*, [2001] developed
a preliminary method to link intermediate-term earthquake prediction methods with
disaster preparedness; such procedures may be quite useful for earthquake-triggered
landslide hazards. Several probabilistic methods have recently been proposed, based
on *Newmark's* [1965] sliding-block model, to assess and spatially map earthquake-
induced landslide hazards [*Jibson et al.*, 2000; *Refice and Capolongo*, 2002; *Carro et
al.*, 2003].

MULTI-FACTOR, EMPIRICAL LANDSLIDE HAZARD ASSESSMENT

General Characteristics and Types

Empirical landslide hazard assessments share some common attributes with terrain
hazard mapping but generally differ with respect to the number of factors considered, their
relationship to past landsliding, and how the factors are evaluated in the context of the
assessment. In empirical landslide analysis, the factors contributing to landslide initiation
are typically established based on characteristics of existing landslides. The end product is
focused on producing maps, or at least useable decision tools, that relate landslide hazards
to measurable environmental attributes—similar to terrain hazard analysis. Both bivariate
and multivariate statistical analyses can be employed to examine each factor or several

factors together, respectively, in combination with the presence or absence of landslides [*Soeters and van Westen*, 1996]. Most approaches assume that landslides are more likely to occur under conditions similar to those of previous failures [*Brabb*, 1984; *Varnes*, 1984]. Numerous examples of these empirical landslide analyses can be found; herein only a few representative examples are discussed.

Multi-factor methodologies typically estimate landslide hazard using the relationships between past landslide patterns with various site characteristics. In such cases, the weighting of site attributes affecting slope stability is important. Factors typically considered include topography, geology, vegetation cover or land use, hydrology, and soil properties. Trigger mechanisms, such as rainfall and seismic patterns, are usually not included because such hazard assessments focus on conditions predisposing hillslopes to failure. Some hazard assessments have been developed using factor weightings based on past experience or professional judgment [e.g., *Stevenson*, 1977; *Newman et al.*, 1978; *Anbalagan*, 1992; *van Westen et al.*, 1993]. While such qualitative hazard mapping based on professional judgment can be very effective if high-quality distributed data and adequate expertise are available [e.g., *Newman et al.*, 1978; *Nilsen et al.*, 1979], the derived factor weighting estimates may vary considerably and lack objectivity. For example, the same criteria and weightings are probably not applicable to different types of landslides in the same region. In other cases, individual factors are simply equally weighted [e.g., *Guillande et al.*, 1995] or methodologies are inadequately specified [*Pachauri and Pant*, 1992; *Perotto-Baldiviezo et al.*, 2004]. Use of GIS allows for accurate and unbiased development of weighting factors typically used in such analysis [e.g., *Carrara et al.*, 1991; *Soeters and van Westen*, 1996; *Dhakal et al.*, 1999]; however, it must be remembered that such weighting factors are only as good as the data bases from which they are derived as well as the errors in the cause–effect relationship implicit in such generalizations.

Statistical approaches to landslide analysis involve developing relationships among factors affecting landslides based on a priori statistical analyses; these relationships are then applied to sites with similar characteristics [*Soeters and van Westen*, 1996]. In bivariate statistical analysis, each factor is evaluated separately in conjunction with landslide density (or volume). This bivariate approach has been widely used to identify factors that are significantly related to landslide occurrence and, if significant, relative weightings are assigned [e.g., *Gupta and Joshi*, 1990; *Mehrotra et al.*, 1996; *Ayalew and Yamagishi*, 2005]. Various statistical methods have been used to calculate factor weightings [*Soeters and van Westen*, 1996]. Multivariate approaches to empirical landslide hazard analysis consider the interrelationships amongst factors in terms of selection and weighting. After sampling all relevant factors at appropriate scales, the presence or absence of landslides is determined. Multiple regression or discriminate analysis is then typically used to analyze the resulting matrix [*Mulder and van Asch*, 1988; *Carrara et al.*, 1991; *Rollerson et al.*, 1997; *Dhakal et al.*, 2000]. In some recent cases, neural network methods have been applied to weight causative factors [e.g., *Lee et al.*, 2004; *Yesilnacar and Topal*, 2005].

The concept of uncertainty or probability has been incorporated into landslide models [e.g., *Ward et al.*, 1981; *Sidle*, 1992; *Popescu et al.*, 1998[1]], particularly related to triggering

---

[1] *Sidle and Wu*, 1999; *Haneberg*, 2004; *Iida*, 2004.

mechanisms, but also to site factors. *Dai and Lee* [2003] introduced probabilistic analysis in a GIS-based multi-factor landslide hazard model. Static dependent variables in the logistic regression model were geology, slope gradient, slope aspect, elevation, slope shape, and land cover, and rolling 24-h rainfall was used as a dichotomous independent variable to denote the presence or absence of a landslide [*Dai and Lee*, 2003]. The model was then applied to rainfalls of varying return periods to predict the probability of landsliding on natural slopes of Lantau Island, Hong Kong, in space and time. Another approach to probabilistic assessment of landslides incorporates updating of risk based on accumulating information or changing concepts using principles in Bayesian decision analysis [e.g., *Wu et al.*, 1996; *Allison et al.*, 2004].

Examples of Multi-Factor Assessment

Early research that predated sophisticated GIS utilized similar concepts as incorporated in the more advanced processing systems of today. In the 1970s, a conceptually advanced slope hazard assessment was developed for the San Francisco Bay region using information derived from maps of slope gradient, previous landslide deposits, and surficial and bedrock geology [*Nilsen et al.*, 1979]. At the scale of 1:125,000, slope stability was assessed for the entire 19,200 km^2 region, including outlying rural areas. Six slope stability categories were designated based on detailed assessments of the site characteristics and previous landslide history: Category 1, generally stable, gentle (0–3°) slopes underlain by bedrock or alluvium; Category 1A, similar to Category 1 but underlain by unconsolidated deposits and thus subject to ground failure during earthquakes; Category 2, relatively stable slopes (3–8.5°) not underlain by landslide deposits or other deposits susceptible to slope failure; Category 3, largely stable hillsides and upland areas (>8.5°) that are not underlain by landslide deposits or deposits susceptible to failure, although small landslide areas may be included; Category 4, hillsides and upland areas (>8.5°) underlain by bedrock that is highly susceptible to failure but not underlain by landslide deposits— little evidence of landsliding exists but these areas are susceptible to landslides, particularly during earthquakes and following human activities; and Category 5, any areas underlain by old landslide deposits—these areas are the most susceptible to future landsliding [*Nilsen et al.*, 1979]. The success of this hazard assessment is largely attributed to the investment in expertise and the intensity of field data compiled over a number of years in this unstable, highly populated region.

In stark contrast to the comprehensive data acquisition system employed in the San Francisco region is the case of stability hazard assessment in the Himalayas of Nepal and northern India. In this region landslides are triggered by both earthquakes and rainfall. However, because of the paucity of spatially distributed data in mountainous regions, these causative factors have not been directly included in multi-factor analyses of landslide hazard potential. A very general approach to landslide hazard assessment in the Ramganga catchment in the Lower Himalayas involved mapping recent and old landslides from aerial photos and overlaying this information on geologic maps, remotely sensed maps of land use, and maps of major faults and thrust zones [*Gupta and Joshi*, 1990]. Four geoenvironmental criteria were used to evaluate landslide hazard: (1) five general lithology groups; (2) five categories of land use; (3) eight distance ranges from major tectonic features; and (4) eight classes of slope aspect. The distribution of landslides on various slope aspects is a surrogate for the influence of rock structure and physiography on landslide occurrence; landslides tend to be more frequent in the direction of the dip of the bedrock. Since the hazard assessment incorporated only recent and older failures (i.e., not potential failures), slope gradient was not used in the analysis [*Gupta and Joshi*, 1990]. This important parameter would obviously improve the GIS-based hazard zonation, especially if inferences related to future land use changes are desired. Percentages of landslide occurrence in all the geoenvironmental subcategories were computed individually and compared to the average landslide incidence; subcategory values >33% higher than the overall average were deemed high risk (weighted as 2), values <33% lower than the average were low risk (weighted as 0), and values in the range of ±33% of the mean were moderate risk (weighted as 1) [*Gupta and Joshi*, 1990]. Each of the four geomorphologic criteria (lithology, land use, distance from tectonic features, and slope aspect) were equally weighted in this preliminary analysis [*Gupta and Joshi*, 1990] (Figure 5.4). More detailed hazard analysis should consider the potential of weighting each of the criteria based on local knowledge and relationships to landslide intensity. The overall methodology is illustrated in Figure 5.4 along with suggestions for improvement. Although forest harvest is not explicitly addressed in this hazard mapping, it can be seen that areas where vegetation is either sparse or totally removed are most susceptible to landslides.

An empirical landslide hazard assessment at the medium scale (1:25,000–1:50,000) based on multivariate statistical analysis within a GIS framework was conducted for the

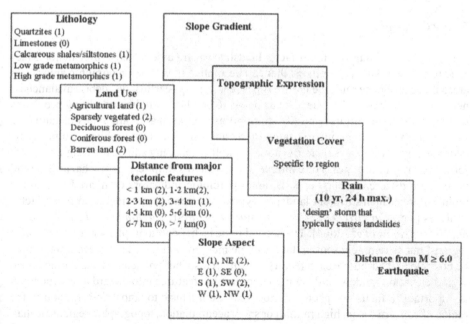

Figure 5.4. Multi-factor empirical landslide hazard assessment for the Ramganga catchment in the Lower Himalayas [based on *Gupta and Joshi*, 1990]. Recommendations for a more detailed, process-focused analysis are shown in the five shaded "boxes".

124 km² Kulekhani catchment in the central Nepalese Himalaya [*Dhakal et al.*, 1999]. Distributions of landslide and non-landslide areas in the catchment were mapped from aerial photos followed by extensive field checking. To determine the factors affecting landsliding, layers of various topographic attributes derived from a DEM along with geology and land use/cover were analyzed by discriminant (Quantification Scaling Type II; QS-II) analysis [*Dhakal et al.*, 2000]. The landslides that were surveyed in the field primarily guided the selection and classification of these factors. Eight terrain factors were included: slope gradient, slope aspect, elevation, drainage basin order, distance from the ridge, distance from the valley, land use/cover, and geology. Each of these factors was categorized into three to seven subclasses. For each analysis unit (grid-cell) of landslide and non-landslide groups, class codes of the eight terrain factors are assigned for the Q-S II analysis. The remaining landslides are then used for the evaluation of the hazard maps. This method is suitable for landslide hazard assessment, because nominal factors such as geology or land use/cover are often most important to discriminate between landslide and non-landslide groups. The quantification of classes of the factors is done in such a way that the proportion of variance between the groups to the total variance is maximized. The calculated "class score" indicates the relative importance of each factor; in this case geology was most important. For the data that are not sampled, the factor-classes are measured and the group to which they belong is predicted from the score of the classes. The scores can then be used to produce a map showing different relative classes of hazard [*Dhakal et al.*, 2000].

Future Challenges

To significantly improve multi-factor landslide hazard assessments, three major issues need to be overcome: (1) Methods that can be applied in broader geographic areas or in areas that experience multiple failure types (e.g., slump–earthflows, debris avalanches) need to be developed; (2) a clear focus needs to be placed on the underlying processes that relate to slope failure; and (3) temporal as well as spatial attributes of landslide susceptibility need to be incorporated in the analysis. The first problem can partly be addressed by developing separate criteria, as well as mapping and factor-weighting rules for different failure types. For example, deep-seated slump–earthflow hazards would necessitate different geologic, soils, and hydrologic indicators compared to shallow rapid failures. Such different landslide types are often not considered in multi-factor analyses [e.g., *Carrara et al.*, 1991; *Varnum et al.*, 1991; *Guillande et al.*, 1995; *Lee et al.*, 2002; *Ayalew et al.*, 2004]. The second issue is difficult to implement in data-sparse regions, but as remotely sensed data become more available and more accurate, it may be possible to incorporate such factors as rainfall distribution, detailed vegetation cover, digital elevation models, and detailed geomorphic features into hazard assessments. It is important to focus on specific factors that are intrinsic to landslide initiation in the region of concern—e.g., high rainfall or snow accumulation, topographic regions of shallow groundwater accretion, steep slopes, hummocky topography, and areas with weak or decaying root strength. Such data that focus on improving process understanding will also benefit the application of landslide hazard assessments to multiple failure types and larger regional applications, as well as the inferences related to forest operations and other land uses. Thirdly, addressing temporal aspects has typically fallen only into the realm of physically based models; however, a recent probabilistic application in Hong Kong indicates one method of combining static and dynamic variables to address both spatial and temporal aspects of the landslide hazards [*Dai and Lee*, 2003]. Probability estimates have also been included in more subjective landslide analyses, including issues related to timber harvesting and forest roads, although specific temporal timeframes were not addressed [*Tsukamoto et al.*, 1982; *Rollerson and Sondheim*, 1985; *Rollerson et al.*, 1986].

It appears that with increasing sophistication of GIS, remote sensing, and statistical/analytical tools, there is a tendency to focus more on new methods rather than trying to understand causal linkages for specific types of landslides [e.g., *Varnum et al.*, 1991; *Guillande et al.*, 1995; *Lee et al.* 2002, 2004]. An advantage of these analytical methods (unbiased factor selection and weighting), can also be a disadvantage, in that geoscience and geotechnical expertise may be ignored in such assessments [*Rollerson et al.*, 1997]. The ability to update hazard assessments based on accumulating information and new concepts needs to be considered [*Wu et al.*, 1996]. Sometimes the simplest analytical methodology may produce the best landslide hazard assessments when thorough field-mapping, good background information, and professional expertise are combined [e.g., *Nilsen and Brabb*, 1977; *Hicks and Smith*, 1981; *Maharaj*, 1995; *Ayenew and Barbieri*, 2005].

DISTRIBUTED, PHYSICALLY BASED MODELS

As a theoretical advance from empirical landslide models based solely on rainfall characteristics, numerous infiltration-based landslide models have been developed for individual sites in both two and three dimensions [e.g., *Anderson et al.*, 1988, *Sammori and Tsuboyama*, 1990; *Haneberg*, 1991a[1]]. Such models (discussed in Chapter 3) offer the advantage of a physics-based approach to assessing the dynamic changes in positive and negative (suction) pressure heads in the soil mantle during the infiltration process and thus are valuable to predict the timing of slope failure relative to rainfall inputs at individual sites with simple slope configurations. However, given the difficulties in applying these theoretical models in topographically complex catchments, they are not discussed in detail in this chapter.

Physically based landslide models [e.g., *Skempton and Delory*, 1957; *Okimura*, 1982; *Sidle* 1992[2]] assess stability in terms of a factor of safety (F_s); i.e., for F_s values $\gg 1$, the slope is stable; as F_s approaches 1, it becomes unstable; and for $F_s \leq 1$, slope failure theoretically occurs. If single parameter input values are used to calculate F_s, this method is most suitable for smaller areas because in larger areas the variations in terrain and soil parameters inherent in the analysis are typically too large [*Lumb*, 1975; *Burton et al.*, 1998; *Haneberg*, 2004]. Distributed, physically based landslide models

---

[1] *Iverson and Reid*, 1992; *Reid and Iverson*, 1992; *Reid*, 1994; *Sammori*, 1995; *Iverson*, 2000; *Baum et al.*, 2002.

[2] *Montgomery and Dietrich*, 1994; *Terlien et al.*, 1995.

have two unique requirements: (1) Spatially and, in some cases, temporally (e.g., rooting strength) distributed model parameters are necessary; and (2) the model output must be spatially and temporally explicit because of the need to know the locations and timing of landslides. Recent advances in incorporating advanced GIS and DEM technology into distributed, physically based modeling has facilitated the prediction of landslides at the catchment scale [*Montgomery and Dietrich*, 1994; *Wu and Sidle*, 1995; *Dhakal and Sidle*, 2003; *Iida*, 2004]. Earlier versions of distributed landslide models could assess spatially distributed aspects but not the dynamic response to rainstorms [e.g., *Ward et al.*, 1982; *Hammond et al.*, 1992].

The SHALSTAB Model

A physically based model for shallow landslide analysis (SHALSTAB) combines digital terrain data with near surface throughflow and the infinite slope model [*Montgomery and Dietrich*, 1994; *Dietrich and Montgomery*, 1998; *Montgomery et al.*, 1998]. For simplicity, the model generally assumes that soils are cohesionless, slope-parallel subsurface flow occurs, unit weights of soils in the saturated and unsaturated zones are equal, and ignores the effects of vegetation root strength [*Dietrich et al.*, 2001]. As such, unconditionally unstable slopes are defined as those where the slope gradient equals the internal angle of friction of the soil. As soil mantles begin to saturate, the critical angle for failure decreases accordingly. Effective steady-state event precipitation drives the shallow groundwater model [i.e., TOPOG, *O'Loughlin*, 1986], allowing for prediction of spatial patterns of equilibrium soil saturation based on analysis of upslope contributing area (a_c), soil transmissivity (t_r) and local slope gradient (β) by the simple relationship

$$\frac{h}{H} = \frac{q}{t_r} \frac{a_c}{b' \sin \beta} \qquad \text{(Equation 5.8)}$$

where q is precipitation minus evaporation, h is the vertical depth of saturation, H is the vertical soil depth, and b' is the cell width across which drainage occurs [*Montgomery and Dietrich*, 1994]. By combining equation 5.8 with the simplifying assumptions used in the infinite slope model and solving for the drainage area per outflow boundary length (i.e., the grid size), the coupled hydrology–slope stability equation is derived that forms the basis of SHALSTAB:

$$\frac{a_c}{b'} = \frac{\gamma_t}{\gamma_w}\left(1 - \frac{\tan \beta}{\tan \phi}\right)\frac{t_r}{q}\sin \beta \qquad \text{(Equation 5.9)}$$

where γ_t is the moist unit weight of soil, γ_w is unit weight of water, and ϕ is the internal angle of friction of the soil [*Montgomery and Dietrich*, 1994; *Dietrich et al.*, 2001]. An underlying assumption of SHALSTAB is that sites with the lowest ratios of effective precipitation to transmissivity (q/T) are the least stable; this relationship holds well in many areas where SHALSTAB has been applied in northern California, Washington, and Oregon [*Montgomery and Dietrich*, 1994; *Dietrich et al.*, 2001]. Further applications reveal that SHALSTAB frequently overpredicts landslides and performs best in steep catchments underlain by shallow bedrock and worst in less steep catchments underlain by thick glacial deposits [*Montgomery et al.*, 1998; *Borga et al.*, 2002; *Fernandes et al.*, 2004]. Due to the steady-state nature of rainfall inputs, SHALSTAB has not been tested for conditions where actual landslides were triggered during specific rain events; rather, effective rainfall was used.

The dSLAM and IDSSM Models

Based on the non-distributed landslide model by *Sidle* [1992], a distributed, physically based slope stability model (dSLAM) was developed to analyze shallow rapid landslides at the catchment scale within a GIS framework [*Wu*, 1993; *Wu and Sidle*, 1995; 1997; *Sidle and Wu*, 1999]. This model assesses the spatial and temporal effects of timber harvesting on slope stability. The distributed model incorporates (1) infinite slope analysis; (2) continuous temporal changes in root cohesion and vegetation surcharge; and (3) stochastic influence of actual rainfall patterns on pore water pressure. A root strength model developed by *Sidle* [1991] that simulates combined root decay and regrowth following timber harvest is used together with a vegetation surcharge model [*Sidle*, 1992] to simulate removal of tree weight and subsequent regrowth. The spatial aspect of the slope stability problem related to timber harvesting, such as locations of failures and changes in the spatial distribution of safety factor due to harvesting strategies in a forest rotation, was previously not investigated. *Moore et al.'s* [1988] TAPES-C model was adapted in the topographic analysis to partition the catchment into relatively homogeneous elements because the stream-tubes in this model are consistent with subsurface hydrologic and geomorphic processes. Rainfall can be applied as synthetic sequences or individual events and this water is routed into the soil mantle in the time increment during which it fell (generally 1 h intervals) [*Wu and Sidle*, 1995; *Sidle and Wu*, 1999]. A one-dimensional form of the kinematic wave equation is then used to route subsurface water through the stream-tubes as slope-parallel flow [*Takasao and Shiiba*, 1988]. Calculated groundwater levels are then incorporated into the infinite slope equation to determine the dynamic factor of safety for individual elements. Recent improvements to the model (now IDSSM) include the ability to simulate multiple harvesting cycles, more efficient handling of rainfall inputs, and an updated distributed shallow groundwater model (DSGMFW) [*Dhakal and Sidle*, 2003, 2004a,b; *Sidle and Dhakal*, 2003].

The model was tested in two steep, forested basins (1.18 and 1.12 km²) in Cedar Creek drainage of the Oregon Coast Ranges. Clearcutting of Douglas-fir forests within the basin was documented back to 1953. The hyetograph from the 29–30 November 1975 storm (178 mm) that triggered many landslides in the region was selected for landslide simulations. Simulated volumes (733 and 801 m³) and numbers (4 and 7) of landslides in the two basins agreed closely with values (734 and 749 m³ and 3 and 6, respectively) measured in the field after the 1975 storm [*Wu and Sidle*, 1995, 1997]. Although the precise locations of the actual 1975 landslides are not available, simulated landslides occurred in the general locations of these failures (Figure 5.5b). All simulated and actual

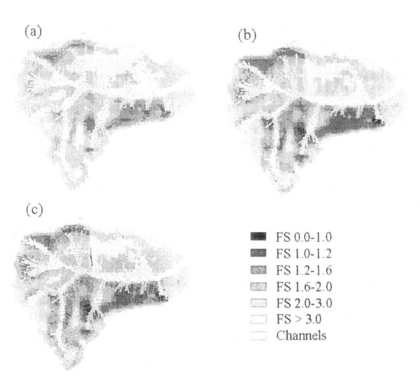

(a)

(b)

(c)

- FS 0.0-1.0
- FS 1.0-1.2
- FS 1.2-1.6
- FS 1.6-2.0
- FS 2.0-3.0
- FS > 3.0
- Channels

Figure 5.5. Spatial distribution of Fs simulated during the November 1975 storm in Cedar Creek, Oregon: (a) at the onset of the storm; (b) at 22 h (just at the end of rainfall); and (c) at 62 h (40 h after the storm) [after *Wu and Sidle*, 1995].

landslides occurred in sites that were clearcut in 1968, 7 yr prior to the November storm. At the onset of the storm, simulated spatial patterns of factor of safety (F_s) were controlled primarily by topography and patterns of past timber harvesting (Figure 5.5a). Areas with initially low F_s values expanded dramatically during the storm, reaching a maximum near the end of the event (Figure 5.5b) and then slowly recovering to pre-storm values (Figure 5.5c) [*Wu and Sidle*, 1995]. Substantial declines in F_s occurred around channel heads and hollows (Figure 5.5b). Although landslide numbers and volumes were accurately simulated, changes in input parameter values for soil cohesion, root cohesion, and soil depth in the range of ±20–30% altered predicted landslide volumes by several-fold [*Wu and Sidle*, 1995, 1997]. Using a Monte Carlo approach, *Sidle and Wu* [1999] simulated 39,000 rainstorms in one catchment that was randomly clearcut over 50% of its total area. Most of the 82 simulated landslides were clustered in a period of about 3–15 yr after clearcutting. By designating vegetation leave areas (uncut areas) on slopes steeper than 40° and in geomorphic hollows within the 50% clearcut catchment, landslide production declined significantly compared to a random 50% clearcutting with no leave areas [*Sidle and Wu*, 1999]. Recent applications of the model in Carnation Creek, British Columbia, indicate that vegetation leave areas reduced landslide volume by 2- to 3-fold [*Dhakal and Sidle*, 2003]. Partial cutting reduced simulated landslides by 1.4–1.6-fold compared to clearcutting. Also, landslide occurrence was influenced by storm characteristics, including mean and maximum hourly intensity, duration, total rainfall and the temporal distribution of short-term intensity [*Dhakal and Sidle*, 2004a]. In addition to the ability of dSLAM and IDSSM to utilize actual rainfall hyetographs as inputs, and thus predict temporal changes in F_s during storms, these models can explicitly analyze complicated scenarios of timber harvesting, making them useful tools for forest planning in potentially unstable terrain.

Other Physically Based Distributed Landslide Models

TRIGRS is one of the few distributed models to incorporate the effects of rainfall infiltration on dynamic pore pressure response in soils [*Baum et al.*, 2002]. This model extends *Iverson's* [2000] infiltration-based landslide model to the catchment scale by adding a solution for an impermeable basal boundary at a finite depth and including a simple runoff routing scheme. Major assumptions include nearly saturated soil conditions, well-documented flow field, and relatively isotropic and homogeneous soil hydrological properties. The steady seepage component, and thus the accuracy of landslide predictions, strongly depends on the initial depth assigned to the water table and the steady infiltration rate [*Baum et al.*, 2002].

Iida [2004] developed a hydrogeomorphological model for shallow landslide prediction that considers both the stochastic character of rainfall intensity and duration, and the deterministic aspects controlling slope stability, where the short-term probability of landsliding is defined as the probability that the saturated soil depth exceeds a critical value. The model was applied to an area east of Hamada, Japan, where extensive landsliding occurred during a heavy rainstorm in 1988. A long-term series of rainstorms was randomly generated and landslides were simulated. The spatial extent of landslides associated with each simulated storm was controlled by both the rainfall as well as the historical sequence of landslides; events with return periods <500 yr triggered about half of the landslides in the long-term [*Iida*, 2004].

Another distributed shallow landslide model, SINMAP, is similar to SHALSTAB but employs different algorithms for calculating contributing area and specifying flowpaths within a rectangular grid [*Pack*, 1997; *Tarboton*, 1997]. *Costa-Cabral and Burges* [1994] provide a detailed discussion of strengths and limitations of such directional flow models based on rectangular grid DEM data. *Dietrich et al.* [2001] show that changes in topographic calculations between the two models may lead to different predictions of unstable sites. SINMAP has been used in a long-term model of the interactions of forest vegetation, forest fire, harvesting, and sediment yields in the Idaho Batholith [*Istanbulluoglu et al.*, 2004].

One of the few distributed approaches to deep-seated landslide modeling was developed by *Miller* [1995]. The model incorporates topographic, geologic, and soils data from 1:24,000 maps with Bishop's modified method of slices stability analysis [*Bishop*, 1955] within a GIS framework. The model was applied to the Montague Creek basin in northwest Washington; soils were assumed to be completely saturated. Considerable discrepancies existed between predicted areas of instability and actual failures [*Miller*, 1995], partly due to the constant saturation assumption.

A physically based model was developed to analyze both the effects of seismic acceleration and rainfall-induced groundwater levels on slope failure probability [*van Westen and Terlien*, 1996]. This GIS-based analysis employs a simple two-dimensional groundwater model that calculates relative saturation of the soil mantle on a daily basis. Variations in soil cohesion, internal angle of friction, and soil depth are also considered; however, effects of vegetation rooting strength and related changes in land use are not considered. The model was applied to a mountainous area surrounding Manizales, central Colombia, along the western flank of the Cordillera Central near the Nevado del Ruiz volcano. The final hazard map was checked with general occurrences of landslides, but the model was not tested for specific rainfall and seismic events that generated such failures.

OVERVIEW OF LANDSLIDE HAZARD ASSESSMENT AND PREDICTION METHODS

Many of the hazard assessment and prediction methods outlined in this chapter can be used as general management tools. Most are designed for application at the regional level, however, some can be applied globally (e.g., Caine's model and our modifications) and others can be adapted to local conditions (e.g., certain mapping techniques, dSLAM and IDSSM). Each example presents some challenge related to application or interpretation of results. Many of the multi-factor empirical methods and physically based models have rather large data requirements. Empirical methods that assess landslide trigger mechanisms as well as the multi-factor analyses are specific to the region in which they are developed. Multi-factor analyses that incorporate only rudimentary data where cause and effect relationships are not clear may not be useful for landslide susceptibility assessments. Additionally, most multi-factor empirical methods have difficulties in differentiating risk or susceptibility at detailed scales in areas with high to extreme susceptibility to landslides. Terrain hazard mapping is a very valuable planning tool for land managers and urban planners to identify potentially unstable sites prior to development activities. Quality of the mapping and thus the utility largely depends on the resources (e.g., geology and soil maps) and expertise available, as well as the commitment to on-the-ground investigations. Methods that assess rainfall trigger mechanisms can be useful in developing early warning systems for shallow, rapid landslide hazards, particularly if coupled with antecedent moisture data [*Keefer et al.*, 1987; *Reid et al.*, 1999].

Distributed physically based models are potentially the most powerful tools in landslide hazard analysis, particularly when they incorporate DEM data based on Lidar [*Dietrich et al.*, 2001], actual rainfall inputs [*Baum et al., 2002*; *Dhakal and Sidle*, 2004a], and long-term land use scenarios [e.g., *Dhakal and Sidle*, 2003]. However, widespread application of these models has been limited because they require distributed input data (including DEMs) and expertise with GIS and computer modeling. While some input data can be augmented by remote sensing and extracted from DEMs, to be effective, these geotechnically based models require accurate distributed data on soil depth and other critical soil properties; such data are typically not readily available. The advantages of stream-tube–based flow systems over rectangular grids in distributed models may gradually disappear as detailed Lidar DEM data become available; however, in the near term, careful attention must be paid to flow-routing algorithms when using rectangular grid-based DEM data [*Costa-Cabral and Burges*, 1994].

There is a conspicuous absence of hazard assessment and prediction methods that address the attributes and processes inherent in deep-seated landslides. Empirical methods that focus on rainfall trigger mechanisms apply only to shallow, rapid failures; other empirical multi-factor analyses use the same criteria for all failure types in a region. At present, terrain hazard mapping appears to be the only suitable strategy for deep-seated landslide hazard assessment, albeit of a very general nature [e.g., *Gage and Black*, 1979; *Krauter and Steingötter*, 1983; *Howes and Kenk*, 1988; *Wegmann*, 2003].

One of the major challenges related to landslide hazard assessment and prediction is the need to better link specific land management activities into empirical analyses and models, as well as terrain hazard mapping. Although some of the current terrain hazard mapping programs address practices like forest road construction and timber harvesting [*Howes and Kenk*, 1988; *Oregon Department of Forestry*, 2003a], no clear distinctions are made to discriminate between these very different impacts on slope stability. Additionally, few multi-factor empirical assessments assess land management issues in a meaningful way, except for the inclusion of very general land cover classes [e.g., *Kienholz et al.*, 1984; *Anbalagan*, 1992]. Distributed, physically based landslide models which include various vegetation attributes (e.g., rooting strength, species distribution) have the potential to generate detailed spatial and temporal landslide scenarios [e.g., *Sidle and Wu*, 1999; *Dhakal and Sidle*, 2003; *Istanbulluoglu et al.*, 2004]. Finally, the need to develop more effective spatial coverage of landslide susceptibility and real-time hazard warnings for poor and vulnerable countries remains apparent. Such methods must be practical with respect to local expertise and technology. The increasing availability of low-cost remote sensing products that can support GIS-based hazard analysis will undoubtedly benefit landslide assessments in such areas [*Leroi et al.*, 1992; *Dhakal et al.*, 2002]. Nevertheless, emphasis still needs to be placed on collecting better field data in these mountainous regions, and it would behoove local governments and international donors to support such efforts rather than focusing too much on high technology.

6

Land Use and Global Change

Both widespread land use activities and more concentrated disturbances affect the magnitude, frequency and type of landslides that occur in many parts of the world. These anthropogenic activities can alter the thresholds for both initiation and acceleration of certain landslide types. Such changes may be cascade down through terrestrial/aquatic ecosystems generating cumulative effects related to displaced and transported sediment [e.g., *Sidle and Hornbeck*, 1991; *Dunne*, 1998]. This chapter focuses on the major widespread land uses that globally influence mass wasting (timber harvesting, forest conversion, grazing, recreation, and fire), as well as concentrated human activities and disturbances (roads, urban development, and mining) that locally affect slope stability. The extent to which various aspects of these management activities and land uses affect slope stability are discussed in detail along with effective prevention methods, stressing non-structural avoidance and control measures. An overview of potential effects of climate change on slope stability is also presented.

TIMBER HARVESTING

In steep terrain with relatively shallow soil mantles, timber harvesting can reduce site stability by (1) causing root strength deterioration following cutting of trees and prior to regeneration of planted or successional forests; and (2) increasing soil water due to the lower evapotranspiration after forest cutting and/or changes in the volume and rate of snowmelt [e.g., *Gray and Megahan*, 1981; *Megahan*, 1983; *Greenway*, 1987]. As noted earlier, root strength deterioration appears to be a much more important factor related to initiation of shallow landslides than increases in soil water. In the Pacific Northwest of North America, alder (*Alunus rubra*) corridors along headwater channels are common legacies of landslides and debris flows that initiated 3–15 yr after timber harvesting (Figure 6.1a). Such vegetative features may date back more than 50 yr. In the tropics, increased soil moisture due to timber harvesting may accelerate deep-seated landslides until new forest vegetation establishes; nevertheless, loss of root strength still appears to be the dominant influence of forest removal on slope stability [*Sidle et al.*, 2006]. Temperate mountain forests that receive large snowfall can experience more landslides after timber harvesting due to increases in snow accumulation and possible increases in peak melt rates [*Megahan*, 1983]. In this section, the term *timber harvesting* assumes that some type of secondary forest will be regenerated on the site. The case of forest clearance and conversion to a different cover type is generally more severe in terms of

Landslides: Processes, Prediction, and Land Use
Water Resources Monograph 18
Copyright 2006 American Geophysical Union
10.1029/18WM07

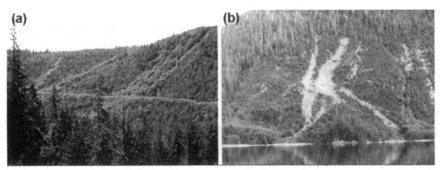

Figure 6.1. Landslides triggered by forest cutting: (a) linear corridors of alder mark the locations of debris flow paths triggered during a large storm in October 1961, about 5–6 yr after clearcutting, in Maybeso Experimental Forest, Prince of Wales Island, Alaska; and (b) numerous landslides in clearcuts in Rennel Sound, Queen Charlotte Islands, British Columbia, during a 5–10 yr return interval storm on October 30 to November 1, 1978, 4–11 yr after most of the clearcutting in the area.

landslide implications and is discussed in later sections of this chapter. The influence of timber harvesting on landslide initiation in all sites also depends on the density of residual trees and understory vegetation, the rate and type of regeneration, site characteristics (e.g., soil water status, nutrients), and water pathways after harvesting. The effects of logging practices (e.g., tractor and cable yarding) on slope stability are discussed independently of various silvicultural practices (e.g., clearcutting, partial cutting).

Silvicultural Practices

Most field studies of the effects of timber harvesting have been in clearcuts; few data are available on effects of other silvicultural practices. Long-term rates of landslide erosion based on field data in steep clearcuts in North America have mostly been complied before the mid-1980's in spite of the current interest in harvesting–landslide relationships. *Sidle et al.* [1985] summarized landslide erosion rates from clearcuts in Pacific Northwest forests in the range from about 0.3 to 3.9 t ha^{-1} yr^{-1}; corresponding rates in undisturbed forests ranged from 0.1 to 1.1 t ha^{-1} yr^{-1}. One of the first studies to directly link the timing of timber harvesting with accelerated landslide initiation was in a large clearcut in Maybeso Valley, Prince of Wales Island, Alaska [*Bishop and Stevens*, 1964] (Figure 6.1a). Landslides persisted for a period of about 9 yr after clearcutting, with more than half of the landslides triggered by a large storm 6 yr after harvesting. The area affected by landsliding during this period was 5 times the estimated area disturbed by landslides during a 100-yr period before logging. An earlier investigation in northern Utah noted an increased tendency for landsliding during major storms following timber harvesting and burning [*Croft and Adams*, 1950].

Some of the more comprehensive field investigations of the effect of clearcutting on landslide erosion rates have been conducted in the Pacific Northwest of North America (Table 6.1). Based on aerial photo interpretation and field surveys covering periods of

TABLE 6.1. Comparison of landslide erosion rates from various land uses and sites based on inventories conducted for ≥10 yr; "roads" denotes forest roads unless otherwise specified.

Study area	Land use	Period of record, yr	Landslide erosion rate, $t\ ha^{-1}\ yr^{-1}$	Management effects relative to undisturbed land	Reference
H.J. Andrews Exp. Forest, Oregon Cascades	Forest	25	1.1	× 1.0	*Swanson and Dryness, 1975*
	Clearcuts	25	3.2	× 2.9	
	Roads	25	34.0	× 30	
Alder Creek, Oregon Cascades	Forest	25	0.6	× 1.0	*Morrison, 1975*
	Clearcuts	15	1.5	× 2.5	
	Roads	15	202	× 337	
Blue River, Oregon Cascades	Forest	34	0.5	× 1.0	*Marion, 1981*
	Clearcuts	22	4.2	× 8.7	
	Roads	25	21.2	× 44	
Klamath Mountains, southwest Oregon	Forest	20	0.56	× 1.0	*Amaranthus et al., 1985*
	Clearcuts	20	3.7	× 6.6	
	Roads	20	36	× 64	
Mapleton, Oregon Coast Ranges	Forest	15	0.25	× 1.0	*Ketcheson and Froehlich, 1978*
	Clearcuts	15	0.9	× 3.6	
Mapleton, Oregon Coast Ranges (all slide-prone soils)	Forest	15	0.4	× 1.0	*Swanson et al., 1977*
	Clearcuts	10	0.8	× 1.9	
	Roads	10	21	× 50	
Coastal southwest, British Columbia	Forest	32	0.14	× 1.0	*O'Loughlin, 1972*
	Clearcuts	32	0.32	× 2.3	
	Roads	32	3.8	× 27	
Capilano River basin, southwest British Columbia	Forest	32	1.3	× 1.0	*Brardinoni et al., 2003*
	Clearcuts	32	5.9	× 4.5	
	Old logging	32	0.9	× 0.7	

TABLE 6.1. *Cont.*

Study area	Land use	Period of record, yr	Landslide erosion rate, t ha⁻¹ yr⁻¹	Management effects relative to undisturbed land	Reference
Chichagof and Prince of Wales Islands, southeast Alaska	Forest	21	1.1	× 1.0	*Swanston and Marion*, 1991
	Clearcuts	21	2.0	× 1.8	
Steep mountain forest area, Puerto Rico	Forest	50	2	× 1.0	*Larsen and Parks*, 1997
	Paved roads	50	10	× 5.0	
Redwood Creek basin, California	1. Roads, 16 basins	26	64	--	*Hagans et al.*, 1986; *Pitlick*, 1995; *Weaver et al.*, 1995
	2. Roads, most erodible basin	26	102	--	
Redwood Creek basin, California	Roads	17	15	--	*Rice*, 1999
New Zealand, mostly granitic mountainous terrain	Background	--	5	× 1.0	*Adams*, 1980; summary of road data by *Fransen et al.*, 2001; *Mosley*, 1980
	New roads	various	266–7600	× 53–1520	
	Older roads	various	30–380	× 6–76	

15–34 yr, landslide erosion from clearcuts in the Cascade Range of Oregon varied from about 1.5 to 4.2 t ha^{-1} yr^{-1}, 2.6 to 8.7 times higher than landslide erosion in forested areas of comparable stability [*Morrison*, 1975; *Swanson and Dyrness*, 1975; *Swanson and Grant*, 1982]. Major storms in December 1964 and January 1965 generated a large number of the total landslides during this period. In the Oregon Coast Ranges, *Ketcheson and Froehlich* [1978] measured about 0.9 t ha^{-1} yr^{-1} of landslide erosion in clearcuts compared to 0.25 t ha^{-1} yr^{-1} in forested areas during a 15-yr period. Most of these landslides occurred during the winter storms of 1975 on slopes >39°. In some of the steepest and most dissected terrain in the Oregon Coast Ranges, landslide losses in clearcuts were about 1.5 t ha^{-1} yr^{-1} compared with 0.8 t ha^{-1} yr^{-1} in clearcuts in all unstable terrain and 0.4 t ha^{-1} yr^{-1} in undisturbed forests [*Swanson et al.*, 1977]. In the Klamath Mountains of southwest Oregon, average landslide erosion over a 20-yr period was 3.7 t ha^{-1} yr^{-1} in clearcuts compared to 0.56 t ha^{-1} yr^{-1} in natural forests [*Amaranthus et al.*, 1985]. *O'Loughlin* [1972] found that landslide erosion in clearcuts in the British Columbia Coast Range averaged about 0.32 t ha^{-1} yr^{-1} compared to 0.14 t ha^{-1} yr^{-1} in undisturbed forest areas. In nearby areas, *Brardinoni et al.* [2003] used aerial photo interpretation and field surveys to assess landslide erosion: Significantly higher rates were found in recent clearcuts (5.7 t ha^{-1} yr^{-1}) compared to older logged sites (0.9 t ha^{-1} yr^{-1}) and undisturbed forests (1.3 t ha^{-1} yr^{-1}). An aerial photo survey with selective ground investigations in Chichagof and Prince of Wales Islands, southeast Alaska, revealed that landslide erosion was twice as high in clearcuts (2.0 t ha^{-1} yr^{-1}) compared to unlogged forests (1.1 t ha^{-1} yr^{-1}); frequency of landsliding in clearcuts during the 21-yr interpretation period was about 4 times higher than in natural forests, but the volume of landslides in clearcuts was only half the size of those in natural forests [*Swanston and Marion*, 1991].

In New Zealand, where beech–podocarp hardwood forests were clearcut and replanted with weaker rooted radiata pine, mass erosion increased from 1.2 t ha^{-1} yr^{-1} before logging to 15–50 t ha^{-1} yr^{-1} during a period from 7 to 10 yr after clear-felling [*O'Loughlin and Pearce*, 1976]. Although this rate is about 10 times the rates experienced in most clearcuts in North America, the data were based on short-term records in which two large storms occurred; nevertheless, these New Zealand data confirm a similar period of landslide susceptibility (i.e., window of low root strength) following harvesting. For all managed forest sites, the effects of timber harvesting on landslide initiation decline significantly after 15 to 25 yr; thus, increased landslide erosion should not persist throughout the entire rotation of regenerating forests, except when stands are harvested at a very young age. Radiata pine plantations in New Zealand as well as fast-growing plantation forests in the tropics may exhibit cumulative declines in site stability since peak rooting strength of trees does not fully recover over rotations with progressively shorter lengths [*Sidle*, 1991; *Sidle et al.*, 2006]. An inventory in forested land in Japan reported higher landslide rates during a 16-yr period after timber harvesting [*Nakano*, 1971]. In Idaho, *Megahan et al.* [1978] found that landslides occurred most frequently 4 to 10 yr after clearcutting, which coincided with the period of minimum rooting strength; by 20 yr after logging, the occurrence of landslides returned to pre-harvesting levels. Many of the North American, New Zealand, and Japanese investigations were based on aerial photo interpretation. This methodology has been shown to underestimate the number of small landslides occurring in forested stands and may exclude as much as 30% of the landslide volume [*Brardinoni et al.*, 2003]. Thus, some of the previously reported increases in landslides due to clearcutting based on aerial photo interpretation could be overestimates.

In addition to the many field investigations in unstable forest terrain already cited, recent catchment-scale modeling studies have confirmed that the period of maximum landslide susceptibility is between 3 and 15 yr after clearcutting [*Sidle and Wu*, 1999; *Dhakal and Sidle*, 2003]. The reported 2- to 10-fold increase in landslide erosion following clear-felling is strongly influenced by the timing of an episodic triggering event (e.g., rainfall or snowmelt) as well as the topographic and geologic attributes of the forested site.

Many studies have assessed the landslide frequency, sometimes including aerial extent and volume estimates, following large storms, where landslide disasters are prevalent. Such event-based data must be interpreted cautiously because without knowledge of past storm and landslide histories, it is possible to obtain different responses related to land use for various antecedent situations. Thus, without prior data, it is difficult to accurately judge the effects of forest harvesting on landslide erosion in the long term. During a large storm in November 1975 in the Oregon Coast Ranges, 77% of the landslides occurred in clearcuts (not associated with roads), and the frequency of landsliding was 23 times higher in clearcuts than in undisturbed forested areas [*Gresswell et al.*, 1979]. An inventory of landslides in four areas of western Oregon following two major storms in 1996 reported that landslide density and erosion were higher in three out of four recently clearcut forests (0–9-yr-old stands: 5.0–8.1 landslides km^{-2} and 9.5–37.3 t ha^{-1}) compared to mature forests (100+-yr-old trees: 2.1–5.2 landslides km^{-2} and 5.2–16.7 t ha^{-1}) [*Robison et al.*, 1999]. However, it is difficult to derive strong inferences from this survey due to intra- and inter-site variability together with small numbers of landslides in certain age classes and the lack of documented landslide history prior to the 1996 storms. Detailed inventories of landslides in two forested areas in Japan revealed that the frequency and area of landslides increased from 3- to 12-fold and from 2.8- to 24-fold, respectively, in clearcut forests compared with natural forests [*Fujiwara*, 1970]. A survey in the Rennell Sound area of the Queen Charlotte Islands, British Columbia, after a large storm (5–10 yr return interval) indicated that landslide frequency and erosion were 14 and 41 times greater, respectively, in clearcuts than in natural forests [*Schwab*, 1983] (Figure 6.1b). As a result of the widespread landslides in clearcuts, the logging company was held liable for damages incurred. Landslides in clearcuts of Rennell Sound were larger (average volume = 520 m^3) than in natural forests (210 m^3), and most of the landslides occurred in sites harvested 4 to 11 yr prior to the storm. Landslide erosion in clearcuts was very high compared to most other surveys (62 t ha^{-1}); average landslide erosion from undisturbed forests was 1.5 t ha^{-1} [*Schwab*, 1983]. Widespread shallow landslides following typhoon 18, which inundated the Fujieda area in Japan in 1982 ($\approx$620 mm of rain in a 60-hr period), were more common in recently clearcut, young-growth forests (47.5 landslides km^{-2}) compared to mature conifer forests (10.9 landslides km^{-2}) and native broadleaf forest (20.2 landslides km^{-2}); however, larger landslides typically occurred in the broadleaf forest [*Omura and Nakamura*, 1983]. Percentage of land area in the land use categories occupied by landslides was 0.12%, 0.30%, and 0.40% for mature conifer forests, young-growth forests, and broadleaf forests, respectively.

Few field studies have assessed the effects of partial cutting and stand tending on landslide occurrence. A recent survey in the northern California Coast Range showed that landslide density was about 5- to 9-fold higher in clearcuts compared to thinned and unthinned second-growth forests on slopes steeper than 31° [*Rollerson*, 2003, personal communition]. Unthinned stands had slightly higher landslide densities compared to thinned stands, possibly attributed to foresters selecting more stable terrain for intensive management [*Rollerson*, 2003, personal communication]. For a rather small sample size of landslides on Prince of Wales Island, southeast Alaska, the average volume of failures in old-growth forests was largest (756 m³), followed by those in second-growth forests (514 m³) and clearcuts (187 m³) [*Johnson et al.*, 2000]. Data on various natural and managed vegetative crown cover conditions in Idaho provide insights on the relationships between partial cutting and landslide initiation [*Megahan et al.*, 1978]. Landslide frequency increased only slightly as overstory crown cover decreased from 100% to 11%; however, for crown covers <11%, major increases in landslides occurred. Such data indicate the importance of understory vegetation in stabilizing slopes.

For sites in coastal Oregon and British Columbia, sequences of Monte Carlo–generated rainfall were used in distributed models (dSLAM and IDSSM) to compare landslide erosion rates for different silvicultural and stand-tending practices; at each respective site the same rainfall sequences were applied for all practices. Most shallow landslides were simulated in a 3- to 17-yr period following the initial timber harvest in a Douglas-fir forest of coastal Oregon [*Sidle and Wu*, 1999]. Based on the sum of root decay and regrowth curves (equations 3.16 and 3.17, respectively) for Douglas-fir [*Sidle*, 1991], the probability of slope failure during a series of five 50-yr rotation clearcuts was about five times higher than for a similar 75% partial cut scenario [*Sidle*, 1992]. Comparisons of landslide erosion were simulated over a 200-yr period for various clearcutting and partial cutting practices in a western hemlock–Sitka spruce forest in Carnation Creek, British Columbia [*Dhakal and Sidle*, 2003]. When the entire 57-ha catchment was harvested, simulated landslide erosion was 0.62, 0.40, and 0.14 t ha⁻¹ yr⁻¹ for clear-cutting, 90% partial cutting, and 75% partial cutting scenarios, respectively (Figure 6.2). Throughout a typical rotation (i.e., 50 yr), landslide erosion was similar whether the entire catchment was clearcut at one time or sequentially or randomly cut throughout the rotation (eventually all trees harvested). Leaving only 10% of the trees on the harvested portion of the site decreased landslide volumes by 1.4- to 1.6-fold compared to sequential clearcutting [*Dhakal and Sidle*, 2003]. Based on root strength simulations, *Sidle* [1991, 1992] found that a two-step shelterwood harvesting system (85% of the stand initially cut, with the remaining 15% cut 12 yr later) generated a lower probability of failure compared to an alternate sequence of thinnings and clearcuts as well as clearcuts alone, but was more prone to landslides than the 75% partial cut scenario for Rocky Mountain Douglas-fir. *Sakals and Sidle* [2004] modeled small-scale changes in patterns of Douglas-fir (*Pseudotsuga mensiezii*) root strength related to different silvicultural practices and found that a random selection cutting (25% of the trees older than 60 yr harvested every 20 yr) caused the smallest decrease in rooting strength (81% of preharvest strength),

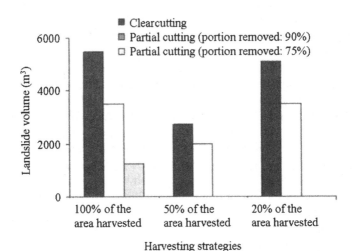

Figure 6.2. Simulated landslide volumes in clearcuts, 90% partial cuts, and 75% partial cuts in Carnation Creek, Vancouver Island, British Columbia. Landslide volumes are for a 200-yr period with four 50-yr harvest cycles; the 20% clearcut/partial cut strategy is applied every 10 yr throughout the 200-yr period [after *Dhakal and Sidle*, 2003].

followed by strip cutting (58% of preharvest root strength) and clearcutting (an average of 47% of preharvest root strength).

Changes in harvesting cycles (rotation lengths) also affect slope stability. Long-term simulations of landslide volume in Carnation Creek, Canada, showed that with sequential decreases in clearcut rotation length, high rates of landsliding persisted through four clearcut rotations due to cumulative declines in rooting strength [*Dhakal and Sidle*, 2003] (Figure 6.3). Such cumulative decreases in simulated root strength were previously noted for radiata pine [*Sidle*, 1991]. Longer intervals between initial and final shelterwood cuttings promote greater rooting strength than short intervals [*Sidle*, 1991]. Although shallow landslide volume is most enhanced by the initial harvest entry [e.g., *Swanson and Fredriksen*, 1982], subsequent clearcutting maintains landslide erosion at levels higher than found in partial cutting or unharvested scenarios, contrary to arguments that clearcutting may only change the timing of natural landsliding and, that over long periods, the total landslide erosion may be independent of timber harvesting [e.g., *Froehlich*, 1978]. This trend of long-term increases in landslide erosion is especially evident when cutting rotations decrease with time (Figure 6.3).

Vegetation leave areas are effective management practices to reduce landslide occurrence. When geomorphic hollows and sites steeper than 40° were protected from harvesting during slope stability simulations, landslide volume decreased by about 1.6- to 2.0-fold in a catchment in coastal Oregon [*Sidle and Wu*, 1999] (Figure 6.4). Greater benefits were realized (1.8- to 2.9-fold decreases in landslide volume) in simulations conducted in Carnation Creek, Canada [*Dhakal and Sidle*, 2003].

Numerous assessments have documented comparative frequencies or areas of landslides amongst different forest management practices or different types of sites. While

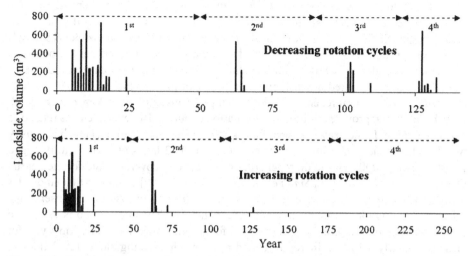

Figure 6.3. Simulated temporal distribution of landslide erosion in Carnation Creek: four clearcut cycles of decreasing rotation length—50, 40, 30, and 20 yr; and four clearcut cycles of increasing rotation length—50, 60, 70, and 80 yr [modified from *Dhakal and Sidle*, 2003].

such surveys provide useful relative information, erosion rates are not available, and thus the issue of magnitude and frequency cannot be addressed. A survey conducted in the Klamath Mountains of northwestern California showed that active forest harvesting (including roads) in vulnerable inner valley gorges and slopes >39° experienced 11 to 26 and 3 to 26 times more landslides per unit area compared to other lands managed for timber [*Wolfe and Williams*, 1986]. In highly unstable terrain near Rennell Sound, Queen Charlotte Islands, British Columbia, *Schwab* [1988] measured an average frequency of 10 landslides km^{-2} (minimum landslide size $\approx$ 400 m^2) in all clearcuts surveyed, with frequencies of 22.4 and 36.7 landslides km^{-2} in clearcuts in the two most sensitive terrain categories. Landslide erosion in Rennell Sound differed greatly between two adjacent

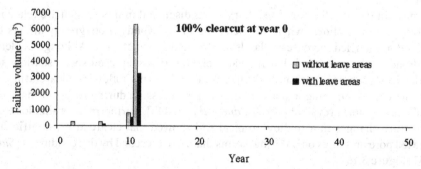

Figure 6.4. Effect of excluding geomorphic hollows and sites steeper than 40° from simulated clearcutting, Cedar Creek, Oregon; at time = 0 the entire 118-ha catchment was clearcut except for these protected areas [after *Sidle and Wu*, 1999].

unstable cutblocks, one which was indiscriminately logged (77% of the most unstable area disturbed by landslides) and the other which allowed some clearcutting but protected previously failed sites (7% of the unstable area disturbed by landslides) [*Schwab*, 1988]. Site productivity on these disturbed sites was projected to recover to about half of the former productivity 80 yr after landsliding [*Smith et al.*, 1986], and costs incurred when landslides impacted downslope second-growth forests reached CAN$1700 ha^{-1} [*Schwab*, 1988]. Several recent landslide aerial photo inventories on Vancouver Island, British Columbia, provide additional information about landslide frequencies related to clearcut logging. In the heavily logged forests around Clayoquot Sound, landslide density was 1.06 landslides km^{-2} in clearcuts compared to 0.22 landslides km^{-2} in unlogged forests [*Jakob*, 2000]. Based on regeneration and canopy coverage data for the nearby Queen Charlotte Islands [i.e., *Smith et al.*, 1983], *Jakob* [2000] estimated that landslides in natural forests would be visible for a period of 40 to 60 yr; using the conservative 40-yr approximation, temporal landslide frequency in Clayoquot Sound is estimated to be 0.0055 landslides km^{-2} yr^{-1} for natural forests and 0.053 landslides km^{-2} yr^{-1} for clearcuts, nearly a 10-fold increase related to timber harvesting alone. Landslides in clearcuts occurred on gentler slopes (more than half on slopes ≤40°) than in unlogged terrain (more than half on slopes >40°), partly because fewer sites >40° were logged [*Jakob*, 2000]. In three other catchments on Vancouver Island, *Guthrie* [2002] found landslide densities ranging from 0.62 to 1.26 landslides km^{-2} in clearcuts compared to 0.17 to 0.75 landslides km^{-2} in natural forests. Applying the same approximations as used in Clayoquot Sound, temporal landslide frequency was estimated as 0.026 landslides km^{-2} yr^{-1} for clearcuts and 0.03 landslides km^{-2} yr^{-1} for natural forests [*Guthrie*, 2002]. While landslides in forested areas of Clayoquot Sound were not significantly larger than those in clearcuts [*Jakob*, 2000], only 25% of the landslides in natural forests in the Vancouver Island catchments were classified as small (0.02–0.2 ha) compared to more than half of those in clearcuts [*Guthrie*, 2002]. The order of magnitude higher landslide densities found in clearcuts in Rennell Sound (Queen Charlotte Islands) compared to nearby sites on Vancouver Island could be related to the inherently greater instability of the Rennell Sound site as well as the more complete documentation of landslides by aerial reconnaissance.

Almost all of the effects of forest harvesting discussed in this section are related to shallow, rapid landslides, where the deterioration of woody roots greatly affects the observed or modeled increases in landslide frequency and erosion. Although evidence has already been presented that indicates clearcutting may predispose shallow soil mantles to higher pore pressures during moderate storms following clearcutting, no field data exist supporting major increases in pore pressure during large storms, when landslides are most prevalent [*Dhakal and Sidle*, 2004b]. Furthermore, a simple model of moisture changes in a shallow tropical soil showed that clearcutting significantly increased pore pressures only during storms that were preceded by dry conditions [*Sidle*, 2005] (Figure 3.16).

The effects of forest harvesting on deep-seated mass movements is less important because root systems generally do not penetrate through the entire regolith. However, little research has been conducted to document the effects of timber harvesting on deep-seated mass movements. Because natural rates of deep-seated mass movement are responsive to seasonal increases in soil moisture [*Swanson and Swanston*, 1977; *Swanston*, 1981; *Wasson and Hall*, 1982; *Iverson and Major*, 1987; *Swanston et al.*, 1995], it follows that higher water contents after timber harvesting should accelerate these processes. In regions where winters are dominated by rain, increases in soil moisture caused by vegetation removal occur primarily in fall and spring and may extend the natural winter period of activity for deep-seated mass movements. Soil creep at a forested site in southwestern Oregon approximately doubled during the second winter after logging [*Swanston*, 1981]. A simple simulation of soil water changes during 1 yr in a tropical rain forest of Malaysia showed that for relatively deep soils (4 m), where an accumulation of 1650 mm of soil water was assumed necessary to initiate or reactivate deep-seated landslides, clearcutting generated a total of 89 more days of movement compared to an uncut forest [*Sidle et al.*, 2006] (Figure 3.16). However, such differences in soil water levels after harvesting are likely short-lived in the tropics due to rapid regeneration.

Increases in deep-seated soil creep induced by logging may also reactivate dormant earthflows by loading downslope areas or reducing lateral support of upslope areas [e.g., *Sharpe and Dosch*, 1942; *Furuya et al.*, 1999]. The loss of root strength after clearcutting could significantly reduce the stability of shallow earthflows and could also permit increases in soil creep. For deep-seated mass movements, the influence of lateral and vertical anchoring by tree root systems is minimal [*Swanston and Swanson*, 1976]. Because slump–earthflows and soil creep affect large portions of the landscape, the specific influence of forest harvesting or other management activities, such as poor road drainage, may be difficult to identify. *Brown and Sheu* [1975] modeled the response of soil creep to clearcutting based on the effects of rooting strength, tree weight, wind stress, and soil moisture. Simulated rates of soil creep decreased immediately after timber harvesting because vegetative weight and wind stress were eliminated, even though evapotranspiration was reduced. Following the onset of root deterioration, creep begins to increase.

Logging Methods

Earlier investigations have clearly shown a strong relationship between logging practices (yarding methods) and the extent and severity of site disturbance [e.g., *Dyrness*, 1965]. While such disturbances affect surface erosion, the impacts on slope stability are likely less, but difficult to assess. If damage occurs to the residual stand (e.g., stems and root systems) during thinning or partial cutting, rooting strength may be affected during the subsequent rotation. Small, shallow landslides occurred in a large clearcut of shallow-rooted Sitka spruce–western hemlock forest in coastal Alaska during the first large storm after logging [*Sidle*, 1984a]. Much of the deep soil organic horizon (including tree roots) was disturbed and displaced in dissected portions of the topography during the ground-based cable yarding. Tractors and rubber-tired skidders destroy much more of the understory vegetation during logging compared to cable yarding and lower impact logging techniques (e.g., skyline yarding, helicopter logging). Retaining a vigorous understory rooting strength has been demonstrated to be beneficial in preventing land-slides following harvesting [*Dhakal and Sidle*, 2003; *Sakals and Sidle*, 2004]. Simulated reductions in landslide volume for both clearcutting and partial cutting scenarios ranged from 3.8- to 4.8-fold when a viable understory cover was retained compared to cases where understory was destroyed during yarding [*Dhakal and Sidle*, 2003]. Ground-based logging also compacts the soil and may reduce the regeneration potential of new trees, thus inhibiting rapid recovery of root strength after logging.

The timing of harvesting operations is more important related to surface erosion than to landslide initiation. Nevertheless, since timber yarding on steep slopes during wet periods produces greater disturbance and compaction, such impacts may channel localized storm runoff onto unstable sites, thus initiating small landslides. From a safety standpoint alone, timber yarding during periods with >100 mm of rain in 24 h is not advised in the Pacific Northwest of North America [*Chatwin*, 1994].

By far the most important issue related to alternative logging (yarding) methods is the amount of roads required to service them. A summary of 16 studies in the USA and Canada showed that both tractor- and ground-based cable yarding require the most roads, more than 10 and 5 times the road disturbance required for skyline and aerial yarding systems, respectively [*Megahan and Schweithelm*, 1983]. Thus, higher rates of landslide erosion would be expected with these two ground-based logging systems compared to skyline and aerial systems when applied in similar terrain (see section on roads).

Effects on Stream Systems and Long-Term Catchment Processes

In steep headwaters where hillslopes are tightly coupled to stream channels, sedi-ment supplied from hillslope landslides is frequently deposited directly in channels where it can immediately initiate a debris flow or where it accumulates until a debris flow occurs sometime in the future [*Gomi et al.*, 2002; *Sidle*, 2005; *Chen*, 2006]. The amount of landslide/debris flow sediment delivered to channels is related to landslide volume, mobility of landslide sediment on hillslopes, types of mass wasting processes, behavior of sediment at hillslope channel junctions, and terrain characteristics [*Fannin and Rollerson*, 1993; *Benda et al.*, 2003a; *Imaizumi et al.*, 2005; *Imaizumi and Sidle*, 2005]. Wide ranges in sediment delivery to channels have been reported, likely attrib-utable to these factors. In the Clearwater National Forest of Idaho, about 23% of the landslide volume was delivered to channels during a 3-yr period [*Megahan et al.*, 1978]. In contrast, *Imaizumi and Sidle* [2005] reported that 64% of the landslide sediment was directly delivered to channels in the very steep Miyagawa Dam catchment in Japan. Intermediate rates of landslide sediment delivery to channels (42%) were estimated in the Sara River basin, Japan [*Nakamura et al.*, 1995]. The implication that little damage is inflicted on downstream habitat, because many landslides either do not reach streams or are confined to the upper channel reaches [e.g., *Ketcheson and Froehlich*, 1978; *Brown*, 1985], is simply not justified in most cases.

In managed forested ecosystems, woody debris in headwater channels plays an impor-tant role in the temporary storage of sediment and the ultimate release via debris flows or hyperconcentrated flows [*Megahan*, 1982; *Gomi et al.*, 2001; *May and Gresswell*, 2004]. Exposed bedrock chutes and low volumes of large woody debris characterize debris flow scour and runout zones in headwater channels; such reaches depend on new recruitment of large wood to reestablish channel structure and store sediment [*Montgomery et al.*, 1996; *Gomi et al.*, 2003; *May and Gresswell*, 2004]. Even when landslide sediment does not reach the channel, it may be transported to channels during later events and can be facilitated by clearcutting [*Swanson et al.*, 1982b; *Fannin and Rollerson*, 1993; *Gomi and Sidle*, 2003]. In some cases, landslide/debris flow deposits can provide longer-term benefits to channels associated with increasing habitat heterogeneity [*Benda et al.*, 2003a].

As headwater streams converge into larger systems, the branched structure of channels becomes important related to the storage and ultimate release of sediment derived from landslides [*Benda and Dunne*, 1997a; *Gomi et al.*, 2002]. Much sediment can be stored at tributary junctions, particularly where the junction angle is sharp; such sites are then susceptible to future debris flows after the accumulation of sufficient material [*Benda and Cundy*, 1990; *Benda et al.*, 2004; *Sidle*, 2005]. Additionally, routing of landslide sediment and debris flows is strongly influenced by changes in valley width and channel gradient [*Benda and Dunne*, 1997a; *Benda et al.*, 2003b; *Sidle and Chigira*, 2004]. At the terminus of debris flows in headwater channels, logjams often form, promoting sediment deposition and subsequent widening and braiding of the channel [*Hogan et al.*, 1995; *Montgomery et al.*, 1996; *Benda et al.*, 2003b; *Gomi et al.*, 2004]. Debris fans formed at such sites are often very unstable features, subject to channel avulsions and frequent inundation by debris flows and hyperconcentrated flows [*Howes and Swanston*, 1994; *Wilford et al.*, 2004]. *Gomi et al.* [2002] noted a shift from predominantly geomorphic-controlled processes in steep headwater streams to largely fluvial-controlled processes in larger streams with branched contributing networks.

Because forest harvesting can alter the timing and amount of sediment delivered to headwater streams, these impacts may also translate into changes in channel structure and biological habitat [*Hogan*, 1987; *Gomi et al.*, 2002; *Wilford et al.*, 2004]. Additionally, forest harvesting affects the loading of large wood into headwater channels [*Froehlich*, 1973; *Hogan*, 1987; *Jackson and Sturm*, 2002]. Following clearcutting, large quantities of small wood are typically left in streams [*Millard*, 2000]; however, the future supply of large wood is greatly reduced for many years [*Hogan*, 1987; *Potts and Anderson*, 1990]. Even if harvesting removes only large trees via partial cutting, the subsequent recruitment of large wood into headwater channels will be reduced for up to 100–150 yr [*Harmon et al.*, 1986]. Accelerated rates of landslides and debris flows following harvesting will have the greatest impact on headwater channels where timber is clear-felled down to the stream (i.e., no riparian leave area). For such cases, recovery of channel structure will take many years because of the lack of large wood recruitment from the riparian corridor [*Hogan*, 1987; *Gomi et al.*, 2003]. Conversely, narrow exposed riparian buffers in steep terrain may be subject to blowdown, thus negating some of the proposed benefits related to a sustained supply of large wood to streams [*Grizzel and Wolff*, 1998; *Martin and Benda*, 2001]. Selectively harvesting large conifers from buffers, while leaving more wind-firm, smaller hardwood trees [e.g., *Froehlich*,

1973] will allow some recruitment of wood to channels, but will decrease the supply of large wood. The role of smaller to moderate-sized logging slash in headwater streams may be beneficial in terms of temporary sediment storage [*Potts and Anderson*, 1990; *Gomi et al.*, 2001].

The "legacy" effects of past timber harvesting activities on landslides and debris flows can be defined as the memory of previous human disturbances in catchments. Such effects constitute one of the succession processes in the continuum from newly disturbed to natural ecosystems (possibly ranging from several decades to >100 yr depending on the ecosystem). Recently, much harvesting is occurring in second-growth forests; thus, understanding legacy effects of past management practices is important for assessing current conditions and impacts of future harvesting. For example, in the Pacific Northwest, most forests are now second-growth conifer and alder, both potentially possessing legacy effects from prior harvest entries. However, past studies have focused on comparisons between old-growth and clearcut forests, with little information generated on how second-growth forests affect slope stability and other catchment processes or on the susceptibility of these forests to future management. Modeling investigations have been conducted on the cumulative effects of successive harvest entries, including different silvicultural practices, on root cohesion and landslide probability [*Sidle*, 1991, 1992], but legacy effects are difficult to quantify in field studies.

Other legacy effects are apparent in channels where large woody debris recruited during earlier logging remains in channels and stores sediment [*Gomi et al.*, 2001]. Additionally, the current condition of riparian stands may result from the legacy of past debris flows, both as a function of magnitude and timing, as well as the ingress of hillslope landslides. These changes in riparian vegetation may be beneficial related to increased habitat heterogeneity [e.g., *Benda et al.*, 2003a] or detrimental, when extensive alder corridors emerge along headwater channels after debris flows or harvesting disturbances near channels [e.g., *Gomi et al.*, 2004].

Control and Avoidance Techniques

Temporal and spatial planning of forest harvest operations, including silvicultural, yarding, and site regeneration practices, can greatly influence the effects of timber harvesting on landslide initiation. Nevertheless, long-term planning of forest cutting operations must carefully weigh the need for roads associated with different logging practices, as roads are the primary cause of landslides and associated sediment delivery to streams (see next section). Effective forest planning should be possible on federal or national lands, but nations endowed with some of the best forest resources have exhibited less than wise and realistic long-term planning during the past few decades. The U.S. Forest Service, for example, jumped from a policy of widespread clearcutting in western USA in the 1960s and 1970s to an almost abject ban on logging by the mid-1980s and 1990s due to somewhat unjustified wildlife concerns. As such, because of the continuing high demands for timber, increasing pressures were placed on private lands and developing nations of Latin America and Southeast Asia, where logging impacts are substantially greater and controls on logging and enforcement of the scant guidelines for forest practices are poor [*Sidle et al.*, 2006]. Because of depressed timber markets and the availability of inexpensive timber from developing countries in Southeast Asia, both the private and public sectors of Japan have opted not to manage and harvest many of their commercial timber supplies, thus relying on imported timber from Southeast Asia. These significant examples indicate that the attitude of "out of sight, out of mind", which has permeated the European continent (where most forests were destroyed centuries ago) for many years, is not in the best interest of global sustainable development of forest lands. As such, what is needed is national governments taking a long-term, practical view of the projected demand for timber (of course, considering the development of alternative products, and weighing their relative global environmental benefits and costs) and recognizing how this can be best realized by planning forest harvesting in the most sustainable manner.

At the site scale, timber harvest planning is also very important to minimize impacts on slope stability. In general, greater slope stability benefits can be derived by implementing appropriate silvicultural practices as opposed to various logging (yarding) procedures. Retaining just a small percentage of forest stands, restricting harvesting on steep slopes and in geomorphic hollows, and retaining viable woody understory vegetation can substantially reduce landslide occurrence [*Sidle*, 1992; *Sidle and Wu*, 1999; *Dhakal and Sidle*, 2003]. Leave areas (cutting restrictions) around hollows or gully headwalls are now being implemented by forest management agencies as a measure to reduce landslide erosion [*Chatwin*, 1994]. Maintaining longer cutting rotations will lessen the probability of landslide occurrence in the long term because rooting strength will recover to nearly the maximum potential in more mature managed forests; conversely, progressively shorter rotations, which are being used in certain plantations, suppress recovery of root strength and thus generate cumulative increases in the probability of landslide occurrence [*Sidle* 1991, 1992; *Dhakal and Sidle*, 2003]. On logged sites which are replanted, selection of trees with higher root strength and rapid regeneration will benefit slope stability. Allowing a thin veneer of logging debris on hillslopes (but not excessive piling) will benefit long-term regeneration by slowly releasing nutrients following decomposition. In steep terrain, the boundaries of clearcut units should be located parallel to the prevailing wind directions whenever possible to minimize the risk of windthrow-initiated landslides [*Chatwin*, 1994].

Partial cutting in steep terrain can be problematic in terms of technology and economics. Skyline logging systems with capabilities of yarding timber laterally as well as along the skyline allow thinning or partial cutting on steep slopes with minimal damage to the residual stand [*Sidle et al.*, 1985]. However, such systems are quite expensive. Partial cutting may also induce some undesirable consequences, such as extensive windthrow, which can directly or indirectly trigger landslides [e.g., *Schwab*, 1983; *Chatwin*, 1994].

Timber harvesting methods that underutilize and/or pile small diameter wood on the hillslope may contribute to stability problems, as this material will migrate into geomorphic hollows and channel heads. As woody debris accumulates, it also traps sediment, thus overloading these unstable sites [*Sidle*, 1980]. Such observations have been made on recently logged hillsides in Nara Prefecture, Japan, where brush piling after clearcutting is used to control potential surface erosion (Figure 6.5).

Figure 6.5. Piling logging slash near ravines and hollows can trap sediment and other debris, thus providing a possible initiation zone for a future debris flow; from southern Nara Prefecture, Japan.

The type of yarding system affects site disturbance and thus the surface erosion potential [*Swanston and Dyrness*, 1973]; however, the impact on slope stability is minor. Skyline logging systems are an economical means of reducing soil disturbance on steep slopes because logs are either partially suspended or totally suspended above the ground [*Sidle et al.*, 1985]. Helicopter and balloon logging allow the site to be harvested with minimal site disturbance in steep isolated areas too hazardous for conventional cable logging systems. *Roberts et al.* [2004] found no statistically significant differences in landslide frequencies between clearcuts that were cable-yarded and helicopter-yarded in British Columbia. Nonetheless, benefits related to landslide prevention can be derived using these more expensive aerial logging methods. For example, they can be used during salvage operations after fires or insect infestations in areas with inadequate road systems [*Sidle*, 1980]. Additionally, they can be used in other remote areas to harvest timber where road construction would create landslide hazards. Combining a logging system that minimizes site disturbance with a reduced level of site preparation by burning, scarification, or herbicide application is probably the most effective way of leaving a viable woody root network at least partly intact during the critical period of forest reestablishment [*Sidle et al.*, 1985]. At the catchment or forest-level scales, planning of the silvicultural and logging systems should proceed directly with planning of the road system (or utilization of existing roads) to achieve maximum benefits related to slope stability and sustainable forest management. The overall timber harvesting plan should consider a variety of yarding options related to terrain conditions, as well as feasible road and landing locations. Minimizing road development and landings in unstable terrain will provide greater benefits for slope stability compared to selecting low-impact yarding systems alone [*Megahan et al.*, 1978; *Chatwin*, 1994].

Unstable Slope Indicators

The identification and avoidance of highly unstable slopes is the most effective and economical method of planning timber harvesting operations in terrain that is prone to landslides [*Swanston*, 1974b; *Sidle et al.*, 1985; *Thomas*, 1985; *Chatwin*, 1994]. Much of the basic hydrogeomorphic and vegetative information needed to identify unstable slopes has been discussed in detail in Chapter 3; however, some specific indicators that are particularly relevant for timber harvest planning (as well as roads, vegetation conversion, residential development, and recreation) are discussed in this section.

Terrain indicators. Many unstable landform features are useful for avoiding or modifying timber harvesting activities in sites with relatively high landslide risk. Hillslope hollows have been discussed related to their high susceptibility to shallow landslides as well as the reoccurrence of mass wasting (Figures 3.5 and 3.6). In glaciated terrain, gully headwalls (similar to hollows but almost always directly linked to channels) are particularly unstable sites that should be avoided in logging operations [e.g., *Brardinoni et al.*, 2003; *Roberts et al.*, 2004]; these sites become more unstable if logging debris accumulates in them [*Sidle*, 1980; *Chatwin*, 1994]. Narrow ridgelines in dissected terrain (between gully headwalls or hollows) may develop retrogressive slumping; such ridgelines are attractive to loggers due to the deflection that can be achieved in yarding operations, but these sites should be very carefully inspected prior to logging [*Chatwin*, 1994]. Linear depressions oriented perpendicular to slope contours are often old landslide scars and represent potential sites of recurrent debris avalanches and debris flows [*Howes and Swanston*, 1994; *Sidle*, 2005]. Evidence of old landslide features can be obtained from both aerial photos and ground reconnaissance [*Howes and Swanston*, 1994; *Hylland and Lowe*, 1997; *Rollerson et al.*, 1997]; such spatially distributed information is useful in avoiding sites that may be destabilized by logging operations. Steep slopes are relatively easy to identify and avoid; based on data from many areas, clearcutting should be restricted to slopes <36° where shallow soil mantles are underlain by bedrock or till unless careful inspection is conducted. However, other site factors must also be considered in slope stability assessments.

Hummocky topography typically characterizes deep-seated landslide terrain (Figure 2.7); it also typifies old landslide deposits. While timber harvesting may not affect these mass movement types as directly as shallow, rapid landslides, care needs to be taken when locating harvesting units and road systems that may change the hydrologic regime or geometry of the site. Aerial photo and remote image interpretation can define general locations of active or dormant slump-earthflows [e.g., *Bovis and Jones*, 1992;

Samarakoon et al., 1993; *van Den Eeckhaut et al.*, 2005]; however, more precise delineations of these features can be gained through ground reconnaissance, particularly where the terrain is obscured by vegetation. Tension cracks, step displacements, and terracettes in the soil around the headwalls, flanks, and toe of slump–earthflows are useful on-the-ground indicators of instability (Figure 6.6) [*Thomas*, 1985; *Howes and Swanston*, 1994; *Bisci et al.*, 1996]. Logging activities and roads should be avoided in such sites. Recent advances in Global Positioning Systems (GPS) have proven useful in monitoring continuous movement of earthflows [*Malet et al.*, 2002; *Coe et al.*, 2003], but such dedicated systems are only available in limited areas managed for timber.

Soil and geological indicators. Soil and geological properties that affect slope stability have been discussed in detail. Surface geology and soils maps, together with field indicators of unstable bedrock and soil conditions, are essential in planning timber harvesting operations. Regolith materials and soil types have been used as localized indicators of slope instability [*Paeth et al.*, 1971; *Durgin*, 1977; *Maharaj*, 1995[1]]. Because such mapping units often contain inclusions, a reconnaissance of geologic and soil conditions should be conducted before harvesting. As noted earlier, highly fractured, faulted, and jointed bedrock and bedrock dipping parallel to the hillslope are very vulnerable to slope failure, as are various sequences of hard and soft rock. Within regional confines, soil and rock color have been successfully used as indicators of unstable soil mantle sequences [*Rockey and Bradshaw*, 1962; *Paeth et al.*, 1971; *Crozier et al.*, 1981]. Red and orange mottles, caused by oxidation of iron, are common in soils that are periodically saturated; the uppermost mottles in the soil profile indicate the approximate height of the maximum seasonal water table [*Thomas*, 1985; *Howes and Swanston*, 1994]. When soils are saturated most of the year, the profile exhibits a

Figure 6.6. Tension cracks and terracettes located near the toe of a slump–earthflow, Oregon Cascades.

---

[1] *Nishiyama and Chigira*, 2002; *D'Amato Avanzi et al.*, 2004.

dark gray or gleyed appearance resulting from the reduction of iron compounds. Such poorly drained sites or sites where subsurface drainage accumulates are highly susceptible to mass failure following timber harvesting. The presence of a relatively impermeable layer, either in the soil or at the soil–bedrock interface, allows the formation of a dynamic water table during storms which triggers many shallow and deep-seated landslides [e.g., *van Genuchten and de Rijke*, 1989; *Terlien*, 1997; *van Asch and Buma*, 1997].

Wet site indicators. Hydrologic and vegetative indicators are useful for identifying seasonally wet sites that may be problematic for timber harvesting in unstable terrain. Seasonally wet areas in steep forested hillsides or hollows are highly susceptible to shallow landslides following timber harvesting. Surface water features, such as ponds, bogs, springs, seeps, and wet glades, in otherwise dry, hummocky terrain are indicators of active deep-seated landslides that should be avoided in timber harvest operations [*Jones et al.*, 1961; *Rockey and Bradshaw*, 1962; *Thomas*, 1985]. Springs and seeps around hollows and on steep slopes are sites of concentrated subsurface water and possibly poor drainage [*Thomas*, 1985; *Howes and Swanston*, 1994]. Such sites are susceptible to shallow landslides following timber harvesting. Patches of windthrow on hillslopes may indicate poorly drained areas or sites with shallow soils (or both), i.e., potentially unstable sites [*Howes and Swanston*, 1994].

Hydrophytes can be used as regional indicators of unstable areas with high groundwater tables or poor drainage. In the Clearwater National Forest of Idaho, patches of ferns were the best indicator of unstable wet areas [*Pole and Satterlund*, 1978]. In the Pacific Northwest, vegetation such as devilsclub, stink currant, cow-parsnip, skunkcabbage, horsetail, alder, and maple have been useful as wet site indicators [*Sidle et al.*, 1985; *Howes and Swanston*, 1994]. The association of horsetail with unstable slopes in Czech Republic was partly attributed to perched groundwater tables and partly to the high silicon and potassium levels of the dominant glauconitic bedrock [*Zaruba and Mencl*, 1969].

Other vegetative indicators. In addition to hydrophytes, several other vegetative indicators are useful in planning timber harvesting operations in potentially unstable terrain. Tilted trees (so-called "jackstrawed trees" or "drunken forests") often grow on slump–earthflows due to unstable soils and moving slopes (Figure 6.7) [*Thomas*, 1985; *Howes and Swanston*, 1994]. Trees that are tipping in generally one direction on steep slopes may indicate a high potential for shallow, rapid landslides. Such unstable headwall areas should be considered as potential leave areas within proposed timber harvest units. Pistol-butted and curved trees may result from active soil creep on hillslopes [*Howes and Swanston*, 1994] (Figure 6.8); however, curved trees should be used together with other indicators of instability, as snow loads will also distort tree form. Active tension and shear cracks in slump–earthflow terrain may be straddled by trees split up the middle by differential ground movement (Figure 6.9). Rates of tree splitting (0.3–9.3 cm yr^{-1}) at ground level agreed well with measured earthflow rates in the Oregon Cascades [*Swanson and Swanston*, 1977] and thus are useful for estimating long-term rates of ground movement. Patches of trees with conspicuous uniform age in otherwise mixed-aged forests may be the result of past landsliding [*Howes and Swanston*, 1994].

Figure 6.7. Differentially tilted (jackstrawed) trees near the head of a rotational slump along the Oregon coast.

Figure 6.8. Pistol-butted or curved trees may indicate active areas of soil creep; from Chichagof Island, Alaska.

Spatial distribution of root systems and degree of root anchoring in the substratum can sometimes be observed along cutslopes. Knowledge of relative tensile strength and biomass of different rooting systems is useful to assess timber harvesting impacts on unstable slopes. For example, Douglas-fir roots are stronger than western hemlock roots, which in turn are stronger than Sitka spruce roots. Many commercial conifer species have lower root strength and biomass than hardwood and native species [*Burroughs and Thomas*, 1977; *O'Loughlin and Watson*, 1979, 1981;

Figure 6.9. A large Douglas-fir tree split by differential earthflow movement; from H.J. Andrews Experimental Forest, Oregon Cascade Range.

Ziemer, 1981[1]]. By using information on relative strength and distribution of rooting systems in conjunction with soil engineering properties, groundwater estimates, and topography, managers can make prudent judgments on alternatives for timber harvesting. For instance, a wet forested site with slopes as steep or steeper than the estimated internal angle of friction of the soil, and evidence of root anchoring into underlying bedrock, would be a likely candidate for shallow, rapid landslides after clearcutting [e.g., *O'Loughlin and Pearce*, 1976; *Schwab*, 1983].

ROADS AND OTHER TRANSPORTATION CORRIDORS

Roads, highways, railways, and trails along hillsides affect slope stability by (1) altering natural hydrologic pathways and concentrating water onto unstable portions of the hillslope; (2) undercutting unstable slopes, thus removing support; and (3) overloading and oversteepening fillslopes, including the road prism (Figure 6.10). The relative importance of these destabilizing factors depends on the design and construction standards of the road and associated drainage system, as well as the natural instability of the terrain. Road fills placed on unstable hillslopes may fail as slumps, debris avalanches, or debris flows [*Swanston*, 1974b; *Burroughs et al.*, 1976; *Coker and Fahey*, 1993; *Wemple et al.*, 2001]. Dormant earthflows and deep-seated slides can be reactivated by cutting or reexcavating transportation corridors through toe slopes [*Swanston and Swanson*, 1976; *Haigh*, 1985; *Jones and Lee*, 1989]. Road fill material placed near the head of dormant slump blocks can reactivate slump–earthflows [*Sidle et al.*, 1985]. Exposed regoliths on cutslope faces and fillslopes are prone to dry ravel and rockfall in steep areas [e.g., *Megahan*, 1978; *Ayetey*, 1991; *Megahan et al.*, 1991]. Improved roads

---

[1] *Sidle*, 1991; *Watson et al.*, 1999.

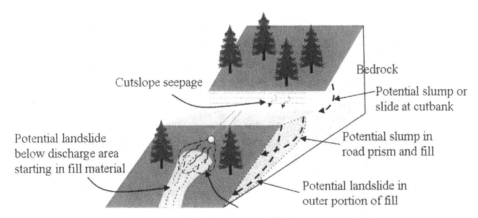

Cutslope seepage

Bedrock

Potential slump or
slide at cutbank

Potential landslide
below discharge area
starting in fill material

Potential slump in
road prism and fill

Potential landslide in
outer portion of fill

Discharge onto fillslope at cross-drain

Figure 6.10. Effects of low-volume roads on slope stability, showing the interactions with hill-slope hydrology.

with careful engineering design and drainage systems will minimize landslide hazards to some extent; however, any road cut into the residual soil mantle or regolith will decrease the stability of the site [e.g., *Megahan*, 1972, 1986; *Burroughs et al.*, 1976]. Large expenditures for slope reinforcement are common along railways in mountainous areas of industrialized countries; however, numerous landslide disasters in the past century are vivid reminders of potential damages associated with such railroad cuts [e.g., *Yamada*, 1970; *Slosson et al.*, 1992; *Schuster*, 1996; *Kiersch*, 2001]. Strategic hill roads in developing countries and low-volume forest roads in developed nations experience frequent landslides due to inadequate road drainage systems, as well as the mechanical destabilization by undercutting and overloading hillslopes [*Bansal and Mathur*, 1976; *Schwab*, 1983; *Haigh*, 1984[1]].

Because of the required excavation into hillslopes, subsurface water may be intercepted by the road and concentrated into drainage systems or on road surfaces [e.g., *Burroughs et al.*, 1972; *Megahan*, 1983; *Wemple et al.*, 1996; *Sidle et al.*, 2004b, 2006] (Figure 6.10). The extent of subsurface flow interception is poorly understood, but depends on (1) the presence of a hydrologic impeding layer below the hillslope soil; (2) the depth of this relatively impermeable layer relative to the depth of the road cut; and (3) the topography upslope of the road cut. If the impermeable layer is exposed at the road cut, then a greater proportion of the subsurface flow will be intercepted by the road compared to the case where the impermeable layer is below the road cut. Even knowing the depth of impeding layers, it is very difficult to estimate the actual amount of subsurface flow intercepted by road cuts because this depends on topography of both the surface and substrate [e.g., *Anderson and Burt*, 1978; *Rice and Lewis*, 1991; *McDonnell et al.*, 1996; *Sidle et al.*, 2001], nature of fractures and weathering in the substrate [e.g., *Komatsu and Onda*, 1996; *Sidle et al.*, 1998; *Noguchi et al.*, 1999], and the influence of

[1] *Hinch*, 1986; *Arnez-Vadillo and Larrea*, 1994; *Sidle et al.*, 2006.

preferential flow pathways above the confining layer [e.g., *Noguchi et al.*, 1999; *Sidle et al.*, 2000a]. Roads excavated through concave slope segments should intercept more subsurface water than those cut through convex or planar slopes, thus roads along concave slopes should be less stable [*Rice and Lewis*, 1991]. Preliminary data from Shiga Prefecture, Japan, suggest that this is the case [*Sidle*, 2006, unpublished data].

In a Malaysia rainforest that was selectively logged, *Sidle et al.* [2006] observed that for storms preceded by rather dry antecedent conditions, subsurface flow intercepted along a road cut comprised <10% of the total road runoff. However, once the site became very wet, cutslope interception of subsurface flow rapidly increased to 60% of the road runoff. In the Idaho Batholith, subsurface flow during snowmelt that was intercepted by a forest road increased surface runoff 7-fold prior to logging and about 18-fold after clearcutting [*Megahan* 1972, 1983]. Given this large influence of roads on local hydrology, it is not surprising that many road-related landslides are caused by inadequate drainage (too few cross-drains or no drainage system), poor maintenance (e.g., plugged culverts and cross-drains), and concentration of road runoff onto unstable downslope sites (e.g., steep slopes and hollows) [*Hinch*, 1986; *Mockler and Croke*, 1999; *Mills*, 1997; *Sidle et al.*, 2004b]. Both shallow, rapid landslides [*Schwab*, 1983; *Montgomery*, 1994; *Wemple et al.*, 2001] and deep-seated failures [*Swanston and Swanson*, 1976] can be triggered or reactivated by concentrated road drainage.

Much of the emphasis herein is on the effects of unimproved roads, as well as nonstructural avoidance and control measures related to such widespread road systems in managed forests, rangelands, farmlands, and recreation areas. Included in this discussion are unimproved roads that comprise major transportation networks in developing nations. Initially, a brief background of the stability problems along major transportation corridors is presented and some final remarks related to structural control of unstable slopes along roads are included; however, both of these topics have been covered extensively in other texts [e.g., *Cleaves*, 1961; *Gedney and Weber*, 1978; *Slosson et al.*, 1992; *Turner and Schuster*, 1996] and are not the main focus of this section.

Highways and Railways

Landslides along major roads, highways, and railways cause considerable damage in steep and unstable terrain and occasionally result in personal injury or loss of life. A 1973 study by the U.S. Federal Highway Administration estimated that annual repair costs related to landslide damage along federally funded highways was $152 million (all costs in this section are converted to US$ at 2000 currency) [*Chassie and Goughnour*, 1976]. A more recent survey estimated annual direct costs of maintenance and repair alone related to landslide damage along U.S. highways (about 20% of the entire road system in the USA) from 1985 to 1990 at $142 million [*Walkinshaw*, 1992]. California reported the highest annual cost for landslide maintenance of all states—more than $20 million during the below-average rain years in the period from 1985 to 1990 [*Schuster and Highland*, 2001]. Several areas in the USA where highway construction has caused major landslide problems include (1) Seattle, Washington [*Palladino and Peck*, 1972; *Highland*, 2003]; (2) eastern Ohio, western Pennsylvania, and West Virginia [*Sharpe*,

1938; *Fleming and Taylor*, 1980; *Delano and Wilshusen*, 2001; *Shakoor and Smithmyer*, 2005]; (3) Minneapolis, Minnesota, freeway [*Wilson*, 1974]; (4) west-central Utah [*DeGraff and Cunningham*, 1982]; and (5) the north coast of California [*Reid et al.*, 1999; *Schuster and Highland*, 2001]. Slope failures along highways in these and other unstable areas have required extensive structural control measures. Such engineering measures are prohibitively expensive for roads with low traffic volumes, which comprise a substantial portion of the road networks in the world.

The high landslide frequency along roads in California has resulted in a comprehensive system for reporting damages. During the winter and spring of 1982–1983, one of the highest damage seasons of recent record in California, approximately $166 million of storm-related damage was sustained by the statewide road system. Most of these costs were associated with landslides. In particular, U.S. Highway 50 was blocked by landslides for an extended period, preventing access into the skiing and gambling resorts of Lake Tahoe. While the cost of highway repairs was $6.2 million, the indirect costs from loss of tourist revenues was estimated at $120 million [*Schuster and Highland*, 2001]. Landslides triggered by heavy rainfall in January 1997 again blocked Highway 50 in northern California, briefly damming the American River. Reopening the highway cost $4.8 million; indirect economic losses from the highway closure exceeded $54 million [*Reid et al.*, 1999]. Several large, active landslides in the same area continue to threaten U.S. Highway 50. Additionally, the Pacific Coast Highway by Big Sur, California, has experienced recurrent landslides.

In 1991, the Pennsylvania Department of Transportation estimated that annual landslide repair costs along road corridors averaged $12.6 million in the State with a similar cost for grading related to landslide mitigation [*Delano and Wilshusen*, 2001]. A number of highway sites in Pennsylvania are also in need of permanent repair at estimated costs of $380,000 to $2.5 million each. Portions of the famous autobahn in Germany are susceptible to earthflow reactivation due to road cuts in relatively gentle topography. Road-related landslides in the Caribbean Islands generate considerable costs, largely related to clearance of debris and repair of roads. The average annual cost of repairing landslide damage to roads throughout the Caribbean is estimated at $21 million, a significant figure in this developing region [*DeGraff et al.*, 1989].

While for most of the examples cited, landslide damages along highways were confined to road repair and maintenance costs, and diversions of traffic, several conspicuous examples of loss of life have been reported. At midnight on 22 June 1988, a failure in colluvial materials upslope of the main Trabzon–Erzurum highway in Turkey, blocked traffic; stranded travelers congregated for the night in a nearby coffee house that was directly in the path of a catastrophic landslide occurring 8 h later [*Jones and Lee*, 1989]. In 1984, the road was widened, producing a 300 m long × 25 m high steep (55°) cutface, with only a 3-m-high retaining wall. Several small failures preceded the 1988 disaster; during the 2-month period prior to the landslide, 233 mm of rain fell, the highest 2-month total in more than 11 yr [*Jones and Lee*, 1989]. A total of 66 bodies were recovered from the landslide debris, with some unaccounted for; traffic was disrupted along this important corridor from Turkey to Iran for 18 days. Recent examples of other landslides along road corridors in which fatalities have occurred are as follows: (1) collapse of a rock cliff in July 1989 above National Highway 305 along the Echizen coast, Fukui Prefecture, Japan, crushing a rock shed that was built over the highway—interestingly, a landslide alarm was issued on the day prior to the disaster but was cancelled due to insufficient rainfall (all 15 passengers in a minibus were crushed to death under the rock shed) [*Yoshimatsu*, 1990]; (2) a landslide origi-nating in a clearcut above Highway 38 in southwest Oregon during a major storm (250 mm in a 2-day period) in November 1996 (1 casualty) [*Achterman et al.*, 1998]; (3) a debris flow in July 1997 along Highway 97A at Swansea Point, Sicamous, British Columbia (1 associated death) [*Singh*, 2004]; (4) the Thredbo landslide, New South Wales, Australia, caused by failure of a road embankment (due to a leaking water main) that impacted a ski resort downslope (July 1997; 18 fatalities); (5) a landslide in September 2004 above Matsuyama Highway, Nihama, Japan (several casualties in a development below the highway; Figure 6.11); (6) a landslide triggered by the October 2004 Chuetsu earthquake along a road near Yamakoshi Village, Niigata Prefecture, Japan (2 fatalities) [*Sidle et al.*, 2005]; and (7) a failure of fill material in Sanyo Highway near Iwakuni, Japan, during a large storm (September 2005; 3 deaths downslope). In some cases, the landslides initiated above the highway; in other cases, the road cuts, fills, or infrastructures within the road prism (e.g., water lines) were the primary destabilizing factors. Highways in such industrialized countries generally have high standards for construction; especially in Japan and Switzerland, consider-able investments are made in structural stabilization of cut and fill slopes. Thus, fewer landslides tend to occur along such engineered highways, but when they do occur, they may have more catastrophic consequences since there is a perception that areas downslope are protected. Also, it is often very difficult to attribute blame to highways related to landslide initiation, especially when stabilization measures are in place.

Railway corridors are typically constructed with due attention to unstable hillslopes; nevertheless, many traverse mountainous areas and are subject to landslide damage. Because of the typically deep cuts into hillsides, rockfall is common along railways [e.g., *Yamada*, 1970; *Evans et al.*, 2002]. In some cases (e.g., the 1903 Frank landslide in Alberta, Canada, and the 1950 Kumanodaira debris flow in Gunma Prefecture, Japan)

Figure 6.11. A landslide originating above Matsuyama Highway in Niihama, Japan, on 29 Sep-tember 2004; several people were killed in the development downslope of the highway, and traffic was disrupted for several days: (a) near headwall of landslide; (b) downslope destruction.

the railroad and nearby inhabitants were simply victims of the hillslope failure; in other cases, the railroad cut oɪ embankment contributed to the landslide disaster [e.g., *Yamada*, 1970; *Schuster*, 1996; *Evans et al.*, 2002]. In April 1958, a small landslide occurred in fill material of the main east–west transcontinental railroad near Baxter, California, which undercut the train tracks and temporarily closed the railway at an estimated cost of $5.9 million per day. The slide eventually triggered larger downslope mass movements impacting US Highway 40 [*Kiersch*, 2001].

In 1928, during a heavy rain, an embankment failure on the Hills Railway Line, near Adelaide, Australia, killed six workers who were demolishing a tunnel to make way for a double rail line. A moderate-sized debris flow triggered by several landslides on the steep slopes above the old Japan Railways Kumanodaira Station in Gunma Prefecture, Japan, demolished the company compound, killing 50 employees and family members in 1950. In 1972, three Canadian Pacific Railway workers were killed when they were struck by a debris flow while maintaining a rail line near Michel, British Columbia [*Evans et al.*, 2002]. More recent examples include the derailment of a passenger train by a small landslide near Granby, Colorado, in April 1985, which injured 26 passengers and caused about $3.5 million in damages [*Schuster*, 1996]. The large Watawala earthflow along the main rail line from Colombo, Sri Lanka, to the central highlands periodically disrupts rail traffic when it is reactivated during rainy periods; extensive movement in June 1992 derailed a freight train [*Bhandari*, 1994]. Several recent landslide incidents occurred along railways near Puget Sound, Washington [*Washington Military Department*, 2003]. During the heavy rains in February 1996, a landslide pushed two locomotives and two rail cars into Puget Sound, resulting in a significant fuel spill. During rain-on-snow events in January 1997, a landslide caused derailment of five freight train cars, interrupting rail traffic for several weeks. A landslide along the Beijing–Kowloon Railway occurred during heavy rains in August 2002; traffic along this important link to Hong Kong was suspended [*People's Daily Online*, 2002].

Several accounts exist of trains narrowly missing large landslides. In 1923 a train traveling southeast from Lexington, Kentucky, was fortuitously stopped in Tallega after a large landslide was discovered by a trackwalker; 2 yr earlier a smaller landslide derailed the night train at the same location, killing the engineer and injuring several others [*Louisville Herald*, 1923]. Along the Canadian Pacific Rail line near Field, British Columbia, a passenger train literally outran a massive debris flow originating from the hillside of Mount Stephen in July 1937 [*Evans et al.*, 2002].

Low-Volume Roads and Trails

Many studies indicate that unimproved roads (largely forest roads) increase landslide erosion by approximately two orders of magnitude compared with undisturbed forest land and by about one order of magnitude compared to clearcuts [e.g., *Morrison*, 1975; *O'Loughlin and Pearce*, 1976; *Gray and Megahan*, 1981[1]] (Table 6.1). For developing nations, mountain roads are sometimes an even greater concern because of the lack of

---

[1] *Marion*, 1981; *Amaranthus et al.*, 1985.

adequate location, planning, and construction measures [e.g., *Haigh*, 1984; *Hinch*, 1986; *Sidle et al.*, 2006]. *Bansal and Mathur* [1976] estimated that an average of 10 small and 0.1 large landslides occur for every kilometer of mountain road in the highly unstable Tehri and Uttar Pradesh areas of the northwestern Himalayas, India. This huge, but approximate, rate of landslide erosion ($\approx$6000 t ha^{-1} of road right-of-way) is several orders of magnitude higher than landslide erosion along forest roads in most developed nations. A survey of road-related landslides in Puerto Rico found that landslide erosion along roads was five times greater (10 t ha^{-1} yr^{-1}) than in the adjacent tropical forest (2 t ha^{-1} yr^{-1}) [*Larsen and Parks*, 1997]. A study in northwestern California measured 21.5 times the erosion (almost all mass wasting) from forest road right-of-ways (170 and 260 t ha^{-1} in two managed forest areas; no timeframe was estimated) compared to the harvested sites (3.4 and 14.3 t ha^{-1}) [*Rice and Lewis*, 1991].

Landslides originating along forest road corridors in western Oregon have been a focal point of discussion since the 1964–1965 winter storms, when extensive damage occurred in both the Cascade and Oregon Coast Ranges. During these earlier events, much of the landslide damage was attributed to poorly compacted and constructed road fills and inadequate or blocked road drains [*Dyrness*, 1967]. Landslide erosion estimated along roads in the western Cascades around this period was 30 to >300 times higher than in undisturbed forests [*Dyrness*, 1967; *Morrison*, 1975; *Swanson and Dyrness*, 1975; *Marion*, 1981]. The greatest landslide losses due to roads in the Cascades was in Alder Creek (about 203 t ha^{-1} yr^{-1}) [*Morrison*, 1975], which was more unstable and had a higher proportion of midslope roads compared to other sites. A summary of landslide erosion data in the Cascade Range from all comparable sources reported average road-related erosion rates of 15 and 128 m^3 ha^{-1} yr^{-1} (20 and 166 t ha^{-1} yr^{-1}) for moderately stable and unstable terrain, respectively, compared with erosion rates of 0.4 and 0.5 m^3 ha^{-1} yr^{-1} (0.5 and 0.65 t ha^{-1} yr^{-1}) for similar undisturbed forest [*Swanson and Grant*, 1982]. In the Mapleton area of the Oregon Coast Ranges, *Swanson et al.* [1977] found that forest roads were responsible for almost 50 times the landslide erosion (20 t ha^{-1} yr^{-1}) compared to natural forests over a 15-yr period. Landslides along roads in the most unstable land unit generated 123 times the sediment compared to forested areas. In 1975, a single storm in the same area caused extensive landsliding. Landslides along roads, however, constituted less than 20% of all reported failures and produced only one-third of the total volume of eroded material [*Gresswell et al.*, 1979]. This decrease in landslides along forest roads was attributed to improved road location, design, construction practices, and maintenance of drainage systems.

More recently, the winter storms of 1995–1996 inflicted severe landslide damage along forest roads in the Pacific Northwest. A study conducted after these storms in the Kilchis River basin of the Oregon Coast Ranges noted that 94% of the larger (>7.6 m^3) road-related landslides were failures in fill material and that most of these occurred on steep slopes (35–39°) [*Mills*, 1997]. More than 80% of these road-related landslides likely entered stream channels. In six areas of the Oregon Coast Ranges comprising 256 km^2, with about 234 km of forest roads, landslide erosion attributed to the storms was 71 t km^{-1} or about 60 t ha^{-1} of road right-of-way; these figures do not include fill "washouts" [*Skaugset et al.*, 1996]. A reassessment of the three most unstable areas in this survey revealed an average landslide erosion of 93 t ha^{-1} from all roads [*Skaugset and Wemple*, 1999]. Landslide volume from midslope roads was more than an order of magnitude higher than from valley bottom and ridge roads; twice as many landslides occurred along valley bottom roads compared to ridge roads, but the volumes were similar [*Skaugset and Wemple*, 1999]. A detailed investigation of landslide erosion sources proximate to forest roads in the western Cascade Range of Oregon following these storms revealed that landslides within the prisms of mid-slope roads produced the highest amount of sediment ($\approx$155 t km^{-1} or about 130 t ha^{-1} of road right-of-way) [*Wemple et al.*, 2001]. Estimated landslide erosion from the prisms of valley bottom and upslope roads was about half of that on mid-slope roads: 71 t km^{-1} (60 t ha^{-1}) and 69 t km^{-1} (58 t ha^{-1}), respectively. *Wemple et al.'s* [2001] detailed mapping indicated that landslide erosion above the road (but which deposited on or passed through the road) was almost 80% of the landslide volume from within the road prism. Most of these landslides that initiated above the road were likely not related to road construction (cutslope failures were included in erosion estimates from the road prism) [*Wemple et al.*, 2001], thus illustrating the difficulties in assigning landslide erosion rates to roads and explaining some of the very high rates reported in other areas, where sources of landslide erosion are not specified.

Results from an aerial photo survey of landslides that entered streams in the Klamath Mountains of southwest Oregon indicated that average mass erosion from roads was 36 t ha^{-1} yr^{-1}; values ranged from 4 to 131 t ha^{-1} yr^{-1} for various geomorphic units [*Amaranthus et al.*, 1985]. These rates were 112 and 16 times higher than landslide erosion from natural forests and harvested sites, respectively.

Several reports that focused primarily on gully erosion from logging roads in Redwood Creek basin, northern California, also present data on landslides along road corridors [*Hagans et al.*, 1986; *Pitlick*, 1995; *Weaver et al.*, 1995]. In 16 catchments surveyed within Redwood Creek (total area 243.9 km^2), about 1.3 t ha^{-1} yr^{-1} of landslide erosion occurred (calculated on the basis of the entire catchment areas) [*Hagans et al.*, 1986; *Pitlick*, 1995]. If it is assumed that roads occupy about 2% of the catchment area (based on *Rice*, 1999, in a subset of these catchments), then landslide erosion from road right-of-ways would be about 64 t ha^{-1} yr^{-1}. *Weaver et al.* [1995] focused on the most gullied basin (246-ha South Copper Creek catchment) in the Redwood Creek survey area; about 17% of the total catchment erosion originated from landslides along roads—constituting about 4.0 t ha^{-1} yr^{-1} for the entire catchment area or 102 t ha^{-1} yr^{-1} from the road right-of-ways based on the road length (10.8 km) and width (8 m) reported elsewhere [*Hagans et al.*, 1986; *Rice*, 1999]. These estimates in Redwood Creek basin are derived from the original reported volumes of soil loss [*Hagans et al.*, 1986; *Weaver et al.*, 1995] modified by the estimated 31-yr period of aerial photo coverage (1947-1978) [*Hagans et al.*, 1986] and using a more realistic value of soil bulk density (1.3 g cm^{-3}). The forest roads in place during this period were largely constructed prior to the full implementation of the Forest Practices Act of 1973 and the Timber Yield Tax Law (AB-1258) [*Martin*, 1989], and although the road system expanded during this period of intense tractor logging [*Best*, 1995; *Rice*, 1999], estimates of "static" road density were used herein because of the lack of road information in the Redwood Basin reports. Some of the forest roads were not used for timber harvesting. A later, more field-intensive, survey focused on logging road erosion in middle Redwood Creek basin during the period from 1980 to 1997 [*Rice*, 1999]; mass wasting from road right-of-ways was estimated at 15 t ha^{-1} yr^{-1}, almost 7 times lower than in nearby Copper Creek during an earlier period. Many small landslides occurred; however, large slumps, while few in number, produced almost half of the landslide mass [*Rice*, 1999]. More than half of all landslide erosion occurred during the winters of 1995–1997, when several large storms occurred. Reasons for the discrepancies between *Rice's* [1999] data and the earlier studies [*Hagans et al.*, 1986; *Pitlick*, 1995; *Weaver et al.*, 1995] may be related to the larger storms included in the earlier but longer studies [*Madej*, 2001]; however, changes in forest practices implemented in 1976 that resulted in improved road drainage and design, reduced road widths and grades, improved alignment, and decreased reliance on tractor logging, likely contributed to the lower landslide erosion reported from 1980 to 1997 [*Rice*, 1999]. Moreover, this latter study focused specifically on road-related erosion and landslides, while the earlier studies did not.

High rates of landslide erosion from forest roads constructed in weathered granite have been documented in several studies in New Zealand. In Dart Valley, South Island, an average erosion rate of 840 t ha^{-1} yr^{-1} occurred along selected road corridors (assuming a 10-m right-of-way influence); however, surface erosion sources were not separated from landslide sources [*Mosley*, 1980]. Only about one-third of this eroded material entered the drainage system. Following four major storms in July and August 1990, 263 landslides were measured along 142 km of forest roads near the same area of South Island; estimated landslide erosion along the road right-of-way was about 1400 t ha^{-1}, equivalent to about 80 yr of sediment yield from surface erosion [*Coker and Fahey*, 1993]. Background erosion rates in the vicinity are estimated at 5 t ha^{-1} yr^{-1} [*Adams*, 1980]. A summary of road-related landslide erosion indicates that landslide erosion decreases on older forest roads: On newly constructed roads mass erosion was 266 to 7600 t ha^{-1} yr^{-1} (assuming a 10-m right-of-way); this rate decreased to 38–380 t ha^{-1} yr^{-1} on roads about 10 yr old [*Fransen et al.*, 2001]. Except for one site (limestone), these high landslide rates all occurred in granitic terrain and were up to three orders of magnitude higher than rates of surface erosion [*Fransen et al.*, 2001]. This observed trend of decreasing mass wasting with road age apparently does not affect ravel; as granitic material weathers on cutslopes, ravel is produced, exposing new granite that in turn weathers and causes ravel [*Megahan and Kidd*, 1972; *Megahan et al.*, 2001] (see Figure 2.13b).

Several recent landslide surveys based on aerial photo interpretation have been conducted for forest roads in British Columbia. In Clayoquot Sound, 20% of the total landslides were associated with forest roads; 63% of these were debris slides, while the remainder were debris flows [*Jakob*, 2000]. In three catchments on Vancouver Island, 26% to 66% of all landslides were caused by forest roads or about 0.35 to 1.33 landslides km^{-2} yr^{-1} along road right-of-ways [*Guthrie*, 2002]. From 38% to 64% of these landslides were directly connected to streams. An earlier field investigation following the 1978 storm in Rennell Sound, Queen Charlotte Islands, noted that 20.8 landslides km^{-2} occurred along forest roads, almost twice the frequency as in clearcuts; however, clearcuts occupied about 90% of the land area modified by logging compared to only 10% for roads [*Schwab*, 1983].

A number of studies have reported that the majority of road-related landslides are small to moderate-sized failures along cutslopes [*Fiksdal*, 1974; *Megahan et al.*, 1978; *Rice*, 1999; *Wemple et al.*, 2001[1]]. Most of these studies indicate that much of this landslide erosion is at least temporarily stored on the road and does not immediately contribute to sedimentation in streams. Nevertheless, such sediment can later be mobilized as surface erosion during subsequent storms and can become a chronic supply to streams [*Sidle et al.*, 2004b, 2006]. Even on relatively gentle slopes, minor retrogressive failures in road cuts may provide a chronic supply of sediment to roads; *Haigh* [1985] found that average recession of road cutslopes in Oklahoma varied from 23 to 27 mm yr^{-1}; tree roots appeared to impart a small reduction in cutslope retreat. A survey of shallow, translational landslides on road cuts in southeastern Ohio also noted the stabilizing role of deep-rooted woody vegetation at many sites [*Shakoor and Smithmeyer*, 2005]. The processes of mass wasting and surface erosion are related to sediment delivery to channels.

Fillslope failures in unimproved roads are often attributed to poor road drainage, inadequate compaction of fills, and inclusion of wood in fills [*Burroughs et al.*, 1976; *Phillips*, 1988; *Douglas et al.*, 1999; *Skaugset and Wemple*, 1999]. While fewer fillslope

Figure 6.12. Landslide in a fill embankment of a forest road resulting from impediment of road drainage, Oregon Coast Ranges.

---

[1] *Shakoor and Smithmeyer*, 2005; *Sidle et al.*, 2005, 2006.

landslides may occur along mountain roads compared to cutslope failures, they are typically larger and travel greater distances, often directly impacting streams. A study in Idaho noted that while only 27% of the road-related landslides were associated with poor road drainage, these fillslope failures contributed more than half of the total landslide sediment directly to channels [*Megahan et al.*, 1978]. Unstable sidecast material and blocked culverts were associated with 58% and 45%, respectively, of road-related landslides along the East Coast of North Island, New Zealand [*Phillips*, 1988].

Poor drainage design (too few or no cross-drains) and plugged cross-drains on insloped, crowned, or ungraded roads can lead to saturation and failure of fill materials [*Megahan et al.*, 1978; *Hinch*, 1986; *Piehl et al.*, 1988] (Figure 6.12). In addition to the greater environmental damage, such failures are generally more costly to repair because the road prism needs to be reconstructed. Loose fill material on midslope roads in steep terrain is very susceptible to shallow failures and may represent a chronic source of sediment along poorly constructed mountain roads [*Burroughs et al.*, 1976; *Mosley*, 1980]. Slumps are also common in road prisms constructed with fills [*Schroeder and Brown*, 1984].

Trails in mountainous terrain support travel by animals, humans, and vehicles, as well as related land management activities. While the major sediment impacts of such trails are associated with surface erosion due to the compacted surfaces, certain trails cut into hillsides may cause small cutslope failures [*Sidle et al.*, 2006]. Additionally, runoff from trails concentrates onto unstable segments of hillslopes or diverts runoff into desiccation cracks, thereby initiating landslides [*Knapen et al.*, 2006]. Mountain trails are ubiquitous in developing regions of Asia, Africa, and South America.

Avoidance Measures

While roads may be appropriately viewed as "necessary evils" in rural mountain settings, substantial reduction in landslide erosion along road corridors can be achieved through wise planning and design measures—notably, road location, construction and drainage practices, maintenance, and consideration of long-term, multiple usage. Where major highways cut through unstable terrain, special care needs to be taken, possibly including warning systems, to avoid injuries. For low-volume mountain roads, issues related to road maintenance and closure need to be carefully considered.

Road location. Probably the greatest benefits related to landslide damage reduction along roads can be achieved through careful road location. Seasonally wet areas on steep slopes should be avoided whenever possible, particularly if associated with geomorphic hollows or other locations of likely landslides. Midslope roads on steep slopes are particularly problematic, due to the large amount of excavation involved. Figure 6.13 illustrates the exponential increase in road right-of-way disturbance for a given road travel width with increasing hillslope gradient. Deep cuts into regoliths may expose unstable sequences of weathered materials as well as removing downslope support. Thus, special attention needs to be paid to soil and geological materials along road cuts as well as joints, fractures, and other unstable structures. Partic-

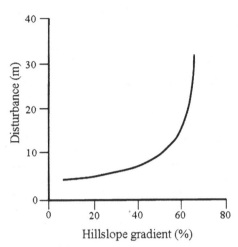

Figure 6.13. Increasing width of road right-of-way disturbance for a fixed road width as a function of slope gradient [after *Megahan*, 1977].

ularly, bedrock dipping parallel to the hillslope may become unstable when roads are cut through these materials. Soils may exhibit relic structures from parent bedrock that indicate areas of potential instability. Areas of old landslides should be avoided. Slump blocks and earthflows can easily be reactivated by locating a road across them. Roads should never cross the head of a rotational failure or undercut the toe. If absolutely necessary to locate roads in such areas, the most desirable location is above the neutral line of the potential slump, with the cutslope uphill of the neutral line and the fill placed downhill (i.e., unloading the head and loading the toe of the potential slump) [*Burroughs et al.*, 1976; *Hutchinson*, 1977]. Such a configuration shifts the center of gravity to a lower, more stable position (Figure 6.14). Poorly consolidated mudstones and clays are subject to slumping when undercut by excavation or when overloaded by embankments [*Zaruba and Mencl*, 1969]. Buried pockets of unconsolidated colluvium also may fail following excavation; such isolated sites are difficult to locate and may represent old valleys or slump escarpments that were filled in by colluvium [*Burroughs et al.*, 1976]. In general, deep cuts into hillslopes should be avoided and alternative routes should be selected if at all possible.

   Valley bottom roads are the least problematic related to landslide initiation, unless they are cut into adjacent hillslopes and remove support or cross debris fans [*Wemple et al.*, 2001]. These roads, however, may generate surface erosion that can easily be delivered to adjacent streams, thus precautionary measures and maintenance need to be undertaken to prevent surface-eroded sediment from reaching channels. Ridgeline roads are also relatively stable locations [*Megahan et al.*, 1978], although landslides have occurred below discharge points along such roads [*Montgomery*, 1994]. Ridgeline roads require little or no excavation and fill material is typically not used, thus eliminating cut and fill slope failures. Midslope roads require extensive excavation and thus may intercept large quantities of subsurface drainage [e.g., *Megahan*, 1972; *Sidle et al.*,

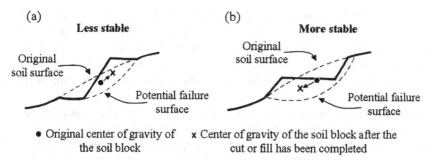

Figure 6.14. The effects of road cuts and fills in relation to reactivating or initiating slumps: (a) Loading the head and/or unloading the toe shifts the center of gravity upwards and increases the potential for slumping; (b) loading the toe and/or unloading the head shifts the center of gravity downwards and decreases the potential for slumping.

2006]; discharge of this additional road drainage on steep and unstable slope segments often triggers landslides. Furthermore any fill placed on steep slopes will potentially create instability problems. Thus, roads located on steep midslope sections are by far the most unstable road locations [*Megahan et al.*, 1978]. A study in managed forest lands of Idaho categorized the percentage of road-related landslides based on slope location as follows: ridgeline, 2%; upper third of the slope, 18%; middle third of the slope, 31%; lower third of the slope, 38%; and slope bottom, 11% [*Megahan et al.*, 1978]. Rolling the road grade of midslope roads allows the road to conform to the natural drainage patterns and thus alleviates some of the instability issues related to concentration of road drainage [*Gardner et al.*, 1978].

The design and layout of mountain road systems is challenging because the same road network may cross various jurisdictions, each with their own set of priorities. In general, the following, not mutually exclusive, considerations should be weighed in laying out mountain road systems from the perspective of slope stability: (1) minimizing road length in steep and unstable terrain; (2) minimizing total road length; (3) minimizing road width, especially on steep, midslope locations; (4) rolling roads to fit hillslope contours and across culverts; (5) minimizing stream and swale crossings; (6) reducing the amount of blasting on steep slopes; and (7) formulating a design that will accommodate the long-term management objectives of the users so that new roads are not necessary. Restricting road grades to <8.5° with short pitches ≤11.3° reduces landslide erosion [*Rice*, 1999]. Adjusting road grade in the field to avoid unstable sites, prior to any excavation, is a cost-effective measure to reduce future landslide damage; if necessary, the proposed road location should be abandoned rather than risking a landslide [*Chatwin*, 1994].

Even for individual land uses, many tradeoffs may be necessary. For example, uphill skyline logging systems generally require less road mileage than ground-based (tractor) systems, but roads wide enough to accommodate skyline towers must be constructed in steep terrain. To minimize the number of crossings over headwater channels and hollows, it may be necessary to build longer roads in potentially steeper terrain. Avoiding wet, unstable hillslopes as well as potential rotational slumps or earthflows may also require diversions in road alignment, thus increasing road length and sometimes requiring adverse grades for short distances. To minimize sediment delivery to streams, disturbances along roads should be located some distance away from channels, possibly necessitating the construction of roads on hillslopes that may be marginally stable [*Megahan*, 1987].

Planning the low-volume road system should also consider future road closure, unless the road is deemed to be a permanent transportation corridor. Road deactivation in steep terrain for the purpose of stabilizing hillslopes is both an expensive and uncertain venture as currently practiced. *Allison et al.* [2004] showed that potentially 98% of the landslide amelioration benefits can be derived by carefully selecting 10% of the most unstable road segments for deactivation and concentrating expenditures on these. The notion held by many reclamation specialists that regrading the landscape at the site of the road will adequately restore hydrogeomorphic functions to natural conditions is naïve at best. Recent studies have suggested that the hydrologic "legacy" of mountain roads will persist decades if not centuries after road closure [e.g., *Schwab*, 1983].

Alternative construction techniques. Low-volume roads constructed on steep hillsides necessitate extensive excavation into potentially unstable soils and bedrock, thus minimizing road width can alleviate some of the slope stability concerns. Road right-of-way disturbance increases rapidly with increasing slope gradient [*Megahan*, 1977] (Figure 6.13). Sidecast or fill material is unstable when placed on slopes >35° [*Chatwin*, 1994], necessitating alternative construction procedures. Thus, proper design of low-volume roads is critical for both function and environmental protection.

Mountain roads are typically constructed according to one of the following designs: balanced cut and fill; variations of cut and fill design; full-benching with end-hauling of waste material; backcasting; and cutting with the waste material pushed downslope (sliver fills) (Figure 6.15). Balanced cut and fill design requires that the volume of

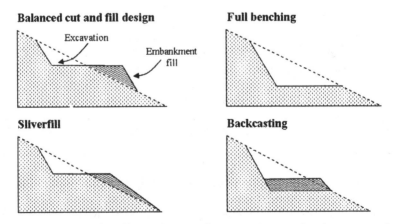

Figure 6.15. Examples of low-volume road construction practices in steep terrain.

excavated material approximately equals that incorporated into the road fill; it is used when possible because it minimizes excavation and hauling costs [*Megahan*, 1976; *Gardner et al.*, 1978]. Full-benching is used in steep sites to minimize the width of the road right-of-way and thus reduce the potential for landslides; however, it is the most expensive construction alternative. A bench is cut into the hillslope equal to the road width and soil and rock materials are end-hauled to a stable disposal site [*Chatwin*, 1994]. Backcasting is a type of full-benching where, following the excavation of the road bench, the bulk of the excavated materials are used to construct a subgrade for the road; such construction will only work if the native materials are competent and well-drained [*Chatwin*, 1994]. Sliverfills are the result of road excavation where the waste soil and rock are simply pushed to the downslope side with little or no engineering design and no attempt to incorporate the materials into the road prism. This practice results in highly unstable sidecast if it is placed on slopes steeper than 31°, especially when materials are not free-draining [*Gardner et al.*, 1978; *Chatwin*, 1994]. If at all possible, sliver fills should be avoided and full-benching should be implemented. Fitting the road alignment to the topography can serve to avoid large cuts and fills [*Gardner*, 1979]. In addition to the road construction design, the method of excavation and the handling of fill materials determine whether the road will create stability problems.

The excavation methods used for cutting roads into hillsides can greatly influence stability as cutslopes are one of the most persistent landslide sites [*Megahan*, 1987; *Arnaez-Vadillo and Larrea*, 1994; *Rice*, 1999]. Generally, the stability of road cuts into residual soil, colluvium, or bedrock decreases with time as weathering proceeds [*DeGraff*, 1978; *Clayton*, 1983]. Road construction with hydraulic shovels or backhoes provides better control of excavated soil and rock compared to bulldozers, decreasing ground disturbance and facilitating more precise cut, fill, and ditch locations [*Bourgeois*, 1978; *Sidle et al.*, 1985]. Where hard bedrock is encountered, blasting is necessary. Controlled blasting techniques such as "pre-splitting" and "smooth wall blasting" control the volume, travel distance, and breakage of the displaced bedrock, thus allowing much of the rock to be removed to more stable sites [*Burroughs et al.*, 1976; *Hinch*, 1986; *Chatwin*, 1994]. During conventional heavy blasting, ground motion can trigger landslides and displaced rocks can overload already steep slopes, eventually initiating debris slides [*Vandre and Swanston*, 1977; *Sidle et al.*, 1985]. Regardless of the excavation method, deep cuts, particularly in unstable regoliths, should be avoided; road realignment must be considered to circumvent such sites [*Gonsior and Gardner*, 1971]. For example, *Gardner et al.* [1978] noted that sloughing occurred on steep cutslopes that exceeded ≈2 m height in weathered granitic soils of Idaho. To minimize road right-of-way disturbance, steep (but not high) cutslopes are desirable, together with controlled excavation and waste rock handling. *Chatwin* [1994] recommended cutslope gradients based on substrate, ranging from 1:1 to 2:1 for lacustrine or marine deposits; from 1:1 to 1.5:1 for non-cohesive alluvium; from 0.25:1 to 1:1 for glacial till; and from 0.5:1 to 1.5:1 for glaciofluvial deposits with boulders. For soft and hard bedrock, benching was recommended with gradients of 0.5:1 and 0.25:1, respectively.

Because landslides in fill materials are typically larger and deliver more sediment to streams compared to cutslope failures [e.g., *Wemple et al.*, 2001; *Sidle et al.*, 2005], construction of road fills must be carefully executed. Prior to incorporating excavated soil and rock into the fillslope, all organic materials should be carefully removed, as these will decay with time, weaken the fills and cause landslides [*Gardner et al.*, 1978; *Douglas et al.*, 1999]. Fill embankments must be thoroughly compacted to increase the strength of the materials and to preclude water entering the road prism [*Gonsier and Gardner*, 1971; *Chatwin*, 1994]. Care should be taken in the field to select ideal soil moisture conditions for optimizing compaction for particular substrates. These conditions can be determined by standard compaction tests for different materials [e.g., *Sidle and Drlica*, 1981] or rules of thumb for different soils [e.g., *Chatwin*, 1994]. For example, in clayey soils, maximum compaction is achieved during slightly wet conditions. Fill material should be compacted in successive layers of 20–30 cm, depending on the compactor used. A roller-type compactor produces a much more uniformly compacted subgrade compared to using a crawler tractor [*Chatwin*, 1994]. Poorly compacted road fills are a major cause of landslides; when combined with saturated to nearly saturated conditions, liquefaction can occur, producing debris flows which travel long distances in steep terrain [*Dyrness*, 1967; *Gonsior and Gardner*, 1971]. In cases where full-benched roads require subgrade material to be placed on the surface (backcasting), and where subsurface water is intercepted by the road cut, either the fill material should consist of coherent coarse aggregates, or rock-filled drainage trenches ("French drains" or "Squamish culverts") should be installed in the subgrade to drain excess water [*Chatwin*, 1994]. Geotextiles can also be used together with such construction practices. Fillslopes should be revegetated as soon as possible after construction [*Gonsier and Gardner*, 1971; *Schwab*, 1994], preferably with the inclusion of deep-rooted woody species to contribute additional stability via rooting strength.

Road drainage. Ideally, road construction should impede natural drainage as little as possible with due attention to avoiding concentration of drainage on unstable slopes. Inadequate road drainage has been cited as the primary cause of damaging landslides in many studies in steep terrain [e.g., *Dyrness*, 1967; *Gonsior and Gardner*, 1971; *Ballard and Willington*, 1975[1]]. Except for lesser used permeable fills, road drainage is typically engineered by one of three road surface configurations: (1) insloped road with an interior ditchline and drainage relief culverts placed at intervals under the subgrade; (2) crowned road surface with similar drainage facilities with some of the water draining off the outer edge of the road onto the slope; and (3) outsloped road surface (Figure 6.16).
    Outsloped roads may effectively divert water from ridgeline roads and other low gradient (3–5%) stretches where a stable road base exists [*Megahan*, 1977; *Sidle et al.*, 1985; *Schwab*, 1994]. Well-designed and maintained outsloped roads may disperse drainage water across the lower hillslope; however, drainage water often will concentrate in ruts on the road surface and discharge at low points, creating potential landslide sites. Outsloped roads are most effective on low-volume sites, where cutbank slumping is likely to obstruct the ditchline [*Schwab*, 1994]. Properly designed insloped and crowned roads allow surface water to drain most effectively when ditchline gradients are > 3%

---

[1] *Phillips*, 1988; *Mills*, 1997; *Skaugset and Wemple*, 1999.

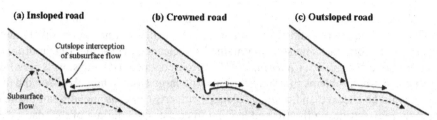

(a) Insloped road   **(b) Crowned road**   **(c) Outsloped road**

Cutslope interception of subsurface flow

Subsurface flow

Figure 6.16. Drainage design for mountain roads: (a) insloped with interior ditchline; (b) crowned road surface with interior ditchline; and (c) outsloped road (no ditch). Dashed arrows indicate water flow paths.

[*Gonsior and Gardner*, 1971]. Adequate sizes and numbers of drainage relief culverts should be installed on both insloped and crowned roads to disperse intercepted subsurface water as well as overland flow from the road surface. Most guidelines for culvert sizing and spacing have been developed for surface erosion control; additional considerations related to slope stability include specific outlet locations and prevention of clogging by woody debris [*Piehl et al.*, 1988; *Mills*, 1997]. A minimum relief culvert diameter of 40 cm is recommended to prevent clogging by organic debris [*Megahan*, 1977]. Inlets of cross-drains can be protected from excessive sediment and debris by installing drop boxes, simple filter structures, or debris racks and by changing the road grade. In addition to conventional relief culvert spacing considerations, culvert discharge should be avoided near natural swales (hollows), deep unstable soils, sites of previous landslides, or fractured bedrock. Culverts should be placed at natural gullies and incipient channels to facilitate natural drainage. In general, desired spacing of drainage relief culverts on roads steeper than 2° vary from about 75 to 260 m, depending on road gradient, soil erodibility, and local climate conditions [*Arnold*, 1957; *Sidle et al.*, 1985; *Chatwin*, 1994]. The topography of upslope areas should be considered, because concave hillslopes will channel more subsurface water to road cuts than planar or convex slopes; current guidelines for spacing of drainage relief culverts do not consider topographic effects. Thus, adherence to guidelines for drainage spacing should be flexible, with due attention paid to the location of drain outlets. Road fills should be protected from culvert discharge by extending cross-drains beyond the fill and riprapping the outfall area with rock. Many road landslides on steep slopes are caused by a combination of oversteepening with fill material and periodic saturation of fills by culvert discharge.

In many mountainous regions of developing countries, as well as some low-volume road sites in developed countries, little or no attention is paid to drainage design. For such roads, storm drainage simply discharges on the hillslope at low points or drainage nodes [e.g., *Sidle et al.*, 2004b; 2006]. This concentrated drainage can trigger landslides in hollows or on steep slopes. In other cases, hollow log culverts used as cross-drains have been observed to decay with time and collapse, leading to landslide initiation [*Douglas et al.*, 1999]. Simple, low-cost measures, such as drainage dips or water bars, can disperse surface runoff from low-volume roads. *Packer* [1967] developed a guide for spacing of surface drains for both insloped and outsloped roads. Off-site impacts can be reduced by placing woody debris downslope of road drainage nodes to trap sediment [*Sidle et al.*, 2004b].

Road maintenance. Substantial benefits related to slope stability can be achieved by properly and frequently maintaining roads and associated drainage systems. Ditches, culverts, and debris racks must be maintained and cleaned to ensure efficient drainage function [*Sidle et al.*, 1985; *Chatwin*, 1994]. Patrols conducted during storms are effective in correcting problems in progress; following large storms, all culverts and debris racks should be checked and cleaned. Blocked culverts can cause ponding of drainage water, thereby saturating the fill and leading to catastrophic failure of the road [*Piehl et al.*, 1988] (Figure 6.12). Road surfaces should be graded during the dry season to ensure proper drainage of surface water. Prior to heavily used periods, roads should be resurfaced with competent crushed rock. Access on unsurfaced or unrocked roads should be limited during the wet season to prevent excessive erosion and rutting on the running surface. When cutslope failures occur, avoid excavating back into the slope when removing failed material, thereby causing additional failures to initiate soon afterwards. Disposal of debris from landslides that accumulates on road surfaces should be placed on stable hillslope locations [*Burroughs et al.*, 1976]. Impending signs of instability, such as subgrade settlement, sagging cut banks, or ponded water on the road prism, should be treated promptly to prevent a major slope failure. Oftentimes the combined efforts related to avoidance (e.g., road location) and maintenance measures can alleviate much of the need for expensive structural controls along low-volume roads.

Warning systems. Landslide warning systems along highways are designed to provide travelers and residents with adequate time to evacuate areas or take alternative routes to avoid impending hazards. As such, these decision systems use real-time data—typically rainfall, pore water pressure, or ground motion in or near known potential failure sites. The U.S. Geological Survey, in cooperation with the California Department of Transportation, installed a real-time system for monitoring landslides along several unstable portions of U.S. Highway 50 in northern California [*Reid et al.*, 1999]. Real-time data on ground motion and pore water pressure are telemetered to Geological Survey computers; graphs of these responses are directly available over the Internet to local officials, geotechnical engineers, and emergency managers. A similar warning system has been installed along a railway corridor in Seattle, Washington [*Reid et al.*, 1999]. Several regions (e.g., Hong Kong, California, Oregon, Japan) that experience widespread landslide problems along roads have established warning systems based on real-time rainfall data. Compliance with these warnings is typically voluntary and based on dissemination of information. For example, in Oregon, there are two warning levels: (1) debris flow advisories to alert landowners and road managers to check road drainage systems and conduct needed maintenance in the event that the predicted rainfall does occur; and (2) debris flow warnings issued when a rainfall threshold is nearly reached so that people can avoid unsafe roadways. In Japan, alarms are issued based on rainfall forecasts in susceptible mountainous areas; following alarms, if a designated threshold of total rainfall occurs (e.g., 140 mm in a single storm for the Echizen coast), the road is closed [*Yoshimatsu*, 1990]. Throughout the railway system in Japan, an automated system shuts down trains during moderately large earthquakes to prevent derailment and potential landslide damages.

Structural Control Measures

As noted earlier, the topic of structural control of landslides has been extensively covered elsewhere. Nevertheless, the several types of structural controls are briefly mentioned as they have been successfully applied as stabilization measures in conjunction with road construction. These controls have been conveniently categorized relative to function as (1) reducing the driving forces and (2) increasing the resisting forces [*Holtz and Schuster*, 1996].

For roads excavated into steep hillsides or other potentially unstable sites, a simple way to increase stability is to reduce the mass of soil in the embankment and cutslope. As mentioned previously, benched cutslopes, reduced excavation depths, excavation to unload the upper part of an unstable mass, and full bench road construction will achieve these objectives. Additionally, incorporating lightweight fill material and removing excess surface and subsurface water via drains are effective measures to reduce the driving force [*Holtz and Schuster*, 1996]. If space permits, excavating the toe of a failure until successive failures generate a stable average slope can be effective. All of these measures are rather costly and should be judiciously applied in sites where alternative road locations are impractical.

Actual structures are more commonly applied in control measures along roads that increase the resisting forces as a method of stabilization. Loading the toe of an unstable cutslope with a rock buttress is a cost-efficient way to stabilize potential slides or slumps; however, often this corrective measure is applied after a problem develops (Figure 6.17). To be effective, the buttress mass should be about 33% to 50% of the potential landslide mass [*Chatwin*, 1994]; thus, this mitigation measure is generally applicable only to smaller potential failures [*Holtz and Schuster*, 1996]. Retaining walls can be constructed where acute stability problems occur and where

Figure 6.17. Retaining wall supporting an unstable road fill in steep terrain; a small rock buttress was constructed to support the unstable cutslope (upper left side of photo).

space is not available to install a buttress or where the size of the potential failure is too large (Figure 6.17). Timber crib walls are commonly used to stabilize relatively small-volume fills along forest roads [Chatwin, 1994]. More expensive retaining structures include steel bin walls, gabions, precast concrete walls, concrete-filled culvert pile walls, and any retaining structures that incorporate either pre-stressed or post-tensioned soil or rock anchors [Schwarzhoff, 1975; O'Rourke and Jones, 1990; Chatwin, 1994]. In addition to these "external" methods of applying a resisting force against potential landslides along roads, soil reinforcement methods have gained popularity in recent years as comparatively cost-effective measures for treating unstable cut and fill slopes along roads [Holtz and Schuster, 1996]. Examples of this "internal" procedure of soil stabilization include the incorporation of polymer or metal strips, geotextile sheets, geogrids, bar and mesh reinforcement systems, earth-anchor reinforcements, and synthetic fibers into backfills. Examples of in situ internal reinforcement procedures in cutslopes and embankments include soil nailing, soil anchors, soil dowels, and micro-piles (so-called "root piles"). Other, less common, methods for stabilizing road embankments include techniques based on chemical treatment (e.g., grouting, ion exchange), electroosmosis, and thermal treatment (e.g., heating or baking clay soils to improve their strength; freezing soils to gain temporary stability). Revegetation with deep-rooted woody species is an attractive low-cost alternative to provide an increment of stability to embankments and excavated cutslopes; wattling is an effective procedure to promote rapid establishment of woody species on such disturbed sites [Schwab, 1994].

Many mountain road systems throughout the world are constructed under different jurisdictions and link activities on public lands to private lands. Thus, it is not surprising that the construction standards as well as the structural and non-structural measures used to control landslides differ greatly from place to place, and investments in such control measures are not always commensurate with resources or people being protected. Additional complications arise because, in some cases, slope stabilization measures along road right-of-ways are funded and implemented through a different government agency than the agency managing the land. Thus, evaluation of the relative costs of the resources at risk and the erosion control measures are not always considered. By utilizing avoidance and non-structural alternatives (e.g., proper road location, road network planning, adapting design standards) for the majority of the road network, and focusing the structural measures only in the most critical areas, more cost-effective solutions for landslide control can be achieved. In Japan, where the Ministry of Construction and Ministry of Forests allocate large budgets each year for slope stabilization along mountain roads, there is no systematic, scientifically based assessment of the relative benefits derived from these expenditures. A recent problem occurred in Miyagawa catchment (Mie Prefecture, Japan), where most of the mountain roads were closed due to landslides incurred during the autumn 2004 typhoons, thus preventing the forest land owners in that region from managing their property (Figure 6.18). Many retaining structures along roads failed during these typhoons; other landslides and debris flows damaged the road system. At present, some of the roads are being reopened

Figure 6.18. An example of one of the many landslides and debris flows that destroyed the mountain road system in Miyagawa dam catchment during the October 2004 typhoon disaster, Mie Prefecture, Japan.

at great expense, including huge expenditures for structural landslide control. Such an extensive disaster in a remote region begs careful consideration of alternative road system planning or even permanent closure of some roads; apparently such alternatives are not being weighed. There is no doubt that a more prudent targeting of such funds would result in huge cost savings and environmental benefits, albeit with some changes in the road system and land use.

CONVERSION OF FORESTS TO AGRICULTURAL LANDS AND PLANTATIONS

Management Effects

Although widespread forest conversion on steep hillslopes is occurring in developing countries of Southeast and East Asia, Africa, and Latin America, a long legacy of permanent forest conversion to agricultural persists in Europe and to a lesser extent in North America, albeit in gentler terrain. Conversion of native forests and previously converted grasslands to exotic radiata pine plantations is ongoing in New Zealand and Australia. Forest conversion in steep terrain to cropland, monoculture plantations, agroforestry, and other exotic vegetation types has the potential to accelerate landsliding, largely by reducing rooting strength, but also by modifying the soil moisture regime [*Sidle et al.*, 2006]. Where these plantations and crops are grown as monoculture, there is generally less ground cover, depleted soil nutrients, deteriorated soil structure, and insignificant rooting strength compared to forests that previously occupied the hillslopes. Tropical forest land in mountainous regions is particularly susceptible to mass failure when converted to agriculture or plantations, although processes and rates are poorly documented in these environments.

Swidden (shifting) cultivation has been practiced for centuries in mountain lands of Southeast Asia, Latin America, and Africa as part of subsistence livelihoods [e.g., *Spencer*, 1966]. Traditional swidden agriculture involves the clearing of forest patches in the dry season, subsequent burning prior to the rainy season to release nutrients, cultivation of the cleared patch for a number of years, fallowing, and secondary regrowth through various stages of succession [e.g., *Spencer*, 1966; *Fox et al.*, 2000]. During the cultivated and cleared periods, steep sites are at greater risk of landslides due to the negligible rooting strength [*Sidle*, et al., 2006]. Because certain types of shifting cultivation may clear forested hillsides for long periods and because reforestation on abandoned, nutrient-depleted agricultural soils is very slow, the impact of slash and burn agriculture is longer-lived compared to timber harvesting. Trees included in modern agroforestry systems offer benefits for rooting strength albeit less then the typical rooting strength of mature forests (Figure 6.19).

Compared to surface erosion, the effects of forest conversion on landslides in developing regions have been much less studied. Progressive forest clearing/conversion and deterioration of the land base in developing countries of Africa, Asia, and Latin America were associated with general increases in landslides [*Haldemann*, 1956; *Harwood*, 1996; *Fischer and Vasseur*, 2000], although specific rates of increase were not reported. *Lanly* [1969] estimated that the land base in the rainforest of the Ivory Coast declined by 30% from 1956 to 1966 due to shifting cultivation; these losses were apparently related to landslides. Increases in landslide erosion following progressive forest conversion in Tanzania from the mid- to late nineteenth century are partly attributed to loss of rooting strength after clearing and burning [*Haldemann*, 1956]. During a large storm in 1969 in the Darjeeling Hills of India, where extensive vegetation conversion from natural forests to tea gardens occurred over the previous 130 yr, debris flows destroyed an average of 20% of the tea garden land area compared with only 1% of the forested area on 20° to 40° slopes [*Starkel*, 1972a]. Most of the landslides on forested land were associated with roads. Failures in the tea gardens were generally shallow (<0.5 m), corresponding to the rooting depth of the tea bushes. In contrast, following a major typhoon in 1982 in Shizuoka Prefecture, Japan, landslide density and extent (% area) in tea gardens was 27.9 landslides km^{-2} and 0.25%, respectively, compared to orange groves (49.6 landslides km^{-2} and 0.22%), native broadleaf forests (20.2 landslides km^{-2} and 0.40%), and recently clearcut forests (47.5 landslides km^{-2} and 0.30%) [*Omura and Nakamura*, 1983]. Conversion of vegetation from forest land to intensive potato culture on steep (30°) slopes in northeast India also generated extensive landslide activity [*Starkel*, 1972b]. Following Hurricane Mitch in late October 1998, a series of erosion plots on hillslopes converted to corn in southern Honduras were largely destroyed by landslides; landslide erosion was estimated at 9961 t ha^{-1} compared to average annual erosion rates of 17 t ha^{-1} [*Haigh et al.*, 2004]. *Marden and Rowan* [1993] reported landslide densities 7.5 times higher in young (<6 yr) exotic pine plantations (0.5 landslides ha^{-1}) than in indigenous forest (0.07 landslides ha^{-1}) on the mountainous East Coast of North Island, New Zealand, following Cyclone Bola in March 1988. In contrast, *Crozier et al.* [1981]

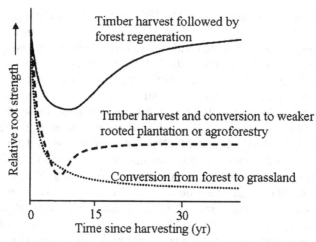

Figure 6.19. Hypothetical changes in rooting strength for scenarios of timber harvesting and subsequent forest regeneration; forest removal and conversion to plantations or agroforestry; and conversion of forest to grassland or pasture.

noted that converted gardens and forests had similar levels of landslide erosion during the large 1980 Cyclone Wally in Fiji, presumably due to the magnitude of this event overwhelming the stabilizing influence of vegetation. In 1979, another storm in Fiji inflicted landslide damage only on hillsides converted to permanent vegetable plots [*Howorth and Penn*, 1979].

The very devastating practice of fuelwood collection by mountain farmers and nomads has destroyed or degraded many forests in the developing world and increased shallow landslide potential in many areas. More than half of all global wood harvested each year is burned as fuel for cooking and heating, mainly in developing countries [*Osei*, 1993]. *Belt and Woo* [1979] attributed the extensive landsliding near Seoul, Korea, partially to past widespread gathering of fuel wood. The Himalaya in Nepal and India are some of the unstable regions most seriously affected by fuelwood gathering [*Amacher et al.*, 1996; *Bhatt and Sachan*, 2004].

While few studies in Southeast Asia have assessed landslide increases in forests that have been converted to cropland or plantations, evidence of such occurrences appear in GIS-based spatial hazard analyses in the tropics [*Gupta and Joshi*, 1990; *Perotto-Baldiviezo et al.*, 2004]. Much speculation exists about the extent and even nature of landslide processes in tropical forests and converted forests of Southeast Asia [e.g., *Diemont et al.*, 1991; *Bruijnzeel*, 2004], largely due to lack of comprehensive data. Reporting of mass erosion in the tropics can also be problematic due to unique interactions between land use on converted hillslopes and erosion mechanisms. Some of the surficial erosion features (e.g., tillage erosion and plow layer erosion) identified on steep cultivated slopes of northern Thailand [*Turkelboom*, 1999] are actually the result of mass erosion processes (i.e., gravity-induced ravel and small-scale flow-type failures, respectively). An example of uncertainties related to landslide–land use interactions in the tropics is apparent from two reports of the November 1988 monsoon-triggered landslides in southern Thailand. *DeGraff* [1990] suggested that forest conversion to weaker rooted rubber plantations was responsible for a higher level of landsliding than would be expected with native forest cover. Conversely for the same storm and general area, *Phien Wej et al.* [1993] noted that landslide density appeared independent of vegetation cover, implying that the storm magnitude overwhelmed the stabilizing influence of the different root strengths. Neither study reported enough data to substantiate these conclusions. Other observations in southern Thailand noted large numbers of landslides and debris flows in converted rubber plantations [*Tang*, 1991; *Harper*, 1993]. *Sidle et al.* [2006] observed three landslide/debris flows in a small headwater catchment in Sumatra that was converted to coffee plantation; average landslide erosion over a 4-yr period was estimated at 27 t ha^{-1} yr^{-1} (Figure 6.20). Slightly higher landslide erosion rates (41 t ha^{-1} yr^{-1}) were found in a larger converted catchment with secondary forest (including bamboo) and paddy fields in northern Thailand [*Sidle et al.*, 2006]. These rates of mass erosion in converted tropical forests are more than an order of magnitude higher than typical rates in clearcut forests of North America.

Terracing steep hillsides for conversion to agricultural production has long been viewed as an effective soil and water conservation practice. However, from the perspective of slope stability, terracing can create problems when water accumulates on terraces and when terrace risers and cutslopes are oversteepened and constructed in unstable soils. Landslides displaced 30% to 50% of the land area occupied by terraced tea gardens in steep terrain (25° to 40°) of the Darjeeling Hills (India) during a major storm in October 1969 [*Starkel*, 1972a]. This finding is consistent with observations in Tanzania [*Haldemann*, 1956] and Uganda [*Knapen et al.*, 2006], where contour cultivation and micro-terracing have contributed to landslides by accumulating and diverting water into potentially unstable portions of the soil mantle. In Southeast and East Asia, terraces constructed on steep slopes have been found to increase landslide

Figure 6.20. Shallow landslides in a recently converted coffee plantation, Sumber Jaya, Sumatra, Indonesia. No landslides were observed in nearby forest sites during the same period.

Figure 6.21. A landslide initiating on a constructed agricultural terrace in the Loess Plateau, Shanxi Province, China.

potential by both concentrating water on the terraces (especially if back-sloped or used for wet paddy fields) and by oversteepening the terrace face [e.g., *Johnson et al.*, 1982; *Billard et al.*, 1993; *Turkelboom*, 1999; *Sidle and Chigira*, 2004; *Sidle et al.*, 2006] (Figure 6.21). *Van Dijk* [2002] attributed 6–50% of the erosion from steep terrace risers in West Java to mass wasting. Out of 290 slumps surveyed in four catchments in the Middle Hills of Nepal, 81% occurred on primarily irrigated rice terraces (khetland); khetland occupied about 30% of the total catchment areas [*Gerrard and Gardner*, 2000]. Only 16% of the debris slides occurred on khetland, while 14% occurred on the less prevalent outward-sloping terraces (17% of the total area) and 20% occurred on abandoned converted land (only 6% of the total area) [*Gerrard and Gardner*, 2000]. A landslide hazard assessment conducted in the Middle Hills of Nepal noted that 78% of the level terraces were very highly susceptible to slumping, although actual measured soil loss from irrigated rice fields was rather low (0.2–0.3 t ha^{-1} yr^{-1}), presumably due to redeposition of slumped materials on lower terraces prior to reaching the catchment outlet [*Shrestha et al.*, 2004].

Control and Avoidance Practices Amidst the Politics

One problem related to erosion control practices in converted lands is that efforts to date have focused on short-term, chronic surface erosion problems and have largely ignored the potential of landslides on steep hillsides [*Sidle et al.*, 2006]. Recent changes in land cover throughout developing regions of Asia, Latin America, and Africa have been driven by incentives from government and international donors that promote high cash value crop production together with poorly coordinated conservation programs, transmigration schemes, rising market prices for certain crops, and the apparent needs of subsistence farmers to generate additional sources of income from the land [*Byron and Arnold*, 1999; *Elmhirst*, 1999; *Cramb et al.*, 2000[1]]. There is a real need to educate not only local farmers, but also international donors and even researchers, that attention must be paid to the interactions between forest conversion to agriculture or plantations and landslide initiation. The unexpected landslides that occurred on surface erosion plots in Honduras during Hurricane Mitch [*Haigh et al.*, 2004] are stark examples of this need to refocus research and management efforts. Certain agroforestry practices may offer socio-economic and biophysical benefits related to sustainability of converted lands, but virtually no information is available on mass erosion process or the cumulative effects of these practices at larger scales [*Steiner*, 1988; *Fischer and Vasseur*, 2000].

The practice of forest conversion to agriculture or plantations in steep and unstable terrain needs to be urgently addressed by national governments and respective land management agencies. Many steep regions of the tropics are experiencing rapid rates of conversion of forests to agricultural lands. Farmers realize short-term gains from high cash value crops, only to have the productivity of the land deteriorate by erosion and landslides to the point where the land is eventually abandoned. Governments need to provide loans or subsidies to encourage farmers and hill tribes to sustainably manage these forests for long-term benefits, thus protecting the land base and downstream resources. International donors may do better to focus on efforts to achieve this difficult task rather than to promote technologies that encourage small-scale surface erosion control measures in unstable terrain. Benefits can be derived by developing both regional and site-specific landslide hazard assessment maps for forest areas that are being considered for conversion. Such maps could serve as a guide to farmers and land managers to avoid unstable sites and concentrate conversion efforts elsewhere.

In areas where forests have been removed by fuelwood gathering, hardwood species could be replanted and harvested on a sustainable basis [*Bhatt and Sachan*, 2004]. Evidence in the mid-hills region of Nepal suggests that, due to widespread forest clearance and high wood prices, families are now starting to use their own lands for fuelwood production [*Amacher et al.*, 1996]; whether this will ameliorate some of the landslide problems by encouraging more sustainable forest management remains to be seen. Lopping of broad-leaf trees for livestock fodder should be discouraged in steep hillslopes [*Bhatt and Sachan*, 2004]. Terracing of steep terrain should proceed cautiously. Large terraces in steep terrain will result in deep cuts into the hillside and high risers—both unstable terrace features (Figure 6.21). Irrigated or ponded terraces should be avoided in steep terrain as they are susceptible to slumping [*Gerrard and Gardner*, 2000; *Knapen et al.*, 2006]. While terrace walls and risers can partially be reinforced with stones [*Johnson et al.*, 1982], such practices are marginal in steep sites. Regrading practices on terraces should not increase the overall height of the terrace bench, thus destabilizing the terrace [*Shrestha et al.*, 2004]. Incorporating trees on potentially unstable terraces and along water courses may lend some stability to these sites [*Johnson et al.*, 1982]. In the Kakani area of Nepal, mountain farmers tend to aban-

---

[1] *Lu et al.*, 2001; *Thapa*, 2001; *Sidle et al.*, 2004a.

don irrigated terraces when impending signs of failure (cracks, small-scale slumping, undercut walls) are noticed; fields are considered stable if no large-scale landslides occur after several years and they subsequently are cultivated again or, if a failure occurs, the site is abandoned or regraded and replaced by an unirrigated terrace [*Johnson et al.*, 1982]. A government-sponsored program to reforest such sites has been met with mixed acceptance. A more desirable approach is to identify such instability problems before they occur and change land use practices accordingly. The practice of excavating temporary ditches to remove excess soil from sites of valley bottom farming or paddy fields, housing sites, and road construction (e.g., ngaguguntur practices in Java) needs to be curtailed [*Diemont et al.*, 1991]. Based on the limited available information, weak rooted plantations and especially agricultural crops should be avoided if at all possible on slopes >20° to minimize landslide potential [*Phien-Wej et al.*, 1993]. Decision models and effective implementation strategies are badly needed in developing countries where conversion of hillslope forests is ongoing. Unfortunately, even the most sophisticated of such models [e.g., *Döös*, 2000] completely ignores the effects and consequences of forest clearing on landslides.

STEEPLAND GRAZING AND GRASSLANDS

Management Effects

Conversion of forest and brush land to pasture in steep terrain has increased landslide erosion in many parts of the world [*Haldemann*, 1956; *Fairbairn*, 1967; *Rice et al.*, 1969; *Temple and Rapp*, 1972[1]] due to the negligible deep rooting strength of grasses compared to that of trees, prompting *Rice's* [1977] characterization of such conversion as "the most hazardous vegetation manipulation that managers can undertake" (Figure 6.22). Animals congregating around streams and water bodies also represent an important cause of mass wasting and sediment transport by disturbing and compacting soils, removing vegetation, and destabilizing streambanks [e.g., *Sidle and Sharma*, 1996].

An example of where conversion of brushland to grassland has accelerated soil mass movement is the San Dimas Experimental Forest in southern California. Following a series of major storms in 1965, both areas and numbers of shallow landslides were 5-fold higher on grass-converted hillslopes compared to brushland [*Corbett and Rice*, 1966]. A more detailed analysis [*Rice et al.*, 1969] of these data revealed a 7-fold increase in landslide erosion during the winter storms of 1965. Based on shallow landslide inventories, landslide erosion was estimated as 188 m^3 ha^{-1} yr^{-1} and 73 m^3 ha^{-1} yr^{-1} for converted grassland and natural brush land, respectively, during a 3-yr period in which several large storms occurred [*Corbett and Rice*, 1966; *Rice et al.*, 1969]. If only slopes >39° are included in the 1965 estimates, converted grassland (820 m^3 ha^{-1}) experienced almost an order of magnitude higher rate of landslide erosion compared to brushland (88 m^3 ha^{-1}) [*Rice et al.*, 1969]. In the San Gabriel Mountains of California, converted

---

[1] *Fransen and Brownlie*, 1995; *Luckman et al.*, 1999; *Gabet and Dunne*, 2002.

Figure 6.22. Extensive grazing by sheep along the south coast of Iceland has altered natural vegetation and contributed to mass wasting.

grassland had similarly high rates (844 m^3 ha^{-1}) of landslide erosion during winter storms of 1969 [*Rice*, 1977].

Widespread conversion of mixed evergreen forests on North Island, New Zealand, during development of mountain pastures from the mid-19th century to 1920 has created ongoing increases in landslide erosion together with reductions in site productivity [*Trustrum et al.*, 1984; *Fransen and Brownlie*, 1995; *Luckman et al.*, 1999]. A total of 9280 km^2 of converted lands on North Island are affected by moderate to extreme mass movement erosion, which accounts for 78.5% of all North Island areas eroded to this extent [*Sidle et al.*, 1985]. Landslides in converted hillslope pastures in the Wairarapa region were estimated to cause an 18–20% decline in land productivity over a 100-yr period [*Trustrum et al.*, 1984; *Luckman et al.*, 1999]. In the southern portion of North Island, New Zealand, *Grange and Gibbs* [1947] measured between 8% and 10% bare ground, mainly attributed to landslide scars in converted pastureland. Landslide density was quite high (1.3% of the area) on two grassland converted catchments (total area 15.7 km^2) in Hawke's Bay hill country, North Island, based on 1943 aerial photos following a large storm in 1938 [*Fransen and Brownlie*, 1995]. From 1971 to 1972, one catchment was afforested with radiata pine; the other catchment continued to be intensively grazed. Following Cyclone Bola in 1988, the overgrazed catchment experienced 130 landslides km^{-2}, while the reforested catchment had 22 landslides km^{-2} [*Fransen and Brownlie*, 1995]; these translate to rates of 33 and 3.1 landslides km^{-2} yr^{-1}, respectively, for the period of aerial photo coverage. *Selby* [1974] estimated that a 100-yr return interval storm on forested slopes of North Island will produce approximately the same level of soil mass movement as a storm with a return period of 30 yr in pasturelands. Benefits of regeneration of tea tree stands on formerly converted pastures were observed on the East Coast, North Island, following Cyclone Bola in 1988 [*Bergin et al.*, 1995]. Although other studies indicate that tea plantations are more susceptible to landslides than native forests, these plantations still had from 65% to 90% less landslide loss compared to converted pasture.

Other evidence of the effects of pastureland conversion and grazing on landslide erosion is documented in various parts of the world. Converted hillslope pastures in Scotland [*Fairbairn*, 1967; *Innes*, 1983], Western Australia [*Kuruppuarachchi and Wyrwoll*, 1992] and Tanzania [*Haldemann*, 1956] have been gradually deteriorating because of soil mass movement. In New South Wales, Australia, increased rates of soil creep have been observed in overgrazed pastures on formerly forested sites that were cleared and burned [*Costin*, 1950]. In the western Uluguru Mountains, Tanzania, *Temple and Rapp* [1972] reported the following distribution of landslides among various land uses after a major storm in 1970: converted grassland, 46.5%; cultivated farm-land, 46.9%; natural forest or brush land, 0.8%; and other, 5.8%. Grasslands were not overgrazed and cultivated soils were not excessively depleted, which indicated that the differences in landslides were mainly caused by the effective rooting strengths. One case study in Ethiopia suggests that the exclusion of animals on pasturelands promotes the re-activation of old slumps by enhancing the infiltration capacity of the soil; however, no processes were investigated at this field site [*Nyssen et al.*, 2002]. Conversely, it has been implied that alterations of subsurface flow pathways and resultant increases in surface runoff in compacted pastures in the Philippines can contribute to the initiation of landslides (due to localized pore water pressure accretion) and debris flows (in wet downslope areas) [*Chandler and Walter*, 1998; *Chandler and Bisogni*, 1999]. Conversion from aspen and brush cover to primarily grassland in high elevation sites of central Utah resulted in an approximately 3-fold increase in shallow landslide movement [*DeGraff*, 1979].

No detailed studies have been conducted that unequivocally assess the role of animal trampling on landslide erosion. While compaction due to grazing reduces the infiltration capacity of soils, and thus potentially reduces the ingress of water to slip surfaces, it has also been noted that slip surfaces may be initiated by animal trampling on hillsides and that preferential flow in compacted soils will be reduced, thus retarding hillslope drainage. Given the other detrimental impacts of overgrazing (increased overland flow and surface erosion), it would be unwise to recommend this as a management measure to reduce the potential of landsliding in steep grasslands.

Avoidance and Control Practices

In steep terrain subject to shallow landslides, grasses provide the least possible protection against slope failure compared to other vegetation cover [e.g., *Rice*, 1977; *Marden and Rowan*, 1993]. Additionally, overgrazing of such hillslopes can further exacerbate landslide and debris flow potential by creating slip planes or terracettes [*Sharpe*, 1938; *Eisbacher*, 1982; *Luckman et al.*, 1999; *Rost*, 1999]. Even though grazing is known to compact soils and thus favor surface runoff at the expense of subsurface flow [e.g., *Jacobsen*, 1987; *Chandler and Walter*, 1998], water recharges into cracks and at breaks in the turf, thus exacerbating small-scale failures at these already susceptible sites. An unusual example of this phenomenon was noted in high

Figure 6.23. Stone collecting by monks (constructing a monastery) in high-elevation grassland, Shanxi Province, China, caused disturbance to the continuous grass cover and promoted small soil slips.

elevation grasslands in China where breaks in the turf, created during hand collection of surface stones by monks for constructing a monastery, initiated small-scale soil slips (Figure 6.23). Given the strong association of grazing intensity with surface erosion [e.g., *LeBaron et al.*, 1979; *Karambiri et al.*, 2003], as well as the isolated evidence for increases in small-scale soil slips, the practice of intense grazing should not be practiced on moderately steep to steep hillsides.

It is clear that conversion of forests and brushlands on hillslopes to pasture should be carefully controlled. Conversion should focus on areas of more gentle slopes and deeper soils where tree and brush roots exert only minor reinforcement of the soil mantle [*Corbett and Rice*, 1966]. Unstable terrain microsites, such as swales and geomorphic hollows, should not be converted to grassland. Other hillsides can be partly stabilized by leaving or replanting some trees in particularly unstable sites, similar to certain agroforestry practices [*Hawley and Dymond*, 1988; *van Noordwijk*, 2000]. Benefits of reforestation of unstable slopes previously converted to pasture are evident in New Zealand [e.g., *Hicks*, 1991; *Bergin et al.*, 1995; *Fransen and Brownlie*, 1995]. Implementing rotational grazing systems may offer some benefits related to slope stability; however, such practices have not been especially effective in New Zealand [*Sidle et al.*, 1985]. Animals can be fenced out of unstable hillslopes to protect natural vegetation. Exclosures around riparian corridors will eliminate cattle trampling and destruction of vegetation along stream banks, thus reducing channel incision, slumping and channel degradation [e.g., *Sidle and Sharma*, 1996]. Detailed economic assessments should be made prior to the conversion of any land to pasture; such assessments should include all indirect costs associated with off-site sedimentation and damage to aquatic ecosystems.

FIRE

Management Effects

Most of the effects of burning on landslide initiation related to establishment of different vegetation cover types have been discussed in the previous two sections; thus, this section focuses only on the unique effects of the fire itself on mass wasting processes, not on changes in the vegetation cover. Fire is widely used as a management tool for clearing land during site conversion, burning residues after cropping or timber harvesting, and aiding in the reestablishment of new crops. Such prescribed fires have widespread consequences for mass erosion, especially in steep terrain; however, large fires attributed to human intervention and ignition are becoming more common as public access into remote areas increases and the wildland–urban interface expands [*Smith*, 1996]. Widespread wildfires caused by lightning strikes will continue to occur regardless of human intervention. Controlled prescribed burning is also used in drier brushland and forest regions to reduce wildfire hazard. In fire-prone regions such as Australia, the Mediterranean, and southern California, native vegetation has evolved with a long history of burning; however, human intervention has complicated the frequency and extent of such fires as well as disrupted the ecological balance of these systems [*Rice et al.*, 1982; *Atkinson*, 1984; *Clark*, 1989; *Smith*, 1996]. On certain sites, fire can impose short-term effects that influence slope stability in a different manner compared to vegetation removal. These fire-induced effects may be additive to other factors accelerating soil mass movement.

Many soils exhibit water-repellent properties after burning, especially after moderate to hot fires [e.g., *DeBano et al.*, 1967; *Morris and Moses*, 1987; *DeBano*, 2000]. The existence and depth of a water repellent layer in the soil depend on the intensity and duration of the fire. Hot fires cause an immediate decrease in the infiltration capacity of surface soils, which translates into increases in surface runoff and surface erosion; these processes can either increase or decrease the probability of landslide/debris flow occurrence. Increases in debris flows can result from accelerated surface erosion that loads unstable slopes or channels, leading to debris flow initiation [*Meyer and Wells*, 1997; *Cannon et al.*, 2001a]. Conversely, it is possible that due to reduced water inputs into the soil mantle after fire (because of hydrophobicity), landslide probability may decrease in the short term in some sites [*Rice*, 1974]. In areas where fire is used in the forest clearing process, additional runoff and erosion (including dry ravel) typically occur during the first few years after burning [*Heede et al.*, 1988; *Scott and Van Wyk*, 1990; *Bruijnzeel*, 2004; *Sidle et al.*, 2004a].

Dry ravel, the downslope movement of individual particles by gravity, is exacerbated by the consumption of organic matter in surface soils during severe fires (Figure 2.13a). Studies in steep sites (>30°) with soils of low bulk density and cohesion indicate that high rates of dry ravel occur after burning [*Krammes*, 1965; *Mersereau and Dyrness*, 1972, *Florsheim et al.*, 1991]. *Rice* [1982] estimated that 25% of the chaparral zone of southern California is eroded by dry ravel at a rate of ≈ 1.4 m^3 ha^{-1} yr^{-1}. Such non-cohesive soils that are susceptible to dry ravel are typically derived from granite, pumice, and certain bedded sediments [*Sidle*, 1980]. Because of the consumption of soil organic matter during burning and the resulting decrease in internal angle of friction, low-cohesion soils that are at or near their natural angle of repose before burning will be subject to dry ravel after burning (Figure 2.13a). Limited research indicates that most of the dry ravel occurs shortly after the fire [*Mersereau and Dyrness*, 1972; *Bennett*, 1982], likely related to the consumption of organic residues in soil that act as binding agents. Thus, long-duration, hot fires generate the greatest potential for dry ravel. As vegetation reestablishes on burned sites, rates of dry ravel decline markedly.

Rice [1982] measured dry ravel erosion of 39 $m^3 ha^{-1}$ during a 3-month period after a wildfire in chaparral. *Krammes* [1965] reported moderate rates of erosion (mostly dry ravel; 2.4 $m^3 ha^{-1}$) from south-facing chaparral slopes in southern California before burning; rates accelerated to 40.6, 6.5, and 10.4 $m^3 ha^{-1}$ during the first three seasons after burning. Ravel erosion on north-facing slopes was an order of magnitude smaller prior to and after burning. In the Oregon Cascades, steep (39°) forest slopes produced an average of 6.0 $m^3 ha^{-1}$ of ravel erosion on south aspects and 1.6 $m^3 ha^{-1}$ on north aspects during a 13-month period after burning [*Mersereau and Dyrness*, 1972]. Dry ravel erosion was inversely proportional to vegetative cover; however, at least 60% of the ravel was estimated to have been missed during the post-fire period because measurements did not begin until 2 months after burning. Measured ravel erosion on these same plots 12 yr later declined to 0.005–0.38 $m^3 ha^{-1} yr^{-1}$, typical of undisturbed forests [*Swanson and Grant*, 1982]. Median ravel rates of 85–221 $m^3 ha^{-1}$ (including some wind erosion) were measured in steep (>31°), unbordered plots in clearcuts in the Oregon Coast Ranges during the first year after slash burning [*Bennett*, 1982]. An average of 65% of this erosion occurred within 1 d of the burn, reflecting the high consumption and/or displacement of the organic horizon and the resulting drying and exposure of the underlying cohesionless mineral soil.

Landslide erosion on steep hillslopes increases several years after burning by the same mechanism as for timber harvesting: gradual decay in rooting strength of vegetation prior to substantial regrowth. As noted, during the first year or two after burning, landslide erosion may actually decline due to hydrophobic conditions in surface soils. However, as rooting systems decay and new vegetation restores infiltration capacity of soils (≈2 to 3 yr), the burned sites become more vulnerable to failure [*Rice*, 1977]. On granitic forest soils in central Idaho, a 20-fold increase in mass erosion occurred after wildfire [*Gray and Megahan*, 1981]. *Rice et al.* [1982] estimated that landslide erosion would increase 3-fold when a prescribed burning interval of 15 yr was used in brushland. With such a burning interval, the age distribution of brush is clustered in the period of minimum rooting strength, thus the site is sensitive to large landslide-triggering storms. Similar increases in landslides were observed in Scotland, where rotational burning of heathland was used to prevent tree regeneration and promote the growth of heather [*Fairbairn*, 1967]. Relatively low landslide erosion rates were reported in two chaparral areas in the San Gabriel Mountains of California burned 1 yr and 50 yr before the 1969 storm season [*Rice*, 1974, 1977]. These areas experienced erosion rates of 10 and 16 $m^3 ha^{-1}$, respectively, compared with approximately 20-fold higher landslide erosion (298 $m^3 ha^{-1}$) on chaparral burned 9 yr before the storm.

Debris flows sites in channels undergo a process of material accumulation. This accumulation may occur over several years or decades and can be greatly enhanced by burning and the resulting surficial transport of sediment and debris into these channels [*Florsheim et al.*, 1991; *Wohl and Pearthree*, 1991; *Meyer et al.*, 1992; *Cannon et al.*, 2001b]. Thus, the accelerated loading of these drainages immediately after fire can increase the probability of a debris flow. The progressive bulking of sediment in the continuum from hillslopes to channels appears to determine the initiation point of debris flows and depends on geomorphic linkages between hillslopes and channels [*Cannon et al.*, 2003]. Additionally, by consuming woody debris in channels, fires may destabilize stored sediment and decrease channel roughness [*Cannon and Reneau*, 2000]. Several studies note that widespread hydrophobic conditions and accelerated dry ravel are not prerequisite for the initiation of debris flows following fire [*Meyer and Wells*, 1997; *Cannon and Reneau*, 2000; *Cannon et al.*, 2001a]. However, a large rain event and subsequent high flow are needed to trigger a debris flow; such failures occur shortly after fires [*Klock and Helvey*, 1976; *Wells*, 1987; *Cannon and Reneau*, 2000] or may occur some time later, after sufficient debris accumulates in channels [*Benda and Dunne*, 1997a; *Meyer et al.*, 2001]. Sporadic increases in sediment supply following fire can lead to widespread channel aggradation, even in large fluvial systems; sediment accumulates at tributary junctions, alluvial fans enlarge, and channel gradients increase downstream and decrease upstream of fans [*Benda and Dunne*, 1997a; *Benda et al.*, 2003b]. Changes in fire regime, such as fire suppression, impact the timing of sediment release from such storage sites [*Miller et al.*, 2003].

Based on the research to date, several processes emerge as the dominant mechanisms of landslide/debris flow initiation following fire. Firstly, dry ravel may occur just after burning and continue for several years, especially if ground vegetation and large portions of the soil organic horizon are consumed. Dry ravel will be very active on slopes steeper than the internal angle of friction of the exposed mineral soils. Secondly, landslide erosion will increase several years after burning due to root strength decay of burned vegetation; the extent of site stability recovery will depend on the woody species that regenerate. Such increases in landslide-related root strength decay will persist for at least 15 yr or permanently if brushland or woodland sites are converted to grassland. Thirdly, landslides may increase following burning due to reduced evapotranspiration and thus promoting higher pore water pressures during storms [*Klock and Helvey*, 1976; *Helvey*, 1980; *Megahan*, 1983]. The second and third processes are not independent and may act together to trigger both landslides and subsequent debris flows. Several other complex geomorphic interactions following fire appear to trigger debris flows. Surface eroded material (both surface wash and dry ravel) after fire may overload channels, creating conditions suitable for the initiation on "in-channel" debris flows or hyperconcentrated flows during one of the first storms following a major fire [*Johnson*, 1984; *Wells*, 1987]. Research following the 1988 wildfires in Yellowstone National Park suggests that the bulking of surface wash materials in lower channel reaches contributed to debris flow initiation after the fire; these in-channel debris flows grew in size as they proceeded downstream by incising channels [*Meyer and Wells*, 1997]. Similarly, fire-affected surface eroded material that accumulates on hillslopes, within zero-order basins, and in headwater channels may fail as a debris flow during a large, but not episodic storm [*Cannon and Reneau*, 2000; *Cannon et al.*, 2001a,b]. Burning of organic debris results in a loss of sediment buffering on hillslopes; similarly, consumption of large woody debris in headwater channels facilitates the release of stored sediment. Both of these processes contribute to more hydraulically efficient delivery of sediment and thus may enhance debris flow activity after fires [*Cannon and Reneau*, 2000]. To understand the complex effects of fire on the timing and extent of debris flows and other landslide processes, it is necessary to evaluate surface wash, dry ravel, channelized debris flows, hillslope debris flows, and landslides separately, as well as their interactions with one another.

Avoidance and Control Practices

Management practices that specifically address the unique problems of fire on slope stability generally focus on controlling the extent, location, severity, and timing of the burn. Prolonged fire suppression increases fuel loads and can create conditions that lead to catastrophic fires, as evidenced by the 1988 Yellowstone fires [*Romme and Despain*, 1989]. Fire must be recognized as a natural ecosystem process [e.g., *Clark et al.*, 1989], and management activities that attempt to suppress fire or control its magnitude by prescribed burning must envision the limitations and consequences of their actions, not only for the fire hazard, but also the subsequent erosion hazard [*Clark*, 1989; *Romme and Despain*, 1989; *Zierholz et al.*, 1995]. It is becoming increasingly difficult to clearly distinguish between natural wildfires and those caused or at least affected by human activities. More frequent fires associated with human intervention result in lower fuel loads; thus the impacts of individual fires on soil and cover are less severe than with a catastrophic wildfire [*Zierholz et al.*, 1995]. Canopy openings created by selective harvesting of rainforests have increased the susceptibility of vegetation to wildfire by facilitating the accumulation and pre-drying of combustible biomass; such pre-drying is further enhanced during droughts [*Stott*, 2000; *Sidle et al.*, 2004a]. Reestablishing fire-adapted vegetation on disturbed sites in areas prone to wildfire can offset some of the adverse consequences of future fires on hillsides [*Brown*, 1998; *Sidle et al.*, 2004a]. Employing longer intervals between burning on brushlands has been demonstrated to reduce landslide occurrence [*Rice et al.*, 1982].

Prescribed fires are used for preparing sites prior to regeneration of new forests, agricultural conversion, and pasture, as well as reducing the probability of catastrophic wildfire near urban and protected areas and in certain brushlands. Hot burns generally occur when fuel loads are high, biomass is pre-dried, and soils are dry; these conditions should be avoided on steep sites during prescribed burning to reduce ravel and surface erosion [e.g., *Florsheim et al.*, 1991; *Sidle et al.*, 2004a]. High intensity fires often consume most of the biomass (including soil organic horizons) leaving the site exposed. Local conditions that exacerbate the impacts of burning (e.g., aspect) must be considered [*Krammes*, 1965; *Mersereau and Dyrness*, 1972]. Burning should be avoided on highly unstable sites. Some of the burning currently practiced on steep sites, including the initial phase of forest conversion and site preparation, could be replaced by manual or mechanical clearing and disposal of slash on the site together with higher utilization of small-diameter wood. Increases in the probability of debris flows caused by burning on slopes need to consider potential impacts on downslope residences. Such off-site impacts are prevalent along the Wasatch Range of Utah, as well as the populated hillslopes of southern California.

URBAN, RESIDENTIAL, AND INDUSTRIAL DEVELOPMENT

Background and Vulnerability

Concentrations of people, structures, and industrial developments in and downslope of unstable sites pose major economic and social risks. The establishment of settlements on and below unstable hillslopes in affluent regions is often due to personal choice, ignorance, and/or complacency related to the potential landslide hazard, or reflects the recognition of the hazard and subsequent investment in structural stabilization measures before construction. The 10 January 2005 landslide at La Conchita, California, which killed 10 people and damaged 36 homes, is a recent example of a disaster where stabilization measures were not in place [*Jibson*, 2005]. As previously mentioned, a number of complex socio-economic factors may promote settlement in unstable sites of developing countries. Vulnerability of residences and commercial structures and the people therein to landslide damage is strongly influenced by the geomorphic (e.g., steep and concave slopes) and atmospheric (e.g., high rates of rainfall) factors that predispose areas to instability, and especially by the potential for moderate to large earthquakes, since many of these structures are built on artificial fill [*Zêzere et al.*, 1999; *Yoshida et al.*, 2001; *Kamai et al.*, 2004] or on naturally unstable deposits subject to liquefaction during ground motion [*Grantz et al.*, 1964; *Highland*, 2003; *Trandafir and Sassa*, 2005].

Changing demographics due to economic, political, and social pressures have concentrated many poor and disadvantaged people in areas and dwellings that are highly vulnerable to landslide hazards [*Templeton and Scherr*, 1999; *Smyth and Royle*, 2000]. Since the 1960s, landslide disasters in the steep hillslopes around Rio de Janeiro, Brazil, have become more frequent due to slum (favela) growth in these areas [*Jones*, 1973; *Smyth and Royle*, 2000; *Fernandes et al.*, 2004]. Such haphazard developments are especially vulnerable because they typically clear forested slopes and lack adequate water supplies, drainage, and waste disposal systems; additionally, the structural quality of buildings is marginal [*Smyth and Royle*, 2000]. Many devastating landslides around Rio de Janeiro occurred during intense summer rainstorms in January 1966, February 1988, and February 1996. From 1986 to 1997, landslides killed 167 people and destroyed 636 houses in the city [*Amaral*, 1997], attesting to the impact of the displacement of poor residents to vulnerable hillsides. This example illustrates the combined effects of social vulnerability as well as land use practices (e.g., cuts into the hillslope, landfills, forest clearance, accumulation of trash, changes in drainage) on landslide disasters

[*Smyth and Royle*, 2000; *Fernandes et al.*, 2004]. In Hong Kong, where on average 11 people are killed each year by landslides, damages are also socially differentiated [*Chen and Lee*, 2004]. The highest percentage of shallow, rainfall-initiated landslides (28%) affects squatter areas in Hong Kong, whereas other buildings and public facilities incur 10% and 5%, respectively, of the landslide impacts [*Chau et al.*, 2004]. In Kingston, Jamaica, where 3.6% of the city consists of pre-historic and newer slope failures, low-income housing is often more seriously affected by landslides compared to other areas [*Ahmad*, 1991]. A recent survey within Kingston identified 2321 landslides, 16.7% of which were considered active and accounted for 4.5% of the total landslide area [*Ahmad and McCalpin*, 1999]. Most of these landslides initiated during large rain storms; however, the M 5.4 and M 6.5 earthquakes of 1993 and 1907, respectively, also triggered widespread landsliding. In Dunedin, New Zealand, pensioner apartments built near an abandoned rock quarry were subject to rock fall hazard; concerns led to the eventual stabilization and annual inspection of the rock walls [*Glassey et al.*, 2003].

Damages incurred by landslides in urban and residential areas, especially in developing countries, are also related to lack of awareness of the impending disaster [*Smyth and Royle*, 2000]. Lack of awareness of landslide risk and lack of preparedness for hazards perpetuate and exacerbate damage in urban, residential, and industrial areas situated in or downslope of unstable terrain. When institutions or governmental organizations fail to plan for hazards or for changing environmental conditions and risks, social vulnerability increases [*Comfort et al.*, 1999; *Adger*, 2000]. *Crozier* [1981] suggests six reasons for lack of community preparedness: (1) insufficient knowledge and understanding of the hazard; (2) inadequate protection measures; (3) acceptance of the risk; (4) lack of community mitigation measures; (5) insufficient resources and options; and (6) false perception of the risk. Some or all of these factors contribute to landslide impacts, depending on the political, economic, and social structure of the community.

Many other examples of how urban development increases landslide damage are noted throughout the world: landslide damages in Anchorage related to the Great Alaska Earthquake of 1964 (M 9.2); the earthquake-triggered (M 7.7) debris avalanche–debris flow in 1970 from Mount Huascaran, Peru, which destroyed the city of Yungay; the 1976 Guatemala earthquake (M 7.5) that impacted Guatemala City; and the 2001 Nisqually, Washington, earthquake (M 6.8) that triggered many landslides and lateral spreads in the greater Seattle metropolitan area have already been discussed. In Japan, earthquakes have triggered extensive landslide damage in many gently sloping residential areas around Sendai (1978 Miyagiken-oki earthquake; M 7.4), Kobe and Nishiomiya (1995 Hyogo-ken Nanbu Earthquake; M 7.2), and Nagaoka (2004 Chuetsu Earthquake; M 6.6) [*Kamai and Shuzui*, 2002; *Okimura and Tanaka*, 1999; *Yoshida et al.*, 2001; *Sidle et al.*, 2005]. Urban areas that have experienced extensive damage due to rainfall-triggered landslides include Wellington, New Zealand; Cincinnati, Ohio; the San Francisco Bay and Los Angeles metropolitan areas, California; Honolulu, Hawaii; Basilicata and Calabria, Italy; Lisbon, Portugal; Vancouver, British Columbia; Nagano and Hiroshima, Japan; Hong Kong; Chongqing, China; Kuala Lumpur, Malaysia; Kingston, Jamaica; Rio de Janeiro, Brazil; and Caracas, Venezuela.

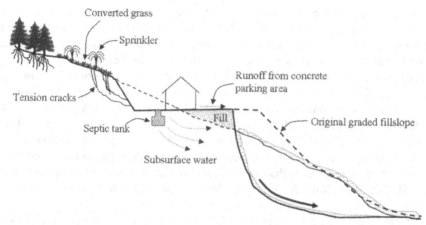

Figure 6.24. A schematic example of how various anthropogenic factors in urban and residential hillslides interact to cause landslides [adapted from *Leighton*, 1966].

Direct Effects

Urban, residential, and industrial development predisposes otherwise stable or marginally stable sites to slope failure by (1) placing poorly compacted fill materials on hillslopes, thus overloading the slope with weak materials; (2) excavating into hillsides and removing support; (3) concentrating water onto potentially unstable sites (e.g., storm drainage); (4) adding additional water via pipe leaks and irrigation; and (5) extensively removing vegetation or converting it to species with weaker root systems. Several of these factors often interact to generate landslides in urban developments (Figure 6.24). The previously described landslides around Rio de Janeiro were affected by all five of these factors [*Smyth and Royle*, 2000; *Fernandes et al.*, 2004]. In Wellington, New Zealand, most of the urban landslides that occurred during the wet winter of 1974 were attributed to artificial cuts and fill materials on recently developed hillsides [*Eyles et al.*, 1978]. The catastrophic 1979 East Abbotsford landslide (18 ha) in a suburb of Dunedin, New Zealand, that destroyed 69 homes and displaced 450 residents, was attributed to a combination of natural topographic and geological conditions as well as a quarry excavation at the toe of the landslide 10 yr earlier, leakage from a water main, and vegetation removal [*New Zealand Commission of Inquiry*, 1980; *Glassey et al.*, 2003]. *Leighton* [1972] estimated that between 25% and 30% of the major landslides in southern California are attributable to construction activities. Many of the shallow landslides in urban Hong Kong are the result of weak fills placed on steep slopes, loss of support from slope excavations, leaks from water conveyance systems, and poor drainage design or malfunction [*Chen and Lee*, 2004]. Hillslope excavation for construction of rural dwellings has contributed to landslide production on the footslopes of the Elgon volcano in Uganda [*Knapen et al.*, 2006]. On more stable hillslopes, industrial and residential developmental activities may accelerate soil creep, which may cause slow but persistent damage to structures and public services [e.g., *Bahar et al.*, 1995].

As with roads, cutslopes in residential developments are sites of instability. Steep cut slopes are superficially attractive to developers since they provide more flat terrain for building sites. Cutslopes were found to be a major cause of shallow debris avalanches over highly weathered, jointed bedrock in Wellington, New Zealand, and Hong Kong, especially when exposed rocks are weathered and clay-rich materials form in joints [*Eyles et al.*, 1978; *Chen and Lee*, 2004]. The removal of lateral support by hillslope cuts is particularly destabilizing where the bedding planes of substrata are parallel to the slope [*Jahns*, 1958; *Leighton*, 1966; *Chen and Lee*, 2004]. The town of Regina Beach, Saskatchewan, is built on a deep earthflow in sensitive marine clay. Various cutting activities together with changes in groundwater due to development have contributed to the overall movement rates that range from <10 mm yr^{-1} to >2 m week^{-1} [*Clifton et al.*, 1986]. The Casal Novo do Brejo landslide in Lisbon, Portugal, is an example of a retrogressive earthflow in a deep clay deposit overlying a very compact clay layer that was triggered during the rainy season of 1989–1990 near the base of a gentle (10°) slope excavated for construction of a building; prior to the cut, the slope was stable [*Zêzere et al.*, 1999]. Excavation into urban hillslopes was at least partly responsible for several other landslides in which lives were lost: Mameyes landslide, Puerto Rico, 1985, at least 129 deaths [*Jibson*, 1992]; Kwun Lung Lau landslide, Hong Kong, 1994, 5 fatalities and 3 injuries [*Wong and Ho*, 1997]; landslides around the developed hillslopes of Rio de Janero, Brazil, during the past two decades, no casualty records [*Smyth and Royle*, 2000]; and Cherry Hills slide, Antipole City, Philippines, 1999, 16 confirmed deaths, 40 people missing, and 100 homes destroyed [*Kamai and Shuzui*, 2002]. This last disaster occurred in a low-income housing development consisting of a series of deep cuts that exposed a weak tuff layer interbedded in claystone. This unstable geologic situation was evident prior to the slope excavation in 1990–1992 for the dense housing project by virtue of an earlier quarry at the site. Additionally, an incipient failure appeared prior to construction at the head of the catastrophic 1999 landslide. A typhoon in 1999 delivered 565 mm of rain in 3 d, triggering the large complex failure: a topple near the head, a deep-seated slide in the main portion, and slumping along the flanks. Field evidence of water pooling in the site of the earlier incipient failure was noted [*Kamai and Shuzui*, 2002]. The manager of the Japanese company in this joint development venture between Philippines and Japan was detained in the Philippines for trial related to this disaster; the Kyoto-based Japanese firm had no prior experience in construction [*Kamai and Shuzui*, 2002].

Residential fillslopes have unique problems related to slope stability, partly because of their design and construction and partly due to the way they respond to external stresses (e.g., earthquakes). Fill material and structures placed on hillslopes load the existing terrain and must be carefully engineered. Compaction of fill material is critically important to increase fill strength and minimize both subsurface water inputs from upslope as well as infiltration of surface runoff [e.g., *Leighton*, 1966; *Chen and Lee*, 2004]. Cases where failures have occurred in deep fillslopes are common; typically these will progressively move in response to rainfall inputs [e.g., *Kwong et al.*, 2004]. Poor compaction and incorporation of organic materials (which will decay and lose strength) into fills exacerbates these stability problems [*Leighton*, 1966; *Wong and Ho*,

1997; *Chen and Lee*, 2004]. Fillslope stability is sensitive to changes or concentrations of subsurface water. In Pasirmuncang, West Java, a meander along the Cisituhiang River was filled to create an area for a residential development. A landslide occurred in this fill in November 1988, destroying 13 homes, probably due to seepage from a spring under the weak fill material [*Santoso*, 1990].

Fillslopes pose rather unique problems related to slope stability during earthquakes. During the Hyogo-ken Nanbu Earthquake in the Kobe region in 1995, about 5100 locations of gently sloping residential land suffered ground damage, with >70% of the damaged slopes located in valley fills [*Yoshida et al.*, 2001]. Older fills incurred more damage than fills constructed after 1961, coinciding with the implementation of residential land development regulations in 1962. Numerous fillslope failures occurred in residential developments in Nagaoka, Japan, during the 2004 Chuetsu earthquake (M 6.6) [*Sidle et al.*, 2005]. Homes and roadways in the Otoyoshi development, built on an old earthflow, were heavily damaged both by reactivation of the earthflow and by failures in fills (Figure 6.25). In another development, Takamachi danchi, about 70 of 522 homes were damaged due to failures in artificial fillslopes [*Sidle et al.*, 2005]. Generally, thicker fills are more subject to damage during earthquakes, but they typically subside rather than fail as slides; major damage usually occurs along

Figure 6.25. Damages in Otoyoshi development, Nagaoka, Japan, during the 2004 Chuetsu earthquake: (a) outline of the old earthflow on which the development was constructed; (b) tension cracks indicating reactivation of the earthflow during the earthquake; and (c) landslide and subsidence damage to homes and roads.

the border of these fills [*Kamai et al.*, 2004]. Thin fills are more dangerous although the frequency of failure is less than in deeper fills during earthquakes; failures typically occur as slides and cause much more damage [*Kamai et al.*, 2004]. Based on neural network analysis of failures during three major earthquakes in Japan (1978 Miyagiken-oki near Sendai; 1993 Kushiro-oki in Hokkaido; and 1995 Hyogo-ken Nanbu near Kobe), the most important predictors of the occurrence of sliding-type failures in valley fills were height of the fill (contributed to 30% of the correct predictions); orientation of the nearest fault (16%); and width of the fill (10.5%) [*Kamai et al.*, 2004].

In residential developments, the quantity of water introduced into the hillslope soil mantle can be greatly increased by irrigation, swimming pools, small artificial ponds, and septic drainage fields (Figure 6.24). Additionally, leakage from water pipes, irrigation ditches, surface drains, and sewer lines can all contribute to instability of fills or hillsides [*Leighton*, 1976; *Thomson and Tiedemann*, 1982; *Smyth and Royle*, 2000[1]]. Leakage from defective or broken water pipes and sewer lines in urban areas has been implicated as a contributing factor in numerous catastrophic landslides—e.g., the 1979 East Abbotsford landslide in Dunedin, New Zealand [*Glassey et al.*, 2003]; the 1985 Mameyes landslide in Puerto Rico [*Jibson*, 1992]; the 1985 Wang Jia Po landslide in Chongqing, China [*Kwong et al.*, 2004]; and the 1994 Kwun Lung Lau landslide in Hong Kong [*Wong and Ho*, 1997]. Water service leakages as well as increased runoff due to a rapid population expansion around 1930 in Basilicata, Italy, are also believed to have contributed to the marked increase in landslide activity in that period [*Gostelow et al.*, 1997]. Changes in hillslope drainage patterns as well as substandard disposal of wastewater have contributed to the numerous landslides around Rio de Janeiro, Brazil [*Smyth and Royle*, 2000; *Fernandes et al.*, 2004]. Breaks in water mains and sewer lines due to earthflow movement can go undetected for some time and lead to accelerated rates of movement, especially in soft clay deposits [*Clifton et al.*, 1986]. Natural rainfall and intercepted drainage, as well as introduced water and leakages, are discharged onto developed hillslopes less uniformly compared to natural rainfall on undisturbed slopes. The myriad of paved surfaces, roof drainage, storm water drainage, and underground drainage systems all concentrate water onto relatively small areas that may be susceptible to landslides because of cutting or filling. Following the incipient failure in the Shum Wan landslide in Hong Kong, blockage of a drainage system contributed to the catastrophic failure that caused extensive damage to industrial facilities and 2 deaths [*Chen and Lee*, 2004]. Irrigation has been implicated as a probable cause of the elevated groundwater levels that triggered a failure in sensitive clay deposits in Edmonton, Canada [*Thomson and Tiedemann*, 1982]. The infusion or concentration of subsurface water, in particular, increases the probability of failure in fill materials and other marginally stable slopes of urban environments.

Residential and industrial development typically involves extensive removal of natural vegetation. The stabilizing influence of deep-rooted brush and trees on hillslopes has been stressed in previous sections of this chapter and in Chapter 3. Leaving forests and

---

[1] *Glassey et al.*, 2003; *Chen and Lee*, 2004.

brushlands in their natural state adjacent to developed structures reduces the probability of shallow, rapid failures in hillslope developments due to the combined effects of rooting strength and transpiration. Shallow-rooted grasses and forbs do little to reduce mass erosion, although they may effectively control surface erosion. The possible destabilizing influence of tall, woody vegetation on very steep slopes (>45°) in Wellington, New Zealand, was attributed partially to vegetation transmitting vibrations to the soil during strong winds, as well as the added weight of vegetation, particularly on oversteepened cut slopes [*Eyles et al.*, 1978]. Although localized failures above cut slopes may be triggered by the weight and/or wind stress transmitted via trees, the general influence of tree cover is to increase the overall stability of developed sites. The recent spread of urban development from the lowlands and valley bottoms to hillsides often provokes extensive woody vegetation removal or conversion, thus increasing the probability and impacts of landslides around many cities [e.g., *Merifield*, 1992; *Olshansky*, 1998; *Fuchu et al.*, 1999; *Smyth and Royle*, 2000; *Fernandes et al.*, 2004].

Urban and industrial areas developed on sensitive clays experience unique slope stability problems. In contrast to non-sensitive soils, sensitive clays exhibit very low strength when remolded and thus the failed materials typically move away from the source area, leaving the new slope unsupported; retrogressive failures are also common in sensitive clay [*Lefebvre*, 1996]. Deposits of sensitive clays typically form gentle topography, thus these areas are superficially attractive for urban and industrial development. Given their inherent instability, any cuts into these materials, concentration of water, or overloading with fills can potentially destabilize these clays. Throughout the sensitive marine clay deposits in southern Canada, slope failures have occurred in urban cut and fill slopes, propagated by groundwater pressures [*Thomson and Tiedemann*, 1982; *Bentley and Smalley*, 1984]. Once failures initiate, retrogressive earthflows can develop and enlarge, causing widespread problems. Urbanization in the Seattle, Washington, area has resulted in numerous slope failures at excavations into overconsolidated clay deposits [*Palladino and Peck*, 1972]; landslide activity at such cutslopes as well as in residential fills increases during earthquakes [*Highland*, 2003].

Avoidance and Mitigation Measures

Many of the avoidance, mitigation, and structural control measures used in conjunction with roads and highways can be applied to urban developments. However, issues related to urban, residential, and industrial planning with respect to landslide hazards are rather unique [*Selkregg et al.*, 1970; *Leighton*, 1976; *Erley and Kockelman*, 1981; *Alexander*, 1989[1]]. Long-range urban planning, coupled with appropriate zoning ordinances, grading codes, and on-site investigations of geologic, geomorphic, soil, and geotechnical properties, can mitigate many potential slope stability problems encountered in hillside residential development [*Leighton*, 1972; *Nilsen et al.*, 1979; *Olshansky*, 1989; *Chau et al.*, 2004]. In cases where high-cost structures or high-risk situations (involving lives and resources) are present, intensive control measures, including structural

---

[1] *Olshansky*, 1998; *Kamai and Shuzui*, 2002; *Glassey et al.*, 2003.

controls, must be considered [*Holtz and Schuster,* 1996; *Bromhead,* 1997; *Chen and Lee,* 2004].

Urban planning and regulation. Detailed investigation of slope stability can often be justified for developing residences on hillsides because of the large capital expenditures incurred in a limited area and the human and environmental consequences realized if a failure does occur. Recognition of factors that predispose sites to slope failure, such as the frequency of episodic rain events, active faults, and locations of ancient landslide deposits, is essential in urban and residential planning [*Nilsen and Turner,* 1975; *McConchie,* 1980; *Zêzere et al.,* 1999; *Kwong et al.,* 2004]. Thus, an extensive data base that includes soils, hydrology, geology, geomorphology, vegetation, seismic, and land-management information is necessary to identify and delineate landslide problems in areas considered for development. These data can be in the form of existing maps at the broad planning scale; however, for detailed site investigations, field observations and sampling are necessary [*Leighton,* 1976; *Nilsen et al.,* 1979]. With the development of detailed digital elevation models from aerial photographs, numerous topographic attributes that affect the stability of mountainous residential and urban landscapes can easily be mapped and assessed using geographic information systems [e.g., *Luzi and Pergalani,* 1999; *Chau et al.,* 2004; *Fernandes et al.,* 2004]. Such digital information is very useful at all planning levels because of the strong influence of topography on slope stability; however, this information needs to be used together with ground investigations, particularly when subtle terrain features such as geomorphic hollows need to be confirmed.

Depending on the governmental structure, land use planning for urban and residential areas can occur at several levels: national, regional, community, and site-specific. As an example of national level planning, the Landslide Prevention Law in Japan designates landslide protection zones in susceptible areas to discourage the following: (1) increases in groundwater or damage to drainage facilities; (2) accumulation of surface water or increases in infiltration; (3) cutting and filling practices; and (4) installation of facilities that may affect landslide movement or initiation (e.g., reservoirs, canals) [*Nakamura,* 1999]. In the USA, a National Landslide Hazards Mitigation Strategy under the leadership of the U.S. Geological Survey outlines a plan for providing landslide hazard maps, guidance for land use planning, and development of public and private policy to reduce damages related to landslides [*Spiker and Gori,* 2000]. This program is currently understaffed, insufficiently funded, and not fully implemented. In New Zealand, planning at the national level related to urban landslides is legislatively administered through the Resource Management Act 1991 and the Earthquake Commission Act 1993, although local communities have the responsibility for collecting information and formulating planning policies [*Glassey et al.,* 2003]. At most national levels of administration, only very general guidelines are typically outlined, and most of the actual planning effort is accomplished at regional and local government levels.

Regional planning for landslide control can be administered at prefectural (State) or multi-county levels. A useful example is the Association of Bay Area Governments (ABAG) in the greater San Francisco–Oakland area of California. This organization provides the policy framework for future urban growth, such as broad guidelines for open space, public safety, and preparedness related to landslide disasters, as well as educational information [*Nilsen et al.*, 1979]. More detailed information at the regional level is now becoming available from the Washington State Department of Natural Resources' Landslide Hazard Zonation Project [e.g., *Wegmann*, 2003]. At the regional level, the basic data (e.g., surface geology, soils, rainfall distribution) needed for landslide hazard analysis is typically assembled and general hazard maps are prepared with information disseminated to communities and the public. University researchers are becoming involved at this level by developing new technologies and predictive capabilities for landslides in urban and residential areas. The U.S. Geological Survey is collaborating with several metropolitan areas to develop tools and predictive measures for both rainfall- and earthquake-induced landslide hazards (e.g., Seattle, Honolulu).

Community-level planning for landslide hazards involves more detailed investigation and mapping sponsored at the local government level, suitable for assessment of stability problems in high-risk areas, and includes regulations for structural design, grading plans, zoning, population density, and specific types of land use [*Leighton*, 1976; *Olshansky*, 1989, 1998]. Community-level maps provide an information base for subdivision and site reviews. Local governments can use these as initial screening tools for development decisions and targeting additional studies or information needs. Hazard designations on these maps serve as "red flags" to alert local governments and developers of high-risk landslide sites. These maps can also form the basis for regulations, in which specific requirements are designated for each hazard zone. An effective way to accomplish these objectives is to attach conservative requirements to potential hazard zones but allow site investigations to relax the regulations. For example, development would be prohibited in zones with high hazard potential unless site-specific studies and design met pre-specified requirements. Such an approach, with a carefully designed hierarchy of hazard zones, is used in several communities in the San Francisco Bay area [*Spangle and Assoc.*, 1988].

Commonly, the Uniform Building Code has been applied throughout western USA as a flexible guideline for grading practices at the community and site-specific levels. However, inputs from geoscience experts and local building officials are required at key points during the grading process [*Erley and Kockelman*, 1981; *Olshansky*, 1998]. Prior to grading, geotechnical investigations, including three-dimensional mapping of landslides and limited subsurface exploratory work, as well as collection of detailed site data (e.g., soils and groundwater information) that can be used as specific guidelines for land use decisions, are required [*Leighton*, 1976]. Requirements of the Universal Building Code (modified for the Los Angeles area in California) include severe restriction of construction on slopes >26°, mandatory setbacks of buildings from the toe of slopes that have been cut and filled, drainage around buildings, curtailment of grading during the rainy season, limitation of fill thickness to <30 m, restrictions on earth moving, and limits on the size and type of equipment that can be used during grading [*Erley and Kockelman*, 1981]. In addition to grading requirements, the Los Angeles code ensures enforcement through a series of seven mandatory inspections throughout the development project: (1) initial inspection prior to grading or brush removal;

(2) toe inspection after clearing, prior to fill placement; (3) excavation inspection before excavation exceeds 3 m; (4) fill inspection before height of fill placement exceeds 3 m; (5) inspection of drainage devices before concrete is poured; (6) inspection of rough grading at completion; and (7) final inspection after completion of all work. Between 1952 and 1962, Los Angeles imposed a semi-adequate grading code prior to the implementation of the Uniform Building Code resulting in a 10-fold reduction in the number of sites that failed. The average damage per failed site more than doubled, however, reflecting the higher capital investments. After the revised grading codes were implemented, which required geotechnical investigations and stricter enforcement, the number of construction sites damaged by landslides between 1963 and 1969 was approximately two orders of magnitude less than during the pre-1952 period, when no codes or restrictions were in place [*Slosson*, 1969; *Erley and Kockelman*, 1981]. Zoning regulations and tax incentives for not developing unstable land can be implemented at the community planning level to minimize landslide risk [*Erley and Kockelman*, 1981; *Schuster and Kockelman*, 1996; *Olshansky*, 1989]. Detailed landslide risk maps can be developed based on statistical analysis of past landslides [e.g., *Luzi and Pergalani*, 1999; *Chau et al.*, 2004]; distributed, physically based models that predict areas of potential instability [e.g., *Montgomery et al.*, 2001; *Fernandes et al.*, 2004]; and empirical landslide hazard models based on site characteristics and soil moisture conditions [e.g., *Newman et al.*, 1978; *Ahmad and McCalpin*, 1999; *Crozier*, 1999].

Site-specific planning for landslide control generally requires a detailed on-site investigation by a geoscientist or geotechnical engineer at the expense of the developer or property owner [*Leighton*, 1976]. This level of investigation is generally appropriate for expensive residential or industrial developments in areas with potential on- and off-site landslide problems. Once appropriate field data are collected, detailed site stability analyses are often warranted [e.g., *Lumb*, 1965; *Fuchu et al.*, 1999]. Such investigations are common in densely populated urban sites where hillslope development is undertaken [*Chau et al.*, 2004] or in sites where strategic or expensive structures are planned [*Clifton et al.*, 1986; *Bahar et al.*, 1995]. Unstable site indicators outlined in the "Timber Harvesting" section of this chapter are useful in delineating problem sites for residential development. At the regional, community, and site levels, it is important to closely integrate geotechnical information with local planning efforts [*Alexander*, 1989].

These four levels at which planning decisions, actions, and regulations are implemented form the basis for effective mitigation of urban landslide disasters. Inherent in such a planning system is good communication amongst all levels of planning. A comprehensive decision-making strategy that should be coordinated through these four planning levels includes (1) problem statement and assessment of overall risk; (2) objectives—i.e., standards for effective landslide disaster mitigation; (3) criteria or benchmarks against which decisions are made (e.g., economic, technical, political, environmental); (4) alternative strategies for achieving objectives; (5) adoption of a detailed plan; (6) implementation of the plan; and (7) performance evaluation, including feedback that can be used to revise objectives and strategies [*UNDRO*, 1991]. An example of how decision-making can be systematically conducted is shown in Figure 6.26. The

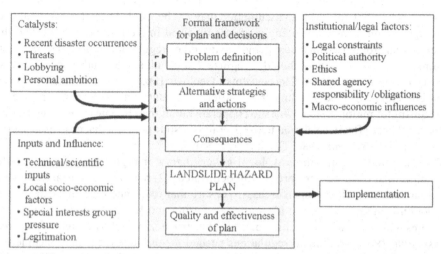

Figure 6.26. Outline of a comprehensive decision-making strategy for urban planning related to landslide hazards.

planning examples and procedures cited in this section are mostly from countries or regions with extensive experience and expertise in dealing with landslide problems (e.g., Canada, Hong Kong, Italy, Japan, New Zealand, USA). In regions with lesser resources and increasing low-income populations—resulting in urban expansion into marginally stable hillsides—implementation of effective planning strategies for the prevention of urban landslides is urgently needed [*Ahmad and McCalpin*, 1999; *Smyth and Royle*, 2000]. For example, from lessons learned during the 1976 Guatemala Earthquake, buildings and infrastructures should be avoided within and near canyon margins in earthquake susceptible terrain [*Harp et al.*, 1981]. The greatest global benefits related to urban landslide hazard reduction can be achieved by focusing international efforts (both donor and research) on feasible solutions in such developing areas. Even more than in industrialized nations, effective solutions in developing regions must focus on non-structural avoidance and mitigation measures [*Erley and Kockelman*, 1981; *Bromhead*, 1997; *Olshansky*, 1998].

A very effective method of avoiding residential and industrial development in unstable areas is the implementation of zoning ordinances and hillside development regulations designed to ensure public safety and land stewardship [*Erley and Kockelman*, 1981; *Olshansky*, 1986, 1989; *Schuster and Kockelman*, 1996]. At the community or even subdivision level, such restrictions include (1) cluster zoning, i.e., concentrating more intense development on stable sites; (2) slope-density regulations; (3) designating a percentage of sites that are to remain in a "natural" state; and (4) development density based on soil/geology characteristics [e.g., *Olshansky*, 1989; *Schuster and Kockelman*, 1996]. All of these land zoning practices attempt to concentrate development activities in more stable landforms. Zoning ordinances must be prepared related to stability problems of specific areas. For example, slope–density zoning assumes that landslide potential is directly related to slope gradient; for urban regions with earthflow and other sensitive clay problems, such an assumption may not be valid. Thus, it is essential that geoscientists and geotechnical engineers are involved in the formation of these ordinances. Also, zoning ordinances include many other regulatory issues, and planning commissions and developers should recognize that not all slope stability issues can be covered by these codes alone. The concept of Geologic Hazard Abatement Districts has been implemented in certain unstable landforms in southern California to provide a mechanism for multiple landowners to cooperate rather than litigate over issues related to the most cost-effective solution to the overall landslide problem that crosses property boundaries [*Kockelman*, 1986; *Olshansky*, 1986].

Prior to subdivision approval, local governments or regulatory bodies should (1) require that unstable sites are clearly delineated; (2) require that unstable sites be designated as low-density open areas; (3) ensure that roads and other infrastructures are not built in unstable areas; (4) minimize cutting and filling operations on potentially unstable sites; and (5) prohibit building construction in high-risk areas [*Erley and Kockelman*, 1981]. Additional regulations needed for unstable areas may include the following: (1) minimize disturbances to vegetation and natural drainages; (2) prohibit filling, irrigation, waste disposal, and cutting of toe areas; and (3) prohibit storage of toxic substances [*Erley and Kockelman*, 1981]. When formulating zoning regulations in unstable areas, communities should develop landslide risk assessments that address the issue of socially acceptable risk [*Fell*, 1994].

State and local governments can offer tax incentives for not developing unstable land, and special tax levies can be imposed on developments existing in landslide-prone areas [*Schuster and Kockelman*, 1996]. Land uses compatible with landslide-hazard areas within residential developments include low-maintenance greenbelts, woodlands, parks, and wildlife refuges. While certain land uses that do not require structures (e.g., golf courses, farming, grazing, parking areas, and storage yards) have been proposed for unstable terrain in developments [*Schuster and Kockelman*, 1996], these may not be desirable since most involve irrigation, concentration of runoff, or alteration in natural vegetation. Land uses that retain natural woody vegetation on such sites are more desirable and will minimize off-site consequences. Septic systems and sand mound filtration systems should be avoided near and above landslide-prone sites. The high cost of landslide insurance can discourage land development in unstable areas [*Erley and Kockelman*, 1981], although *Olshansky* [1996] suggests that insurance coupled with the concept of assessment districts, whereby small communities in potentially unstable areas essentially become self-insured, has merits. In theory, such insurance would provide a more equitable distribution of costs and benefits to property owners in landslide-prone regions and discourage structures and development in particularly unstable sites. Commercial landslide insurance, while available in California prior to 1960, did not cover damage for repair of subsurface foundation areas and allowed insurance companies to cancel policies following a landslide claim. The case of Pfeiffer versus General Insurance Company, related to a landslide that initiated upslope of his property in Orinda, California, was decided in favor of the plaintiff, who was awarded the maximum amount covered in the insurance policy; this award applied not only to the dwelling but also to foundation improvement, and it was determined that the insurance policy could not be cancelled, establishing a precedent for landslide insurers [*Kiersch*, 2001]. However, currently no domestic-based landslide insurer exists in the USA; only high-cost international insurance is available.

Control and mitigation measures. It is difficult to generalize about stable angles for cut slopes required in residential and industrial development because these depend on the properties of the exposed regolith and the types and magnitudes of the inherent landslide triggering processes. In non-sensitive soils, *Leighton* [1966] recommends that cut-slopes be excavated to gradients ranging from 1:1 (45°) to 3:1 (18°). Cuts should be avoided if at all possible in sensitive clays [*Thomson and Tiedemann*, 1982; *Bentley and Smalley*, 1984] and in regoliths where weak layers are interbedded in otherwise stronger materials where a slip surface may be exposed [e.g., *Chen and Lee*, 2004]. Another approach to establishing a design cutslope angle in urban hillsides is to document the gradient at which most hillslopes in the same material fail and then apply a safety factor. For example, in Hong Kong it has been shown that most shallow landslides occur on hillslopes steeper than 30° [*Fuchu et al.*, 1999; *Chen and Lee*, 2004]. Thus, for shallow cuts, a stable angle would appear to be in the range of 25° to 28° ($\approx$1:2). Local experience and landslide inventories can aid in establishing such design criteria. As in the case with roads, deeper cuts into steep sites (i.e., higher cut face) create greater instability. Horizontal drains placed into deep cuts improve stability by draining excess pore water [*Krohn*, 1992]; however, these need to be checked and maintained as they can clog over time and become ineffective [*Chen and Lee*, 2004]. In some cases, steep cuts into bedrock on hillslopes are warranted, but careful inspection of bedrock weathering, jointing, folding, and fracturing at nearby exposed sites should be done prior to excavation [*Wyllie and Norrish*, 1996]. Additionally, the design of cutslopes in regions prone to earthquakes requires special considerations [*Kamai and Shuzui*, 2002].

Design of stable fillslopes is often more difficult than cutslopes because this involves the reshaping of terrain along with changes in the inherent properties of soil materials. Fillslopes normally should be constructed at gradients gentler than cutslopes in similar regoliths. For non-sensitive soils, *Leighton* [1966] recommends gradients ranging from 1.5:1 (34°) to 2:1 (27°) or even flatter for well-designed and compacted fillslopes. Compressible soils and sensitive clays should not be used as fillslope materials. Artificial fills need to be uniformly compacted to increase soil strength and prevent buildup of excess pore water pressure, which may lead to slope failure or subsidence. Prior to placement and compaction, all organic debris, trash, and rocks larger than about 30 cm should be removed from the fill material. During the entire grading process (both cutting and filling), careful inspection should occur to ensure that the contractor is in compliance with existing regulations [*Erley and Kockelman*, 1981]. Surface drains route runoff away from areas of potential instability. While subsurface drains in fills provide an inexpensive means of improving stability by reducing excess pore pressure, drains need to be strategically placed and maintained (to prevent clogging) to ensure long-term effectiveness and thus reduce the risk of rainfall and earthquake-induced fill failures.

In earthquake-prone residential regions, gently sloping, thin valley fills are more susceptible to sliding than deeper fills because of the lower friction along the sides of thin fills; deep valley fills have higher frictional resistance along their sides and thus are not likely to slide as far during earthquakes, but they are susceptible to subsidence [*Kamai et al.*, 2004]. In general, placement of artificial fill on even gentle hillsides in earthquake-prone areas can create many problems related to landslides and subsidence [*Kamai and Shuzui*, 2002; *Yoshida et al.*, 2001; *Sidle et al.*, 2005]. The natural frequency of earthquakes appears important, with high-frequency earthquakes causing the greatest threat of failure in shallow fills, while large, long-travel distance earthquakes are more likely to cause failure in deeper fills [*Kamai and Shuzui*, 2002]. All of these factors should be considered in urban and residential planning in seismically active areas when any fills are proposed.

Establishment of deep-rooted woody vegetation on cleared areas of developed sites reinforces shallow soils or fills, removes some subsurface water via transpiration, and reduces surface erosion [*Erley and Kockelman*, 1981; *Sidle et al.*, 1985; *Merifield*, 1992]. In areas prone to brush fires, selection of fire-resistant brush species is desirable for revegetation. Vegetation that is not well adapted to site conditions and requires irrigation should be avoided on potentially unstable sites. If windthrow of trees is a problem or if cutslopes are very steep (>45°), deep-rooted brush species can be planted. Brush species that exhibit high root shear strengths include Pacific red elder, snowbrush, and huckleberry [*Ziemer and Swanston*, 1977; *Ziemer*, 1981].

Most intensive structural control measures that are effective for slope stability problems in urban areas are very expensive because of the size of the cuts and fills and the level of investment being protected. Some of these measures for roads have been outlined earlier and are not repeated here. Numerous papers and reports have outlined effective structural methods that can be applied to slope stability problems at urban sites [e.g., *Leighton*, 1966; *Rogers*, 1992; *Holtz and Schuster*, 1996]. These intensive control and mitigation measures can be categorized as follows: (1) excavation and recompaction; (2) retaining structures; (3) subsurface drainage; (4) soil reinforcement via geosynthetic materials; (5) mechanically stabilized embankments; and (6) combinations of (5) and (2) [*Rogers*, 1992]. The first method is rather expensive and has liabilities related to handling and drying of soil materials. Methods (2), (5), and (6) are very expensive. Method (4) is moderately expensive and the effective life of geosynthetic materials is unclear [*Rogers*, 1992]. Subsurface drainage, Method (3), is rarely used alone but presents a less expensive solution. Generally, these costly structural measures should be applied only after other avoidance and prevention alternatives have been exhausted.

MINING

Surface mining operations can affect landslide occurrence via deep cuts into regoliths and rocks (e.g., sand and gravel quarries, clay pits), subsequent placement of unstable overburden or waste materials on steep slopes, open pit mines, and disposal of mine wastes (spoil heaps) [*Bell et al.*, 1989; *Bentley and Siddle*, 1996; *Stead and Eberhardt*, 1997]. Underground mining can also contribute to landslides following subsidence [*Thomson and Tiedemann*, 1982; *Shea-Albin*, 1992; *Slaughter et al.*, 1995[1]], although these failures have been poorly investigated. Many similarities exist between stability problems in mining cuts with those related to roads and urban development; however, mining cuts are often deeper and may expose numerous potential weak layers where failures can originate [*Post and Borer*, 2002; *Glassey et al.*, 2003; *Ural and Yuksel*, 2004]. On-site risks may be different for mining-related landslides if operational personnel are aware of the potential failure, as opposed to a less aware general public (in the case of landslides along roads or in urban areas). However, numerous cases of off-site consequences of mining-related landslides (both active and abandoned mines) have been reported [e.g., *Bentley and Siddle*, 1996; *Ghose*, 2003; *Glassey et al.*, 2003]. Other types of mining-related slope failures include ground collapse from underground mines and failures of mine tailings impoundments; these specialized types of slope failures are treated in numerous engineering papers and texts [e.g., *Jeyapalan et al.*, 1981; *Vick*, 1996], and thus are not covered here.

Historically, many landslide and rockfall disasters have occurred related to mining operations due to the lack of regulations and mine safety guidelines. In some cases, these disasters have impacted off-site residences and facilities. Major rockfall disasters have occurred in conjunction with talc pits and slate quarries in the Alps of northern Italy and Switzerland [*Heim*, 1932]. Excavation into terraced clay beds in the Hudson River valley, New York, caused a catastrophic landslide that inflicted extensive damage on adjacent brickyards and killed 20 people [*Newland*, 1916]. *Bentley and Siddle* [1996] noted that prior to the catastrophic October 1966 debris flow in Aberfan, South Wales, which killed 144 people (including 116 children in a junior high school), the effect of mining on frequent landslides in Welsh coalfields was largely dismissed by engineers. This disaster, which originated from a spoil pile above the town of Aberfan, prompted a major investigation, extensive revisions in legislation related to spoil piles, and, surprisingly, the adaptation of well-known geotechnical methods and analyses to coalfields of South Wales for the first time [*Bentley and Siddle*, 1996]. As mentioned earlier, the 1979 East Abbotsford landslide in Dunedin, New Zealand, that destroyed 69 homes was partly attributed to a quarry excavated at the toe 10 yr earlier [*Glassey et al.*, 2003]. Subsidence over an underground coal mine in Slovakia caused a landslide that destroyed part of the village of Podhradie in 1978 [*Klukanová and Rapant*, 1999]. One of the more recent on-site landslide disasters in mines occurred in Filedelfia, Colombia, in November 2001; 51 miners were killed when a slope failure occurred as the result of hydraulic mining (undercutting) [*Hinton et al.*, 2003]. Because of operating improvements, regulations, and mine safety codes, controls have been implemented in many countries that have drastically reduced the risk of mining-related landslides to both operational personnel and surrounding areas [e.g., *Zipper et al.*, 1989; *Bentley and Siddle*, 1996]. Nevertheless, numerous developing nations in Asia, Latin America, Africa, and Eastern Europe, where advanced safety regulations are not in place or

---

[1] *Klukanová and Rapant*, 1999; *Donnelly et al.*, 2001.

are poorly enforced, continue to experience landslide disasters related to mining operations [e.g., *Klukanová and Rapant*, 1999; *Ghose*, 2003; *Hinton et al.*, 2003]. With the increasing worldwide demand for extractable natural resources and fossil fuels, large tracts of land are being disturbed by surface mining. Most surface mining operations have two impacts in common: extensive disturbance of land and the generation of large volumes of disturbed spoil material that must be regraded or otherwise disposed of on the landscape. Overburden material that is removed, as well as processed waste material, undergoes an increase in volume during handling and poses a grading and disposal problem, especially when the overburden-extractable resource ratio is high. In flat to gentle terrain, overburden material is often piled on-site, where slopes form at an angle of repose; these are subject to shallow slides during the rainy season and dry ravel during the drier months, as well as surface erosion [*Russell*, 1979; *Sidle et al.*, 1993]. In steep terrain, overburden waste is either pushed downslope, where it may later fail, or is regraded on the site following resource extraction [*Bell et al.*, 1989]. Water ponded behind spoil bank slopes can initiate failures by subsurface seepage [*Savage*, 1950; *Hoffman et al.*, 1964]. Stability of such slopes can be improved by compacting the spoil material and draining ponded water from behind spoil banks. In many regions, the uncontrolled dumping of mine waste material on hillsides has led to widespread problems of slope stability and sedimentation. In the Himalayan mining region of India alone, more than 60% of the area is covered by mining waste [*Ghose*, 2003]. Many of the earlier landslides in the coal fields of South Wales are attributed to placement of loose spoil material on hillsides [*Bentley and Siddle*, 1996].

Cross-valley fills used in coal and phosphate mining operations in steep, dissected terrain pose potential problems related to long-term stability. Because these large fills typically cover ephemeral or perennial streams, maintaining the drainage through the fill is critical to dissipating the buildup of pore water pressure that could lead to catastrophic failure [*Jeppson and Farmer*, 1980; *Farmer and Peterson*, 1985]. In a well-constructed waste rock fill in southeastern Idaho, subsidence and rock creep decreased significantly over a 7-yr period after construction; periodic saturation occurred in portions of the cross valley fill, but overall, it was well-drained [*Farmer*, 1980; *Sidle et al.*, 1994]. While such large, cross-valley fills require free internal drainage to maintain stability, smaller fills on steep slopes should be compacted to increase shear strength and preclude water ingress [*Bell et al.*, 1989].

The practice of "approximate original contour" placement of overburden waste has been enforced in the USA since the enactment of the Surface Mining and Control Reclamation Act (SMCRA) in 1977. Although this reclamation practice may be aesthetically desirable, the concept has shortcomings when applied in steep terrain, where minespoil fillslopes are significantly less stable than natural slopes of similar steepness. In addition, the natural slope–regraded spoil interface may provide an area for positive pore water pressures to develop and may increase landslide probability. *Bell et al.* [1989] noted that many of these backfills cause long-term slope stability problems in the coal regions of central Appalachian because of (1) oversteepened and/or convex fill configurations (>30°); (2) excess seepage in fills on moderately

Figure 6.27. Landslides in a multi-benched, open-pit gold mine, Borealis mine, Hawthorne, Nevada.

steep (<25°) slopes; (3) placement of the fill toe beyond the edge of the mining bench; and (4) inaccurate estimations of spoil shear strength and fill safety design factors. Thus, rather than adopting inflexible regulations for slope regrading (e.g., maximum angles of fills), it is better to locate, design, and construct fills in a manner sensitive to specific site conditions.

Landslides and rockfalls along cut faces of mines (e.g., highwalls, footwalls) are controlled by slope steepness, joint structure, faults, orientation of bedding, location of weak regolith zones, and ingress of groundwater [*Stead and Eberhardt*, 1997; *Sonmez et al.*, 1998; *Bye and Bell*, 2001; *Post and Borer*, 2002]. External factors that affect cutslope stability include blasting [*Bauer and Calder*, 1971], earthquakes [*Moreiras*, 2004], geometry of the pit or cut face [*Sjöberg*, 1996], and the history of seismic and tectonic activity. Benching techniques are typically employed for deep cuts into hillsides or for deep, open-pit mines to limit the extent of rockfall or landslides; however, circular failures can occur throughout multiple benches in discontinuous rock/regolith masses [*Bye and Bell*, 2001]. Such multiple-bench failures typically occur on slopes of 40–55° [*Sjöberg*, 1996] (Figure 6.27). At smaller scales, simple planar and wedge failures are common on cutslopes, but as scale increases, more complex rotational and toppling failures dominate [*Sjöberg*, 1996; *Bye and Bell*, 2001]. Slab-type failures have occurred on slopes as gentle as 30°, while other failures more typically occur on slopes >33° [*Sjöberg*, 1996]. Because stability of cutslopes is determined by complex interactions of slope gradient, jointing, regolith properties, and groundwater, it is difficult to rely on cutslope gradient criteria alone; as with fillslopes, a detailed assessment of site properties is required prior to designing cutslopes [e.g., *Sonmez et al.*, 1998; *Bye and Bell*, 2001]. Loading tops of cutslopes with spoil can induce landslides [*Sonmez et al.*, 1998]. If blasting is used to remove in situ overburden, the resulting ground motion weakens regoliths and exerts a dynamic loading that can trigger landslides, especially in weathered regoliths [*Bauer and Calder*, 1971; *Bye and Bell*, 2001]. Mining of peat deposits generally involves shallow excavations, but these can trigger landslides on rather gentle slopes due to removal of support and changes in subsurface drainage patterns [*Warburton et al.*, 2004].

The effect of underground mining on landslides is difficult to determine due to other contributing factors and thus has not been studied in detail. Subsidence associated with underground mining may initiate rock creep and reactivate geologic faults [*Klukanová and Rapant*, 1999; *Donnelly et al.*, 2001]. Where geological faults intersect valley sides or where they have developed subparallel to slopes, regolith strength may be reduced and preferential pathways may develop that allow groundwater to flow into soils [*Donnelly et al.*, 2001]. Subsidence causes cracking in regoliths, which can facilitate the movement of water into or out of potentially unstable slope segments [*Mather et al.*, 1969; *Sells et al.*, 1992; *Sidle et al.*, 2000b]. Rock creep often increases after the onset of underground mining, which, in turn, opens subsurface pathways for infiltrating water, leading to deep-seated slope failures; this process is believed to have contributed to the catastrophic landslide in Slovakia in 1960 that destroyed 180 homes [*Klukanová and Rapant*, 1999]. A survey along a headwater channel in central Utah that experienced 0.3–1.5 m of subsidence after underground mining of coal revealed no increase in new landslides in the first 1–2 yr after subsidence [*Sidle et al.*, 2000b].

Few inventories have been conducted that assess the incidence of mining-related landslides over large areas. A detailed landslide mapping program for the areas surrounding seven urban centers of West Virginia (3367 km² total area) revealed that approximately 8% of the landslides mapped were associated with mine spoil banks [*Lessing and Erwin*, 1977]. Failures associated with surface mining of coal in West Virginia were mainly caused by overloading slopes with poorly consolidated spoil material, oversteepening slopes, and inhibiting drainage. A survey of mine spoil slides in eastern Kentucky in 1967 noted that landslides affected 6% of the mined outcrop area [*Williams*, 1973]. Nearly half of these spoil slides occurred in valleys, near or in natural drainages. After the Aberfan, South Wales, landslide disaster, a comprehensive, but general landslip survey [*Conway et al.*, 1980] was conducted for the entire region. A qualitative inventory of abandoned tailings from silver, lead, and zinc mines in the Coeur d'Alene mining district of northern Idaho indicated that there was a low potential for landslides in more than 80% of the sites surveyed (including both reclaimed and unreclaimed spoils) [*Gross et al.*, 1979]. It is clear that better surveys of landslides on active and, especially, abandoned mine lands are needed to define the magnitude and risk of the hazard.

Most surface mine reclamation efforts have focused on regrading and revegetation. While such practices offer aesthetic benefits and likely reduce surface erosion, they not very effective in addressing slope stability [*Bell et al.*, 1989]. For surface mines where slope stability is a concern, reclamation objectives must consider: (1) whether deep-rooted woody vegetation will help stabilize reclaimed slopes; (2) if surface or subsurface drainage is necessary to remove excess water from critical portions of the site; and (3) whether structural control measures are needed to ensure site stability

Figure 6.28. Examples of effects of recreational activities on slope stability: (a) shallow land-slides and active dry ravel along a mountain road into a cross-country ski area, South Island, New Zealand; and (b) initiation of a shallow landslide from a hiking trail (circled), Nagano Prefecture, Japan.

[*Sidle and Brown*, 1992]. For such unstable sites, more focus needs to be placed on physical and geotechnical properties of the disturbed materials and sites as well as hydrological processes.

TOURISM AND RECREATION

Tourism has increased in many mountainous regions of the world during the past few decades, placing new anthropogenic pressures on fragile landscapes due to access roads, hiking trails, lodging facilities, camping areas, and development of related infrastructure and support facilities. In more developed nations, pressure from expansion of ski resorts in steep terrain (Figure 6.28a) has contributed to landslides [*Watson*, 1985]. After more than 40 yr of ski and hiking development in the Upper Guil River Valley of the French Alps, a 30-yr return period flood event caused extensive landsliding along the channelized river, not to mention the effects of development on peak flows [*Arnaud-Fassetta et al.*, 2005]. Similarly, channel changes in the Beas River, Himachal Pradesh, India, resulting from recreational development in the floodplain, caused peak flows to undercut adjacent hillslopes, thereby triggering numerous landslides [*Sah and Mazari*, 1998]. Reductions in channel capacity caused by such recreational development and related urban expansion could exacerbate damages from debris flows. Additionally, erosion from hillslope recreational areas can load headwaters or hollows, thus triggering landslides and debris flows. Within the past few decades there has been a rapid increase in the number of recreational and permanent dwellings in steep forest terrain. Such development along the Wasatch Front in Utah has exacerbated landslide and debris flow hazards [*Sidle and Dhakal*, 2002]. The incidence of forest fires has increased in recent years, partly as the result of recreational activities and even arson; these widespread events have caused increases in landslide and debris flow activity, as noted earlier [e.g., *Cannon and Reneau*, 2000]. Widespread development of tourism infrastructures and sites (e.g., golf courses) in formerly forested mountains of Asia likely contributes to increased landslide erosion [e.g., *Lu et al.*, 2001; *Sidle et al.*, 2006].

Recreational pressure has caused trails formerly used only by people and animals to be widened for vehicles [*Sidle and Dhakal*, 2002]. While no studies have documented the effects of trails on mass erosion, based on observations in many parts of the world, it is obvious that hiking, animal riding, and off-road vehicles generate shallow landslides by displacing soil on steep slopes and by channeling surface runoff onto unstable areas (Figure 6.28b). Additionally, especially in dry or poorly vegetated areas, extensive dry ravel can occur as the result of these disturbances. While the direct effects of these various recreational activities on landslide erosion have not been systematically assessed, in places where vegetation conversion, terracing, hydrologic changes, and other site alterations occur, mass erosion would be expected to increase at rates similar to those reported for other land disturbances.

New technology has brought to bear a series of unforeseen impacts on unstable hillslopes in many regions of the world, not the least of which are developing nations with fragile mountain environments. Most of these impacts are in the privy of affluent residents and tourists or even ecotourists, but are nevertheless creating environmental damage as more and more people have access to steep and remote terrain [*Sidle and Dhakal*, 2002]. While certain land management agencies claim to be addressing these recreational impacts, the current philosophy of "customer satisfaction" and "responsiveness to user needs" practiced by the US Forest Service and many other land management agencies in developed nations is avoiding this important environmental issue. Therefore, to expect better guidelines for protection against recreational impacts in developing countries, where income from tourism is an essential part of local economies, is wishful thinking at best.

GLOBAL CLIMATE CHANGE

Changes in the world's climate due to increases in atmospheric carbon dioxide and other gases can potentially affect landslide initiation in numerous ways. Greenhouse gases in the atmosphere are vital controls on the temperature of the Earth. There is strong evidence that the levels of several greenhouse gases have increased during the past 150 yr [*Ciesla*, 1995], and climate models predict a global warming of about 2°C in the 110-yr period from 1990 to 2100 [*Acosta et al.*, 1999]. Regional estimates of changes in precipitation can be made from general circulation models (GCMs), but storm intensity information is not available or is unreliable. The most common global climate change scenarios that may affect landsliding are increasing mean air temperature and changes in regional annual and seasonal precipitation [*Zimmermann and Haeberli*, 1992; *Sidorova*, 1998; *Buma*, 2000]. Although certain types of landslides may be influenced by these regional climate changes, the initiation and persistence of slope failures are more related to the timing and short-term perturbations of rainfall and snowmelt. Thus, GCMs provide only background information and general climatic trend data that may be useful in framing scenarios of landslide response to long-term climatic change.

The incorporation of climatic change scenarios into landslide hazard models is a complex task because of the lack of spatial resolution of climate change predictions.

Prediction complexity further increases when we try to relate the influence of projected temperature rises to environmental factors such as species composition changes, rooting depth, rooting strength, evapotranspiration, soil depth, soil cohesion, and vegetation regrowth. In climate-controlled studies of root biomass response for different crops, root biomass increased with elevated levels of atmospheric CO_2 [*Rogers et al.*, 1994]; thus, higher CO_2 may promote greater root cohesion due to denser and possibly stronger root systems. However, longer term changes in vegetation community structure may be a more significant factor related to root strength and landslide susceptibility. Temperature change also affects weathering rates and the distribution of plant communities. Changes in vegetation in turn affect weathering rates, soil characteristics, and freeze–thaw cycles. Nevertheless, precipitation patterns are the most critical factors affecting landslides, and predictions of rain intensity and duration and of snowmelt timing and magnitude are weak at best [*Loaiciga et al.*, 1996; *Buma and Dehn*, 1998]. Thus, assessment of effects of global climate change on landslide processes is largely relegated to scenario evaluations.

The response of shallow, rapid landslides to climate change is the most difficult mass wasting process to assess because these failures depend on uncertain projections of rainfall intensity or rates of snowmelt. The occurrence and rates of deep-seated mass movements may increase in some regions characterized by winter rainfall and rain-on-snow events in response to predicted increases in winter precipitation in northern latitudes (above 45°N) [*Sidle and Dhakal*, 2002]. However, with predicted shorter winters, the period of deep-seated mass movement may decrease. Increased temperatures may accelerate weathering processes and increase the susceptibility and rate of slow, deep-seated mass movement. The amount and timing of water released from snowpacks will affect deep-seated mass movements. In mid-latitude regions where the snow levels are expected to decrease, slow, deep-seated mass movements will occur earlier but for a shorter period of time and probably at a reduced seasonal movement rate [*Sidle and Dhakal*, 2002]. In northern latitudes where snowpacks increase, mass movements will probably occur earlier (due to climate warming), have increased seasonal rates of movement, and experience a longer period of activity. Glacial thinning and retreat in western Canada associated with climate warming during the last 100–150 yr triggered landslides due to debuttressing effects; additionally, outburst floods and debris flows initiated from moraine- and ice-dammed lakes [*Evans and Clauge*, 1994]. *Bovis and Jones* [1992] used recent climate change records, dendrochronological data, and stratigraphic records to show that movement of large earthflows in British Columbia responded to Holocene climatic changes. Dating of lichens on debris flow deposits in Scotland indicated that climate change during the past half millennium did not increase debris flow activity; instead, these increases were related to burning and overgrazing during the 19th and 20th centuries [*Innes*, 1983]. Similarly, *Innes* [1985] observed no indication of climate change effects on debris flows in southwest Norway, nor were anthropogenic increases detected in the past 500 yr.

Any climate change scenario in steep terrain that exacerbates wildfire would probably increase the rates of dry ravel. Such effects would be most pronounced in arid to semi-arid regions [*Heede et al.*, 1988]. Since higher temperatures related to climatic change are likely to affect high-elevation permafrost distribution, thawing of permafrost may also induce ravel and dry creep at higher elevations [*Sidle and Dhakal*, 2002]. Climate changes that promote more frequent freezing–thawing or wetting–drying cycles would induce greater dry ravel and dry creep on steep, disturbed or partially vegetated hillslopes. Conversely, fewer of these cycles would tend to reduce dry ravel and dry creep.

Overall, it is difficult to predict the effects of potential long-term climate changes on landslide activity. Long-term changes in average climate conditions (temperature and precipitation) as well as possible shifts in the frequency of extreme events are expected as a result of climate change. Depending on the degree to which such changes are realized, they may or may not play important roles in influencing landslide scenarios in various regions of the world [*Evans and Clauge*, 1994; *Wyss and Yim*, 1996; *Sidorova*, 1998; *Buma*, 2000]. The effect of climate change on other environmental factors, such as vegetation and soil, may introduce more complex interactions and scenarios related to landslide occurrence. Warming in glaciated terrain will likely increase landslide activity [*Evans and Clauge*, 1994], but such effects typically occur in relatively unpopulated regions and are unlikely to cause disasters, except for cases of moraine- or ice-dammed lake collapse [e.g., *Hewitt*, 1998]. Given the difficulties in associating past climate change scenarios with landslide activity [e.g., *Innes*, 1997], as well as the widely recognized effects of land use and land cover change on these geomorphic processes [e.g., *Singh*, 1998; *Slaymaker*, 2000; *Sidle et al.*, 2006], a higher priority must be placed on understanding land use–landslide interactions and applying this knowledge to management of mountainous and unstable terrain [*Slaymaker*, 2001].

7

Summary

During the past two decades, knowledge of the complex nature of processes that predispose hillslopes to landslides and initiate slope failure has advanced. Recent investigations of chemical weathering in various hillslope materials provide insights into susceptibility to landsliding. However, such findings need to be better related to hydrologic processes within hillslopes to fully appreciate their influence on landslide initiation. It is now clear that roots of woody vegetation impart a significant strength in relatively shallow soil mantles; much of this strength is derived from the vertical anchoring of roots into more stable substrate, although recent research has suggested that in some cases lateral root strength may provide an equal or even greater stabilizing influence. The spatial manifestation of root strength needs further study to better assess issues such as cumulative effects of previous vegetation management schemes and effects of different forest harvesting, stand tending, and vegetation conversion practices on the distributed nature of landsliding. Likewise, better knowledge of how variable site factors and soil properties interact to affect soil strength will provide insights into why landslides occur in certain sites and not in others of similar terrain. In this regard, recent hydrologic studies have elucidated certain mechanisms related to slope failure, but have also uncovered new challenges. While progress has been made in predicting reductions in soil shear strength via suction and pore water pressure dynamics during and following rainfall infiltration, we now realize that many other hydrologic controls in unstable hillslopes may invalidate some of the assumptions in models based on Darcian flow and Richards' equation. For example, extensive preferential flow paths, which may exchange water between the soil mantle and weathered bedrock and route the subsurface flow rapidly through selected portions of the regolith, complicate the spatial and temporal nature of pore water pressure dynamics in unstable hillslopes. Similarly, complex surface and subsurface topography can significantly affect pore water dynamics; these features can only be captured in distributed models. Such distributed models for shallow groundwater have now been developed, but they generally employ simplifications that ignore the infiltration phenomena and thus cannot simulate the effects of the suction decreases that occur during rainfall infiltration. The long-term rates of soil accretion in geomorphic hollows are now better understood in certain environments, but more studies are needed on the spatial and temporal linkages of landslides and debris flows in steep terrain related to soil accretion. Additionally, issues related to the contrasting recurrence periods for landslides in hollows that have been estimated by different methods need to be resolved, especially related to specific infilling or weathering processes. Knowledge of the dynamic loading that causes landslides during earthquakes

Landslides: Processes, Prediction, and Land Use
Water Resources Monograph 18
Copyright 2006 American Geophysical Union
10.1029/18WM08

has progressed, as has knowledge of relations of landslides with earthquake magnitude and distance from the epicenter, but real-time predictions remain elusive. Improved methods have been developed to predict landslides, both at individual sites and throughout catchments. Many of the distributed landslide models demand extended spatial coverage of soil and site data that are typically unavailable. Such models will benefit greatly from algorithms or pedotransfer functions that relate easily measured or remotely sensed soil and site properties to input data that are more difficult to obtain (e.g., soil shear strength parameters, pore water pressure). However, in many cases, there is no substitute for accurate field-based information, underlying the need for such data collection by government organizations. One major advantage of distributed landslide models that include rooting strength and other vegetation effects is that they facilitate long-term simulations of forest harvesting and vegetation manipulation; such temporal scenarios are virtually impossible to investigate by other means and are important for forest and other land use planning. New analytical methods are available to assess the effects of ground motion on landslide initiation; however, until better earthquake prediction methods are available, such analyses can predict scenario consequences of earthquakes only in potentially unstable terrain. At present, real-time landslide warning systems are mostly in the privy of a few industrialized nations with applications in selected, densely populated regions. Landslide warning systems based on rainfall forecasts, real-time rainfall data, and ground motion could save lives and reduce damages if properly implemented in populated areas of developing nations. Nevertheless, most mountainous terrain in both the developed and developing world is far too remote for accurate, real-time warning systems to be employed in the next few decades. Thus, it is necessary to continue to improve terrain hazard mapping techniques and multi-factor empirical analyses by better incorporating processes that directly contribute to specific types of landslides. Such methods continue to provide a low-cost form of landslide hazard assessment that is useful in land use planning. However, improvements related to specific types of land uses that have different impacts on slope stability need to be considered in these analyses, rather than treating all land uses similarly. A major challenge, especially in poor nations, is to incorporate landslide hazard maps and analyses into effective regional or community development planning.

As indicated by the depth of discussion and numerous references cited in this book, much new information is now available on the interaction of various land uses with landslide processes. Interestingly, in spite of the spirited debate concerning effects of timber harvesting and other logging practices on landslide erosion, few data on actual erosion rates from landslides have been reported in the past two decades; i.e., much of the new research has focused on comparisons of numbers (and possibly areas) of landslides for different logging practices, often following large storm events. While such data are interesting, they can paint a somewhat biased picture of long-term effects of forest practices on landslide erosion if not placed within the context of prior landslide episodes. Thus, it is critical that government agencies invest in continuing landslide inventories, or implement new long-term inventories that measure actual erosion fluxes for various forest practices, and ensure that rainfall data are available for such areas. Granting agencies typically do not fund these important long-term investigations at the necessary spatial scales. Low-volume road construction continues to be the most conspicuous source of landslide erosion per unit area affected. While some reductions in landslides along forest roads in steep terrain have been apparent in recent years due to improved location, construction, drainage, and maintenance, landslide erosion from road right-of-ways is still typically more than one to three orders of magnitude higher than in undisturbed forests and often at least an order of magnitude higher than in clearcuts. Developing nations such as China and India, with extensive networks of poorly constructed roads in mountainous areas, experience very high rates of landslide erosion and considerable economic loss and environmental damage related to these road failures. Knowledge gained from the numerous site- or area-specific studies on road-related landslides needs to be used by planning authorities together with emerging socio-economic analyses to better evaluate the design and location of road *systems*, especially low-volume roads, where it is difficult to justify expensive landslide control measures. Considerable insights have emerged in the past 20 yr related to effects of fire and mining practices on landslide erosion. Grazing and recreational impacts, however, remain poorly documented.

Most of the landslide research and prediction/avoidance applications have taken place in highly industrialized nations. While these nations all experience landslide problems, they typically have the management infrastructure to deal with landslide disasters and related environmental issues. While one could argue that such landslide prevention strategies related to land use are not optimally implemented in many of these nations, nonetheless the basic knowledge and access to resources is generally available. Thus, "mistakes" made in developed nations related to land use in potentially unstable terrain generally have small consequences compared to management mistakes in developing countries. The low loss of life in landslide disasters in developed compared to that in developing nations is testament to these differential consequences. Much of the mountainous terrain in Africa, South America, Southeast Asia, East Asia, and the Middle East is in nations that lack the resources and possibly the knowledge to deal with the interactions between land use and landslides. Furthermore, many such areas are pushing the limits of resource extraction, often spurred on by internal corruption or external multi-national corporations. Included are land uses that cause some of the highest rates of landslide erosion: forest conversion to agriculture or monoculture plantations, mining, steepland grazing, and the necessary road and trail systems for all land uses. Thus, the greatest global challenges facing government and international agencies, as well as landslide researchers and land managers, are the acquisition of better fundamental data on long-term landslide erosion in mountainous regions of developing nations and implementation of existing landslide knowledge in current land use planning. International donor organizations have typically ignored landslide problems in the developing world and have taken the more myopic approach of focusing on surface erosion issues at smaller scales. Whilst such information is valuable for managing individual field plots, a larger scale approach is needed to integrate the complete spectrum of erosion processes in time and space. This dilemma is epitomized by the ongoing use of and international support for large terraces in developing nations, which require substantial excavations and roads in steep terrain. While these terraces may retard surface erosion, they supply large amounts of sediment via landslide erosion. From the limited data available, plus our current knowledge on interactions amongst natural processes and landslide initiation, it appears that land use in the tropics may exacerbate landslide erosion to a greater extent than in similarly steep, temperate terrain. The most damaging land uses, in addition to roads, are conversion of natural forests to cropland, pasture, or monoculture plantations (e.g., coffee, rubber), where root systems are weak compared to pre-existing forest cover, soil structure is destroyed by cultivation, and evapotranspiration is lower year-round. The interactions of these land cover changes deserve the immediate attention of international donors supporting soil and water conservation and related research in the tropics. Current remote-sensing investigations focus primarily on land cover change and are simply not detailed enough to address many important aspects related to landslides. Because of the rapid rates of land cover change in Southeast Asia and Latin America, effects on landslide erosion are most pressing in mountainous areas of these regions. Instead of reacting to landslides only when major disasters occur (e.g., lives are lost or major property damage occurs), a more systematic and process-based approach to assessing and avoiding landslide hazards in the tropics is urgently needed. As learned from examples in developed nations (e.g., New Zealand), landslides can cause significant reduction in the productivity of crops, pastures, and plantations in steep, previously forested lands.

Typical of many areas of applied earth sciences, landslide investigations have increasingly relied on remotely sensed data and existing maps in developing land management plans. Also, many studies heavily employ geotechnical investigations, laboratory simulations, and modeling exercises at the exclusion of field process studies. Certainly breakthroughs in remote-sensing technology and GIS processing of these data have facilitated landslide hazard analysis at larger scales. Nevertheless, an overreliance on such methods is dangerous because they often preclude critical understanding of the inherent processes operating in various sites [*Sidle*, 2006]. Modeling and simulation studies are useful to understand specific mechanisms or to examine long-term landslide scenarios that are not possible in field investigations. In many ways, such studies are safer—they produce attractive results that are difficult to contest and often impossible to verify. However, the concern is that such computer- and laboratory-based investigations are gaining favor to the exclusion of field investigations, the sites where landslides actually occur [*Sidle*, 2006]. Funding agencies in developed countries appear to increasingly support development of new experimental equipment, remote-sensing research, GIS applications, and other "high-tech" methods related to landslide hazards at the exclusion of field-oriented process investigations. Field investigations may be less attractive these days because of the difficulties of field research in steep terrain and the uncertainties involved. Many younger scientists and students are more attracted to high-technology approaches to landslide investigations than to dealing with the heterogeneities and difficulties of working in the real world. Without being exposed to process investigations at an early stage in their education, there is little hope to attract good students into these areas of geosciences and engineering. Thus, in the past few decades, fewer and fewer promising young scientists appear motivated to pursue careers that focus on field-process studies related to landslides. This is indeed a shame; partly this may be due to funding availability, but it also partly reflects directions taken by many government land management agencies, which have also relegated a low priority to field process investigations. Surely, a balance of approaches is needed, but we must continue to include and expand our knowledge of processes contributing to landslides into analyses and development of land management plans. To provide realistic results and meaningful applications, this combined approach must include detailed field investigations.

Notation

a_1 $H_\infty - H_0$

a_c upslope contributing area

a_i average cross-sectional area of roots in size class i

A total soil cross-sectional area

A_f point in the stress diagram (σ vs. τ) describing the stress state at failure on the sliding surface

A_h point in the stress diagram describing the stress state corresponding to a water table of height h above the sliding surface

A_P constant in Philip's infiltration equation which is related to K

A_p point in the stress diagram describing the stress state at failure corresponding to the peak strength on the sliding surface

A_r point in the stress diagram describing the stress state at failure corresponding to the residual strength on the sliding surface

A_R proportion of the soil cross-sectional area occupied by roots

A_0 point in the stress diagram describing the initial stress state on the sliding surface

A' potential area affected by landslides (km^2)

b width of the slice in Bishop's modified method

b_I inflection point of the soil depth recovery curve (after a landslide)

b' cell width across which drainage occurs

c soil cohesion

c' effective soil cohesion

C slope of the water retention curve

$\Delta c'$ apparent soil cohesion due to matric suction

ΔC root cohesion

D decay rate of dimensionless root cohesion (actual root cohesion divided by maximum root cohesion)

D_s storm duration (h)

E Young's modulus of elasticity

f infiltration capacity or rate at some time t

f_c "final" infiltration capacity (at $t \to \infty$)

f_0 initial infiltration capacity (at $t = 0$)

F cumulative depth of infiltration

ΔF difference between shear stress and shear strength when slope failure occurs at peak strength

F_s Factor of safety for slope stability (S/T)

h vertical depth of groundwater above the failure plane

h_p	pressure head (always positive for saturated flow)
h_z	elevation head
H	vertical soil depth
H_o	vertical soil depth remaining just after a landslide
H_∞	upper limit of vertical soil depth or accretion (for the case of infilling after a landslide)
H_t	vertical soil depth at some time t after a landslide
$\hat{H}$	total hydraulic head (sum of elevation and pressure heads)
I	mean rainstorm intensity (mm h^{-1})
k_h	horizontal earthquake acceleration coefficient
k_v	vertical earthquake acceleration coefficient
K	hydraulic conductivity of the soil or regolith
K_e	earthquake load
K_{sat}	saturated hydraulic conductivity
l	unit travel length for subsurface flow
L_a	arc length AC for the Fellenius ordinary method of slices
M	earthquake magnitude (Richter scale)
n	soil porosity
N	normal force on the slip surface within a slice
Q	precipitation minus evaporation
R	regeneration of dimensionless root cohesion (actual root cohesion divided by maximum root cohesion)
R_e	maximum distance from the earthquake epicenter for which landslides occur (km)
s	soil shear strength
S	resistant force along the sliding surface within a slice
S'	sorptivity of the soil (a parameter relating to the penetration of the wetting front)
S_f	effective suction at the wetting front
S_R	shear strength of individual roots
t_r	soil transmissivity
T	tangential force acting along a sliding surface within a slice
T_i	tensile strength of roots in size class i
T_R	average tensile strength of all roots or of individual roots
u	pore water pressure
Δu	pore water pressure increment
u_a	soil air pressure
v	macroscopic or so-called Darcy velocity for flow in porous media
W	total weight acting on a slice or the slope
W_e	sum force of the earthquake load K and weight (W)

W_s total weight of a circle-slip slice including vegetation

α angle between the weight vector (W) and the sum vector W_e

α_r angle of root deformation (under stress)

β slope gradient (in degrees)

γ soil unit weight (average)

γ_d dry soil unit weight

γ_{sat} saturated soil unit weight

γ_t moist unit weight of the soil above the water table (not saturated)

γ_w unit weight of water

$\gamma_{1/3}$ soil unit weight at 1/3 atmosphere moisture tension

θ angle between the weight (W) and earthquake load (K_e) vectors

σ total normal stress

σ' effective normal stress

σ_0' effective normal stress at the potential failure plane

τ shear stress

τ_0 shear stress on the failure plane

$\Delta\tau$ shear stress increment

τ_R average tensile strength of roots per unit area of soil

ϕ internal angle of friction of the soil

ϕ' effective internal angle of friction

ϕ_p' internal angle of friction at peak strength

ϕ_r' internal angle of friction at residual strength

χ index of soil moisture condition; 0 for completely dry soils and 1 for saturated soils

ψ_M matric suction

References

Abe, H., and H. Kawakami (1987), Collapse phenomena and negative pore water pressure of unsaturated soil, *Proc. of Symp. on Mechanical Properties of Unsaturated Soil*, Jpn. Geotech. Soc., pp. 45–54 (in Japanese).

Abe, K., and M. Iwamoto (1986a), Preliminary experiment on shear in soil layers with a large direct-shear apparatus, *J. Jpn. For. Soc.*, *6*(2), 61–65.

Abe, K., and M. Iwamoto (1986b), An evaluation of tree-root effect on slope stability by tree-root strength, *J. Jpn. For. Soc.*, *68*(12), 505–510.

Abe, K., and M. Iwamoto (1990), Simulation model for the distribution of tree roots—application to a slope stability model, *J. Jpn. For. Soc.*, *305 72*(5), 375–387.

Abe, K., and R. R. Ziemer (1991), Effect of tree roots on a shear zone: Modeling reinforced shear stress, *Can. J. For. Res.*, *21*, 1012–1019.

Abernethy, B., and I. D. Rutherford (2000), The effect of riparian tree roots on the mass-stability of riverbanks, *Earth Surf. Process. Landforms*, *25*, 921–937.

Abramson, L. W., T. S. Lee, S. Sharma, and G. M. Boyce (2002), *Slope Stability and Stabilization Methods*, 736 p., John Wiley & Sons, New York.

Achterman, G., B. Kintigh, V. Tarno, J. Welsh, C. Wooten, W. Krieger, S. Foster, H. MacPherson, J. Beaulieu, and K. Cubic (1998), Joint Interim Task Force on Landslides and Public Safety, *Report to the 70th Legislative Assembly*, State Flood Control Plan Task Force, Salem, OR.

Acosta, R., M. Allen, A. Cherian, S. Granich, I. Mintzer, A. Suarez, and D. von Hippel (1999), *Climate Change Information Sheets*, edited by M. Williams, 62 p., United Nations Environment Programme (UNEP), International Environment House, Geneva, Switzerland.

Adams, J. (1980), Contemporary uplift and erosion of the Southern Alps, New Zealand, *Geol. Soc. Am. Bull.*, *91*(Part II), 1–114.

Adger, W. N. (2000), Institutional adaptation to environmental risk under the transition in Vietnam, *Ann. Assoc. Am. Geogr.*, *90*, 738–758.

Ahmad, R. (1991), Landslides triggered by the rainstorm of May 21–22, 1991, *Jamaica J. Sci. Tech.*, *2*, 1–13.

Ahmad, R., and J. P. McCalpin (1999), Landslide susceptibility maps for the Kingston metropolitan area, Jamaica with notes on their use, Dept. of Geog., Univ. of the West Indies, UDS Publ. No. 5.

Ahnert, F. (1970), Functional relationship between denudation, relief and uplift in large mid-latitude drainage basins, *Am. J. Sci.*, *268*, 243–263.

Ahuja, L. R., and S. A. El-Swaify (1979), Determining soil hydrologic characteristics on a remote forest watershed by continuous monitoring of soil-water pressures, rainfall and runoff, *J. Hydrol.*, *44*, 135–147.

Åkerman, H. J. (1993), Solifluction and creep rates 1972–1991, Kapp Linne, west Spitsbergen, in *Solifluction and Climatic Variation in the Holocene*, edited by B. Frenzel, J. A. Matthews, and J. A. Gläser, pp. 225–250, Gustav Fischer Verlag, New York.

Aleotti, P. (2004), A warning system for rainfall-induced shallow landslides, *Eng. Geol.*, *73*, 247–265.

Alexander, D. (1989), Urban landslides, *Prog. Phys. Geogr.*, *13*(2), 157–191.

Alexander, D. (1992), On the causes of landslides: human activities, perception, and natural processes, *Env. Geol. Water Sci.*, *20*(3), 165–179.

Allison, C., R. C. Sidle, and D. Tait (2004), Application of decision analysis to forest road deactivation in unstable terrain, *Environ. Manage.*, 33, 173–185.

Amacher, G. S., W. F. Hyde, and K. R. Kanel (1996), Household fuelwood demand and supply in Nepal's Tarai and Mid-Hills: choice between cash and outlays and labor opportunities, *World Dev.*, 24(11), 1725–1736.

Amaral, C. (1997), Landslide disasters management in Rio de Janeiro, *2nd Pan-American Symp. on Landslides* (II PSL/2 COBRAE), ABMS, ABGE e ISSMGE, Rio De Janeiro, pp. 209–212.

Amaranthus, M. P., R. M. Rice, N. R. Barr, and R. R. Ziemer (1985), Logging and forest roads related to increased debris slides in southwest Oregon, *J. Forestry, 83*, 229–233.

Ambers, R. K. R. (2001), Relationships between clay mineralogy and hydrothermal metamorphism, and topography in the Western Cascades watershed, Oregon, USA, *Geomorphology, 38*, 47–61.

Amen, B. (1990), The hydrologic role of the unsaturated zone of a forested colluvium-mantled hollow, Redwood National Park, California, *Redwood National Park Res. and Dev. Tech. Report No. 26*, pp. 1–50.

Anbalagan, R. (1992), Landslide hazard evaluation and zonation mapping in mountainous terrain, *Eng. Geol., 32*, 269–277.

Anderson, C. J., M. P. Coutts, R. M. Ritchie, and D. J. Campbell (1989), Root extraction force measurements for Sitka spruce, *Forestry, 62*, 127–137.

Anderson, D. M., R. C. Reynolds, and J. Brown (1969), Bentonite debris flows in northern Alaska, *Science, 164*, 173–174.

Anderson, M. G., and T. P. Burt (1977), Automatic monitoring of soil moisture conditions in a hillslope spur and hollow, *J. Hydrol., 33*, 27–36.

Anderson, M. G., and T. P. Burt (1978), The role of topography in controlling through flow generation, *Earth Surf. Process. Landforms, 3*, 331–344.

Anderson, M. G., and P. E. Kneale (1980), Pore water pressure and stability conditions on a motorway embankment, *Earth Surf. Process. Landforms, 3*, 347–358.

Anderson, M. G., and P. E. Kneale (1982), The influence of low-angled topography on hillslope soil-water convergence and stream discharge, *J. Hydrol., 57*, 65–80.

Anderson, M. G., M. J. Kemp, and D. M. Floyd (1988), Applications of soil water finite difference models to slope problems, *Proc. 5th Int. Symp. on Landslides*, Vol. 1, pp. 525–530, Lausanne, Switzerland.

Anderson, S. A., and N. Sitar (1995), Analysis of rainfall-induced debris flows, *J. Geotech. Eng, 121*(7), 544–552.

Aniya, M. (1985), Landslide-susceptibility mapping in the Amahata River basin, Japan, *Ann. Assoc. Am. Geogr., 75*, 102–114.

Ariathurai, R., and A. Kandiah (1978), Erosion rates of cohesive soils, *J. Hydraul. Div. Am. Soc. Civ. Eng., 104*, (HY2), 279–283.

Arnaez-Vadillo, J., and V. Larrea (1994), Erosion models and hydrogeomorphological function on hill-roads (Iberian system, La Rioja, Spain), *Z. Geomorph. N.F., 38*(3), 343–354.

Arnaud-Fassetta, G., E. Cossart, and M. Fort (2005), Hydro-geomorphic hazards and impact of man-made structures during catastrophic flood of June 2000 in the Upper Guil catchment (Queyras, Southern French Alps), *Geomorphology, 66*(1–4), 41–67.

Arnold, J. (1957), Engineering aspects of forests soils, in *An Introduction to Forest Soils of the Douglas-fir Region of the Pacific Northwest*, pp. 1–15, Western Forestry and Conservation Assoc., Portland, OR.

Asare, S.N., R. P. Rudra, W. T. Dickinson, and W. T. Wall (1997), Frequency of freeze-thaw cycles, bulk density and saturation effects on soil surface shear and aggregate stability in resisting water erosion, *Can. Agric. Eng., 39*, 273–279.

Atkinson, G. (1984), Erosion damage following bushfires, *J. Soil Conserv. N.S.W., 40*, 4–9.

Atkinson, T. C. (1978), Techniques for measuring subsurface flow on hillslopes, in *Hillslope Hydrology*, edited by M. J. Kirkby, pp. 73–120, John Wiley, New York.

Aubertin, G. M. (1971), Nature and extent of macropores in forest soils and their influence on subsurface water movement, *Res. Pap. NE-192*, 33 p., For. Serv., U.S. Dep. of Agric., Broomall, PA.

Auer, K., and A. Shakoor (1993), A statistical approach to evaluate debris avalanche activities in central Virginia, *Eng. Geol., 33*(4) 305–321.

Ayalew, L., and H. Yamagishi (2005), The application of GIS-based logistic regression for landslide susceptibility mapping in the Kakuda-Yahiko Mountains, Central Japan, *Geomorphology, 65*, 15–31.

Ayalew, L., H. Yamagishi, and N. Ugawa (2004), Landslide susceptibility mapping using GIS-based weighted linear combination, the case in Tsugawa area of Agano River, Niigata Prefecture, Japan, *Landslides, 1*(1), 73–81.

Ayetey, J. K. (1991), The extent and effects of mass wasting in Ghana, *Bull. Int. Assoc. Eng. Geol., 43*, 5–19.

Aynew, T., and G. Barbieri (2005), Inventory of landslides and susceptibility mapping in the Dessie area, northern Ethiopia, *Eng. Geol., 77*, 1–15.

Ayonghe, S. N., G. T. Mafany, E. Ntasin, and P. Samalang (1999), Seismically activated swarm of landslides, tension cracks, and a rockfall after heavy rainfall in Bafaka, Cameroon, *Nat. Hazards, 19*(1) 13–27.

Bahar, R., B. Cambou, and J. J. Fry (1995), Forecast of creep settlements of heavy structures using pressuremeter tests, *Computers and Geotechnics, 17*(4), 507–521.

Bailey, R. G. (1971), *Landslide hazards related to land use planning in Teton National Forest, northwest Wyoming*, 131 p., For. Serv., U.S. Dep. of Agric., Ogden, UT.

Ballard, T.M., and R.P. Willington (1975), Slope instability in relation to timber harvesting in the Chilliwack Provincial Forest, *The Forestry Chron., 51*, 59–62.

Balteanu, D. (1976), Two case studies of mudflows in the Buzau Subcarpathians, *Geograf. Ann., 58A*, 165–171.

Bansal, R. C., and H. N. Mathur (1976), Landslides—the nightmare of hill roads, *Soil Conserv. Digest, 4*, 36–37.

Baruah, U., and A. C. Mohapatra (1996), Slope mass movement and associated soils in East Khasi and Jaintia Hills of Meghalaya, *J. Indian Soc. Soil Sci., 44*(4), 712–717.

Basher, L. R., and P. J. Tonkin (1985), Soil formation, soil erosion and revegetation in the central South Island hill and the mountain lands, in *Proc. Soil Dynamics and Land Use Seminar*, edited by I. B. Campbell, pp. 154–169, New Zealand Soc. Soil Sci and N.Z. Soil Conserv. Assoc.

Basile, A., G. Mele, and F. Terribile (2003), Soil hydraulic behaviour of a selected benchmark soil involved in the landslide of Sarno, 1998, *Geoderma, 2026*, 1–16.

Bauer, A., and P. N. Calder (1971), The influence and evaluation of blasting on stability, in *Stability in Open Pit Mining*, edited by C. O. Brawner and V. Milligan, pp. 83–93, AIME, New York.

Baum, R. L., and R. W. Fleming (1996), Kinematic studies of the Slumgullion, Hinsdale County, Colorado, in *The Slumgullion Earth Flow: a Large-Scale Natural Laboratory*, edited by D. J. Varnes and W. Z. Savage, pp. 9–12, U.S. Geol. Surv. Bull. 2130.

Baum, R. L., and M. E. Reid (1995), Geology, hydrology and mechanics of slow-moving clay-rich landslide, Honolulu, Hawaii, in *Clay and Shale Slope Instability*, edited by W. C. Haneberg and S. A. Anderson, pp. 79–105, Rev. in Eng. Geol., vol. 10, Geol. Soc. Am., Boulder, CO.

Baum, R. L., and M. E. Reid (2000), Ground water isolation by low-permeability clays in landslide shear zones, in *Landslides in Research, Theory and Practice*, edited by E. N. Bromhead, N. Dixon, and M.-L. Ibsen, Vol. 1, pp. 139–144, Thomas Telford, London, UK.

Baum, R. L., W. Z. Savage, and J. W. Godt (2002), TRIGRS—A Fortran program for transient rainfall infiltration and grid-based regional slope-stability analysis, *Open-file report 02-424*, 35 p. (2 appendices), U.S. Geol. Surv.

Baver, L. D., W. H. Gardner, and W. R. Gardner (1972), *Soil Physics*, 4th ed., 498 p., John Wiley, New York.

Beaulieu, J. D. (1974), Geologic hazards of the Bull Run watershed, Multnomah and Clackamas Counties, Oregon, 77 p., *Bull. 82*, Oregon Dep. of Geol. and Miner. Ind., Portland.

Beaumont, C., P. Fullsack, J. Hamilton (1992), Erosional control of active compressional orogens, in *Thrust Tectonics*, edited by K. R. McKay, pp. 1–18, Chapman and Hall, New York.

Bechini, C. (1993), Natural conditions controlling earthflows occurrence in the Eden Canyon area (San Francisco Bay region, California), *Z. Geomorph. N.F.*, Suppl.-Bd 87, 91–105.

Bell, J. C., W. L. Daniels, and C. E. Zipper (1989), The practice of "approximate original contour" in the Central Appalachians. I. Slope stability and erosion potential, *Landscape Urban Plan.*, *18*, 127–138.

Belt, G. H., and B. M. Woo (1979), An analysis of landslide damage and slope stabilization near Seoul, Korea, *Bull. 31*, 9 p., Forest, Wildlife and Range Exp. Stn., Univ. of Idaho, Moscow.

Benda, L. (1990), The influence of debris flows on channels and valley floors of the Oregon Coast Range, U.S.A., *Earth Surf. Process. Landforms*, *15*(5), 457–466.

Benda, L., and T. Dunne (1997a), Stochastic forcing of sediment routing and storage in channel networks. *Water Resour. Res.*, *33*, 2865–2880.

Benda, L., and T. Dunne (1997b), Stochastic forcing of sediment supply to the channel network form landsliding and debris flow, *Water Resour. Res.*, *33*, 2849–2863.

Benda, L., K. Andras, D. Miller, and P. Bigelow (2004), Confluence effects in rivers: Interactions of basin scale, network geometry, and disturbance regimes, *Water Resour. Res.*, *40*, W05402, doi:10.1029/2003WR002583.

Benda, L., D. Miller, P. Bigelow, and K. Andras (2003b), Effect of post-wildfire erosion on channel environments, Boise River, Idaho, *For. Ecol. Manage.*, *178*, 105–119.

Benda, L., C. Veldhuisen, and J. Black (2003a), Debris flows as agents of morphological heterogeneity at low-order confluences, Olympic Mountains, Washington. *Geol. Soc. Am. Bull.*, *115*, 1110–1121.

Benda, L. E., and T. W. Cundy (1990), Predicting depositions of debris flows in mountain channels, *Can. Geotech. J.*, 27, 409–417.

Bennett, K. A. (1982), Effects of slash burning on surface soil erosion rates in the Oregon Coast Range. M.S. thesis, 70 p., Oregon State Univ., Corvallis.

Bentley, S. P., and H. J. Siddle (1996), Landslide research in the South Wales coalfield, *Eng. Geol.*, *43*, 65–80.

Bentley, S. P., and I. J. Smalley (1984), Landslides in sensitive clays, in *Slope Instability*, edited by D. Brunsden and D. P. Prior, pp. 457–485, John Wiley & Sons, Chichester, UK.

Berggren, B., J. Fallsvik, and L. Viberg (1991), Mapping and evaluation of landslide risk in Sweden, in *Landslide*, edited by Bell, pp. 873–878, Balkema, Rotterdam.

Bergin, D. O., M. O. Kimberley, and M. Marden (1995), Protective value of regenerating tea tree stands on erosion-prone hill country, East Coast, North Island, New Zealand. *N.Z. J. For. Sci.*, *25*(1), 3–19.

Berris, S. N., and R. D. Harr (1987), Comparative snow accumulation and melt during rainfall in forested and clear-cut plots in the western Cascades of Oregon, *Water Resour. Res.*, *23*, 135–142.

Best, D. W. (1995), History of timber harvest in the Redwood Creek basin, Northwestern California, in *Geomorphic Processes and Aquatic Habit at in the Redwood Creek Basin, Northwestern California*, edited by K. M. Nolan, H. M. Kelsey, and D. C. Marron, U.S. Geol. Surv. Prof. Pap. 1454-C, pp. C1-C7, U.S. Govt. Printing Office, Washington, DC.

Beven, K. J. (1984), Infiltration into a class of vertically non-uniform soils, *Hydrol. Sci. J., 29*(4), 425–434.

Beven, K. J. (1987), Towards the use of catchment geomorphology in flood frequency prediction, *Earth Surf. Process. Landforms, 12,* 69–82.

Beven, K., and P. Germann (1982), Macropores and water flow in soil, *Water Resour. Res., 18*(5), 1311–1325.

Bhandari, R. K. (1994), Watawala earthslide in Sri Lanka, *Landslide News (Jpn. Landslide Soc.), 8,* 28–31.

Bhandari, R. K., and K. M. Weerasinghe (1996), Pitfalls in subrogating slope maps for landslide hazards maps, *Proc. 17th Asian Conf. on Remote Sensing,* Colombo, Sri Lanka, Asian Assoc. of Remote Sensing.

Bhatt, B. P., and W. S. Sachan (2004), Firewood consumption along an altitudinal gradient in mountain villages in India, *Biomass and Bioenergy, 27,* 69–75.

Billard, A., T. Muxart, E. Derbyshire, J. T. Wang, and T. A. Dijkstra (1993), Landsliding and land use in the loess of Gansu Province, China. *Z. Geomorph. N.F.,* Suppl. Bd. 87, 117–131.

Bisci, C., F. F. Burrattini, F. Dramis, S. Leoperdi, F. Pontoni, and F. Pontoni (1996), The Sant'Agata Feltria landslide (Marche Region, central Italy): a case of recurrent earthflow evolving from a deep-seated gravitational slope deformation, *Geomorphology, 15,* 351–361.

Bishop, A. W. (1955), The use of the slip circle in the stability analysis of slopes, *Geotechnique, 5,* 7–17.

Bishop, A. W., L. Alpan, G. E. Blight, and I.B. Donald (1960), Factors controlling the strength of partly saturated cohesive soils, *ASCE Research Conf. on Shear Strength of Cohesive Soils,* pp. 503–532, Univ. of Colorado, Boulder, CO.

Bishop, D. M., and M.E. Stevens (1964), Landslides on logged areas, southeast Alaska, *Res. Rep. NOR-1,* 18 p., For. Serv., U.S. Dep. of Agric., Juneau, AK.

Bisson, P. A., and R. E. Bilby (1982), Avoidance of suspended sediment by juvenile Coho salmon, *North Am. J. Fish. Manage., 2*(4), 371–374.

Blaschke, P. M. (1988), Vegetation and landscape dynamics in eastern Taranaki, North Island, New Zealand, Ph.D. thesis, Victoria Univ. of Wellington, N.Z.

Blikra, L. H. (1990), Geological mapping of rapid mass movement deposits as an aid to land-use planning, *Eng. Geol., 29,* 365–676.

Blong, R. J., and D. L. Dunkerley (1976), Landslides in the Razorback area, New South Wales, Australia, *Geograf. Ann., 58A,* 139–147.

Bodri, B. (2001), A neural-network model for earthquake occurrences, *J. Geodynamics, 32,* 289–310.

Bogucki, D. J. (1976), Debris slides in the Mt. Le Conte area, Great Smoky Mountains National Park, USA, *Geograf. Ann., 58A,* 179–191.

Bonilla, M. G. (1975), A review of recently active faults in Taiwan, *Open-file report 75–41,* 43 p., U.S. Geol. Surv.

Boore, D. M. (1972), A note on the effect of simple topography on seismic SH waves, *Bull. Seismol. Soc. Am., 62,* 275–284.

Boore, D. M. (1973), The effect of simple topography on seismic waves: Implications for the acceleration recorded at Pacoima Dam, San Fernando Valley, California, *Bull. Seismol. Soc. Am., 63,* 1603–1609.

Borchardt, G. A. (1976), Clay mineralogy and slope stability, *Special Report,* 41 p., Calif. Div. of Mines and Geol., Sacramento, CA.

Borga, M., G. D. Fontana, C. Gregoretti, and L. Marchi (2002), Assessment of shallow landsliding by using a physically based model of hillslope stability, *Hydrol. Process., 16,* 2833–2851.

Bormann, B. T., and R. C. Sidle (1990), Changes in productivity and distribution of nutrients in a chronosequence at Glacier Bay National Park, Alaska, *J. Ecol., 78,* 561–578.

Bormann, B. T., D. Wang, F. H. Bormann, G. Benoit, R. April, and M. C. Snyder (1998), Rapid, plant-induced weathering in an aggrading experimental ecosystem, *Biogeochemistry, 43,* 129–155.

Botschek, J., S. Krause, T. Abel, and A. Skowronek (2002), Hydrological parameterization of piping in loess-rich soils in the Bergisches Land, Nordrhein-Westfalen, Germany, *J. Plant Nutr. Soil Sci., 165,* 506–510.

Bouchen, M. (1973), Effect of topography on surface motion, *Bull. Seismol. Soc. Am., 63*(2), 615–632.

Bourgeois, W. W. (1978), Timber harvesting activities on steep Vancouver Island terrain, *Proc. of 5th North American Forest Soils Conf.,* pp. 393–409, Colorado State Univ., Fort Collins.

Bovis, M. J., and M. Jakob (1999), The role of debris supply conditions in predicting debris flow activity, *Earth Surf. Process. Landforms, 24,* 1039–1054.

Bovis, M. J., and P. Jones (1992), Holocene history of earthflow mass movements in south-central British Columbia: the influence of hydroclimatic changes, *Can. J. Earth Sci., 29,* 1746–1754.

Brabb, E. E. (1984), Innovative approach to landslide hazard and risk mapping, *Proc. 4th Intern. Symp. on Landslides,* Vol. 1, pp. 307–324, Toronto, Ontario, Canada.

Brand, E. W., M. J., Dale, and J. M. Nash (1986), Soil pipes and slope stability in Hong Kong, *Q. J. Eng. Geol., 19,* 301–303.

Brardinoni, F., O. Slaymaker, and M. A. Hassan (2003), Landslide inventory in a rugged forested watershed: a comparison between air-photo and field survey data, *Geomorphology, 54,* 179–196.

British Columbia Ministry of Forests (1995), Coastal Watershed Assessment Procedure Guidebook (CWAP), *Forest Practices Code of British Columbia,* 66 p., British Columbia Min. of For., Canada.

Bromhead, E. N. (1997), The treatment of landslides, *Proc. of the Inst. Civil Eng., Geotech. Eng., 125,* 85–96.

Bronders, J. (1994), Hydraulic conductivity of a tropical residual soil prone to landslides, *Q. J. Eng. Geol., 27,* 375–382.

Brooks, S. M., and M. G. Anderson (1995), The determination of suction-controlled slope stability in humid temperature environments, *Geograf. Ann., 77A,* 11–22.

Brooks, S. M., M. J. Crozier, N. J. Preston, and M. G. Anderson (2002), Regolith stripping and the control of shallow translational hillslope failure: application of a two-dimensional coupled soil hydrology-slope stability model, Hawke's Bay, New Zealand, *Geomorphology, 45,* 165–179.

Brown, C. B., and M. S. Sheu (1975), Effects of deforestation on slopes, *J. Geotech. Eng. Div. Am. Soc. Civ. Eng., 101,* 147–165.

Brown, G., and G. W. Brindley (1980), X-ray diffraction procedures for clay mineral identification, in *Crystal Structures of Clay Minerals and Their X-Ray Identification,* edited by G. Brown and G. W. Brindley, pp. 305–359, Monograph No. 5, Mineralogical Society, London, UK.

Brown, G. W. (1985), Landslide damage to the forest environment, Proc. of a Workshop on Slope Stability: Problems and Solutions in Forest Management, *General Tech. Rep. PNW-180,* pp. 26–29, For. Serv., U.S. Dep. of Agric., Portland. OR.

Brown, N. (1998), Out of control: fires and forestry in Indonesia, *Trends Ecol. Evol., 13*(1), 41–44.

Brown, S., J.M. Anderson, P.L. Woomer, M.J. Swift, and E. Barrios (1994), Soil biological processes in tropical ecosystems, in *The Biological Management of Tropical Fertility,* edited by P. L. Woomer and M. J. Swift, pp. 15–46, TSBF, Wiley-Sayce, UK.

Bruijnzeel, L. A. (2004), Hydrological functions of tropical forests: not seeing the soil for the trees? *Agric. Ecosys. Environ., 104*(1), 185–228.

Brunsden, D. (1985), Landslide types, mechanisms, recognition, identification, in *Landslides in the South Wales Coalfield*, edited by C. S. Morgan, pp. 19–28, Proc. Symp. April 1–3, 1985, Polytech. of Wales, UK.

Brunsden, D., and D. B., Prior (Eds.) (1984), *Slope Instability*, 620 p., John Wiley & Sons, Chichester, UK.

Buchanan, P., and K.W. Savigny (1990), Factors controlling debris avalanche initiation, *Can. Geotech. J., 27*, 659–675.

Buchanan, P., K. W. Savigny, and J. de Vries (1990), A method for modeling water tables at debris avalanche headscarps, *J. Hydrol., 113*, 61–88.

Buma, J. (2000), Finding the most suitable slope stability model for the assessment of the impact of climate change on a landslide in southeast France, *Earth Surf. Process. Landforms, 25*, 565–582.

Buma, J., and Dehn, M. (1998) A method for predicting the impact of climate change on slope stability, *Environ. Geol., 35*, 190–196.

Buma, J., and T. van Asch (1996), Slide (rotational), in *Landslide Recognition—Identification, Movement, and Courses*, edited by R. Dikau, D. Brunsden, L. Schrott, and M.-L. Ibsen, pp. 43–62, John Wiley & Sons, Chichester, UK.

Burke, T. J., D. N. Sattler, and T. Terich (2002), The socioeconomic effects of a landslide in Western Washington, *Environ. Hazards, 4*, 129–136.

Burns, D. A., R. P. Hooper, J. J. McDonnell, J. E. Freer, C. Kendall, and K. Beven (1998), Base cation concentrations in subsurface flow from a forested hillslope: the role of flushing frequency, *Water Resour. Res., 34*(12), 3535–3544.

Burns, S. F. (1996), Landslides in the West Hills of Portland: a preliminary look, *Oreg. Geol., 58*(2), 42–43.

Burroughs, E. R., and B. R. Thomas (1977), Declining root strength in Douglas-fir after felling as a factor in slope stability, *Res. Pap. INT-190*, 27 p., For. Serv., U.S. Dep. of Agric, Ogden, UT.

Burroughs, E. R., G. R. Chalfant, and M. A. Townsend (1976), *Slope Stability in Road Construction*, 102 p. U.S. Dep. of the Inter., Bur. of Land Manage., Portland, OR.

Burroughs, E. R., M. A. Marsden, and H. F. Haupt (1972), Volume of snowmelt intercepted by logging roads, *J. Irrigation and Drain. Div. Am. Soc. Civ. Eng., 98*(IR1), 1–12.

Burton, A., T. J. Arkell, and J. C. Bathurst (1998), Field variability of landslide model parameters, *Environ. Geol., 35*(2–3), 100–114.

Buttle, J. M., and D. J. McDonald (2000), Soil macroporosity and infiltration characteristics of a forest podzol, *Hydrol. Process., 14*, 831–848.

Bye, A. R., and F. G. Bell (2001), Stability assessment and slope design at Sandsloot open pit, South Africa, *Int. J. Rock Mech. Min. Sci., 38*, 449–466.

Byron, N., and M. Arnold (1999), What futures for the people of the tropical forests? *World Dev., 27*, 789–805.

Caine, N. (1976), Summer rainstorms in an alpine environment and their influence on soil erosion, San Juan Mountains, Colorado, *Arct. Alp. Res., 8*(2), 183–196.

Caine, N. (1980), Rainfall intensity-duration control of shallow landslides and debris flows, *Geograf. Ann. 62A*, 23–27.

Calder, I. R., I. R. Wright, and D. Murdiyarso (1986), A study of evaporation from tropical forest-West Java, *J. Hydrol., 89*, 13–31.

Campbell, A. P. (1966), Measurement of movement of an earthflow, *Soil Water, 2*(3), 23–24.

Campbell, G. S. (1985), *Soil Physics with Basic: Transport Models for Soil-Plant Systems*, Developments in Soil Science 14, Elsevier, Amsterdam.

Campbell, R. H. (1975), Soil slips, debris flows, and rainstorms in the Santa Monica Mountains and vicinity, southern California, *Prof. Pap. 851*, 51 p., U.S. Geol. Surv.

Campy, M., J. F. Buoncristiani, and V. Bichet (1998), Sediment yield from glacio-lacustrine calcareous deposits during the postglacial period in the Combe D'Ain (Jura, France), *Earth Surf.
Process. Landforms, 23*, 429–444.

Cannell, M., and M. Coutts (1988), Growing in the wind, *New Scientist, 117*(1596), 42–43.

Cannon, S. H., and S. D. Ellen (1985), Rainfall conditions for abundant debris avalanches, San
Francisco Bay Region, California, *Calif. Geol., 38*, 267–272.

Cannon, S. H., and S. L. Reneau (2000), Conditions for generation of fire-related debris flows,
Capulin Canyon, New Mexico, *Earth Surf. Process. Landforms, 25*, 1103–1121.

Cannon, S. H., E. R. Bigio, and E. Mine (2001b), A processes for fire-related debris-flow initiation,
Cerro Grande fire, New Mexico, *Hydrol. Process., 15*(15), 3011–3023.

Cannon, S. H., J. E. Gartner, C. Parrett, and M. Parise (2003), Wildfire-related debris-flow genera-
tion through episodic progressive sediment-bulking processes, western USA, in *Debris-Flow
Hazards Mitigation: Mechanisms, Prediction, and Assessment*, edited by D. Rickenmann and
C.-L. Chen, pp. 71–82, Millpress, Rotterdam.

Cannon, S. H., R. M. Kirkham, and M. Parise (2001a), Wildfire-related debris-flow initiation
processes, Storm King Mountain, Colorado, *Geomorphology, 39*(3–4), 171–188.

Canuti, P., P. Focardi, and C. A. Garzonio (1985), Correlation between rainfall and landslides, *Bull.
Int. Assoc. Eng. Geol., 32*, 49–54.

Carey, S. K., and M.-K., Woo (2000), The role of soil pipes as slope runoff mechanism, Subarctic
Yukon, Canada, *J. Hydrol., 233*, 206–222.

Carrara, A., M. Cardinali, R. Detti, F. Guzzetti, N. Pasqui, and P. Reichenbach (1991), GIS technique and statistical models in evaluating landslide hazard, *Earth Surf. Process. Landforms,
16*, 427–445.

Carro, M., M. De Amicis, L. Luzi, and S. Marzorati (2003), The application of predictive modeling
techniques to landslides induced by earthquakes: a case study of the 26 September 1997 Umbria-
Marche earthquake (Italy), *Eng. Geol., 69*, 139–159.

Carson, M. A., and M. J. Kirkby (1972), *Hillslope Form and Process*, 475 p., Cambridge Univ.
Press, London.

Carson, M. A., and D. J. Petley (1970), The existence of threshold hillslopes in the denudation of
the landscape, *Trans. Inst. British Geogr., 49*, 71–95.

Chandler, D. C., and J. J. Bisogni (1999) The use of alkalinity as a conservative tracer in a study of
near-surface hydrologic change in tropical karst, *J. Hydrol., 216*, 172–182.

Chandler, D. G., and W. F. Walter (1998), Runoff responses among common land uses in the uplands
of Matalom, Leyte, Philippines, *Trans Am. Soc. Agric. Eng. 41*(6), 1635–1641.

Chang, J., and O. Slaymaker (2002), Frequency and spatial distribution of landslides in a mountainous drainage basin: western foothills, Taiwan, *Catena, 46*, 285–307.

Chang, J.-B., and S.-H. Lin (1995), Root strength characteristics on dominant plants at landslide area
in central cross-island highway, Taiwan, *J. Chinese Soil & Water Conserv. 26*(4), 235–243.

Chappell, N. A., J. L. Ternan, A. G. Williams, and B. Reynolds (1990), Preliminary analysis of
water and solute movement beneath a coniferous hillslope in mid-Wales, UK, *J. Hydrol., 116*,
201–215.

Chassie, R. G., and R. D. Goughnour (1976), National highway landslide experience, *Highway
Focus, 8*, 1–9.

Chatwin, S. C. (1994), Measures for control and management of unstable terrain, in *A Guide for
Management of Landslide-Prone Terrain in the Pacific Northwest*, edited by S. C., Chatwin, D.
E. Howes, J. W. Schwab, and D. N. Swanston, pp. 91–171, Land Management Handbook No. 18,
B.C Ministry of Forests, Crown Publications Inc., Victoria, Canada.

Chau, K. T., Y. L. Sze, M. K. Fung, W. Y. Wong, E. L. Fong, and L. C. P. Chan (2004), Landslide
hazard analysis for Hong Kong using landslide inventory and GIS, *Computers & Geosci., 30*,
429–443.

Chen, H. (2006), Controlling factors of hazardous debris flow in Taiwan, *Quaternary Int., 147*, 3–15.

Chen, H., and C. F. Lee (2004), Geohazards of slope mass movement and its prevention in Hong Kong, *Eng. Geol., 76*(1–2), 3–25.

Chigira, M. (1990), A mechanism of chemical weathering of mudstone in a mountainous area, *Eng. Geol., 29*(2), 119–138.

Chigira, M. (1993), Dissolution and oxidation of mudstone under stress, *Can. Geotech. J., 30*(1), 60–70.

Chigira, M. (2001), Micro-sheeting of granite and its relationship with landsliding specifically after the heavy rainstorm in June 1999, Hiroshima Prefecture, Japan, *Eng. Geol., 59*(3–4), 219–231.

Chigira, M. (2002a), The effects of environmental changes on weathering, gravitational rock deformation and landslides, in *Environmental Change and Geomorphic Hazards in Forests*, edited by R. C. Sidle, pp. 101–121, IUFRO Research Series, No. 9, CABI Publishing, Wallingford, Oxen, UK.

Chigira, M. (2002b), Geologic factors contributing to landslide generation in a pyroclastic area: August 1998 Nishigo Village, Japan, *Geomorphology, 46*, 117–128.

Chigira, M. (2003), Weathering profiles and soil structures in the areas generated many shallow landslides during the time of rainstorm, *Proc. of New Challenges for the Assessment of Hazard by Shallow Landslide Generated by Rainstorm*, pp. 35–48, June 20–21, 2003, Disaster Prevention Res. Inst., Kyoto Univ., Japan.

Chigira, M., and T. Oyama (1999), Mechanism and effect of chemical weathering of sedimentary rocks, *Eng. Geol., 55*(1–2), 3–14.

Chigira, M., and O. Yokoyama (2005), Weathering profile of non-welded ignimbrite and water infiltration behavior within it in relation to the generation of shallow landslides, *Eng. Geol., 78*(3–4), 187–207.

Chigira, M., M. Nakamoto, and E. Nakata (2002), Weathering mechanisms and their effects on the landsliding of ignimbrite subject to vapor-phase crystallization in the Shirakawa pyroclastic flow, northern Japan, *Eng. Geol., 66*(1–2), 111–125.

Chorowicz, J., E. Lopez, F. Garcia, J.-F. Parrot, J.-P. Rudant, and R. Vinluan (1997), Keys to analyze active lahars from Pinatubo on SARS ERS imagery, *Remote Sens. Environ., 62*(1), 20–29.

Churchill, R. R. (1982), Aspect-induced differences in hillslope processes, *Earth Surf. Process. Landforms, 7*, 171–182.

Ciesla, W. M. (1995), Climate change, forests and forest management, an overview, *FAO Forestry Paper 126*, FAO, United Nations, Rome.

Ciolkosz, E. J., G. W. Peterson, and R. L. Cunningham (1979), Landslide-prone soils of southwestern Pennsylvania, *Soil Sci., 128*(6), 348–352.

Clark, J. S. (1989), Effects of long-term water balances on fire regime, north-western Minnesota, *J. Ecol., 77*, 989–1004.

Clark, J. S., J. Merky, and H. Moller (1989), Post-glacial fire, vegetation, and human history on the northern alpine forelands, south-western Germany, *J. Ecol., 77*, 879–925.

Clark, M. J., and M. Seppälä (1988), Slushflows in a subarctic environment, Kilpisjarvi, Finnish Lapland, *Arct. Alp. Res., 20*(1), 97–105.

Clayton, J. L. (1983), Evaluating slope stability prior to road construction, *Res. Pap. INT-307*, 6 p., For. Serv., U.S. Dep. of Agric., Ogden, UT.

Cleaves, A. B. (1961), *Landslide investigations, a field handbook for use in highway location and design*, 67 p., U.S. Dept., Commerce, U.S. Government Printing Office, Washington, DC.

Clifton, A. W., R. T. Yoshida, and R. W. Chursinoff (1986), Regina Beach – a town on a landslide, *Can. Geotech. J., 23*, 61–68.

Close, U., and E. McCormick (1922), Where the mountains walked, *National Geographic, 41*, 445–464.

Coates, D. R. (1977), Landslide perspectives, in *Reviews in Engineering Geology*, Vol. 3, Landslides, pp. 3–28, Geol. Soc. Am., Boulder, CO.

Coker, R.J., and B.D. Fahey (1993), Road-related mass movement in weathered granite, Golden Downs and Motueka Forests, New Zealand: a note, *J. Hydrol. N. Z., 31*(1), 65–69.

Coe, J. A., W. L. Ellis, J. W. Godt, W. Z. Savage, J. E. Savage, J. A. Micheal, J. D. Kibler, P. S., Powers, D. J. Lidke, and S. Debrey (2003), Seasonal movement of the Slumgullion landslide determined from Global Positioning System surveys and fields instrumentation, July 1998–March 2002, *Eng. Geol., 68*(1–2), 67–101.

Collis-George, N. (1977), Infiltration equations for simple soil systems, *Water Resour. Res., 13*(2), 395–403.

Collison, A. J., and M. G. Anderson (1996), Using a combined slope hydrology/stability model to identify suitable conditions for landslide prevention by vegetation in the humid tropics, *Earth Surf. Process. Landforms, 21*, 737–347.

Comfort, L., B. Wisner, S. Cutter, R. Pulwarty, K. Hewitt, A. Oliver-Smith, J. Wiener, M. Fordham, M. Peacock, and F. Krimgold (1999), Reframing disaster policy: the global evolution of vulnerable communities, *Environ. Health, 1*, 39–44.

Commandeur, P.R., and M. R. Pyles (1991), Modulus of elasticity and tensile strength of Douglas-fir roots, *Can. J. For. Res., 21*, 48–52.

Console, R. (2001), Testing earthquake forecast hypotheses, *Tectonophysics, 338*(3–4), 261–268.

Conway, B. W., A. Forster, K. J. Northmore, and W. J. Barclay (1980), South Wales coalfield landslips survey, *Report 80/4*, IGS Special Surveys Div., Eng. Geol. Unit.

Corbett, E. S., and R. M. Rice (1966), Soil slippage increased by brush conversion, *Res. Note PSW-128*, 8 p., For. Serv., U.S. Dep. of Agric., Berkeley, CA.

Cornforth, D. H. (2005), *Landslides in Practice: Investigation, Analysis, Remedial/Prevention Options in Soils*, 624 p., John Wiley & Sons, New York.

Costa, J.E. (1991), Nature, mechanics, and mitigation of the Val Pola landslide, Valtellina, Italy, 1987–1988, *Z. Geomorph. N.F., 35*, 15–38.

Costa-Cabral, M. C., and S. J. Burges (1994), Digital elevation model networks (DEMON): a model of flow over hillslopes for computation of contributing and dispersal area, *Water Resour. Res. 30*(6), 1681–1692.

Costin, A. B. (1950), Mass movements of the soil surface with special reference to the Monaro Region of New South Wales, *New South Wales Soil Conserv. Serv. J., 6*, 73–85.

Coussot, P., and M. Meunier (1996), Recognition, classification and mechanical description of debris flows, *Earth Sci. Rev., 40*, 209–227.

Cramb, R. A., J. N. M. Garcia, R. V. Gerrits, and G. C. Saguiguit (2000), Conservation farming projects in the Philippine uplands: Rhetoric and reality, *World Dev., 28*, 911–927.

Croft, A. R., and J. A. Adams (1950), Landslides and sedimentation in the North Fork of Ogden River, May 1949, *Res. Pap. 21*, 4 p., For. Serv., U.S. Dep. of Agric., Ogden, UT.

Crosta, G. (1998), Regionalization of rainfall thresholds: an aid to landslide hazard evaluation, *Environ. Geol., 35*(2–3), 131–145.

Crozier, M. J. (1968), Earthflows and related environmental factors of eastern Otago, *J. Hydrol. N. Z., 7*, 4–12.

Crozier, M. J. (1969), Earthflow occurrence during high intensity rainfall in eastern Otago (New Zealand), *Eng. Geol., 3*(4), 325–334.

Crozier, M. J. (1981), The character of natural hazards in different physical and social settings, *Proc. 11th New Zealand Geographical Conf.*, pp. 106–109.

Crozier, M. J. (1999), Prediction of a rainfall-triggered landslide: a test of the antecedent water status model, *Earth Surf. Process. Landforms, 24*, 825–833.

Crozier, M. J., and R. J. Eyles (1980), Assessing the probability of rapid mass movement, *N. Z. Inst. Eng. Proc. Tech. Group, 6*, 1(G), 2.47–2.51.

Crozier, M. J., R. J. Eyles, S. L. Marx, J. A. McConchie, and R. C. Owen (1980), Distribution of landslips in the Wairarapa hill country, *N. Z. J. Geol. Geophys., 23*, 575–586.

Crozier, M. J., R. Howorth, and I. J. Grant (1981), Landslide activity during Cyclone Wally, Fiji: A case study of Wainitubatolu catchment, *Pacific Viewpoint, 22*, 69–88.

Cruden, D. M., and D. J. Varnes (1996), Landslide types and processes, in *Landslides—Investigation and Mitigation*, edited by A. K. Turner and R. L. Schuster, pp. 36–75, Special Report 247, Transport. Res. Board, National Res. Council, National Academic Press, Washington, DC.

Cruden, D. M., S. Thomson, B. D. Bomhold, J.-Y. Chagnon, J. Locat, S. G. Evans, J. A. Heginbottom, K. Moran, D. J. W. Piper, R. Powell, D. Prior, and R. M. Quigley (1989) Landslides: extent and economic significance in Canada, in *Landslides: Extent and Economic Significance*, edited by E. E. Brabb and B. L. Harrod, Proc. 28th Intl. Geol. Congr., Symp. on Landslides, Wash., DC., pp. 1–23.

Curry, R. R. (1966), Observation on alpine mudflows in the Tenmile Range, central Colorado, *Geol. Soc. Am. Bull., 77*, 771–776.

Dai, F. C., and C. F. Lee (2002), Landslide characteristics and slope instability modeling using GIS, Lantau Island, Hong Kong, *Geomorphology, 42*, 213–228.

Dai, F. C., and C. F. Lee (2003), A spatiotemporal probabilistic modelling of storm-induced shallow landsliding using aerial photographs and logistic regression, *Earth Surf. Process. Landforms, 28*, 527–545.

Dai, F. C., C. F. Lee, S. J. Wang, and Y. Y. Feng (1999), Stress-strain behaviour of a loosely compacted volcanic-derived soil and its significance to rainfall-induced fill slope failure, *Eng. Geol., 53*(3–4), 359–370.

D'Amato Avanzi, G., R. Giannecchini, and A. Puccinelli (2004), The influence of the geological and geomorphological settings on shallow landslides. An example in a temperate climate environment: the June 19, 1996 event in northwestern Tuscany (Italy), *Eng. Geol., 73*, 215–228.

Davis, L. L., and L. R. West (1973), Observed effects of topography on ground motion, *Bull. Seismol. Soc. Am., 63*(1), 28–298.

DeBano, L. F. (2000), The role of fire and soil heating on water repellency in wildland environments: a review. *J. Hydrol., 231*, 195–206.

DeBano, L. F., J. F. Osborn, J. S. Krammes, and J. Letey (1967), Soil wettability and wetting agents...our current knowledge of the problem, *Res. Pap. PSW-43*, 13 p., For. Serv., U.S. Dep. of Agric., Berkeley, CA.

DeBoer, D. W. (1979), Comparison of three field methods for determining saturated hydraulic conductivity, *Trans. Am. Soc. Agr. Eng., 22*(3), 569–572.

DeGraff, J. V. (1978), Regional landslide evaluation: two Utah examples, *Environ. Geol., 2*(4), 203–214.

DeGraff, J. V. (1979), Initiation of shallow mass movement by vegetative conversion, *Geology, 7*, 426–429.

DeGraff, J. V. (1990), Landslide dams from November 1988 storm events in southern Thailand, *Landslide News (Jpn. Landslide Soc.), 4*, 12–15.

DeGraff, J. V. (1991), Determining the significance of landslide activity: examples from Eastern Caribbean, *Caribbean Geography, 3*(1), 29–42.

DeGraff, J. V., and C.G. Cunningham (1982), Highway-related landslides in mountainous volcanic terrain: an example from west-central Utah, *Bull. Assoc. Eng. Geol., 19*(4), 319–325.

DeGraff, J. V., R. Bryce, R. W. Jibson, S. Mora, and C. T. Rogers (1989), Landslides: their extent and significance in the Caribbean, in *Landslides: Extent and Economic Significance*, edited by E. E. Brabb and B. L. Harrod, pp. 51–80, A. A. Balkema, Rotterdam.

Delano, H. L., and J. P. Wilshusen (2001), Landslides in Pennsylvania, *Educational Series 9*, 34 p., Pennsylvania Geol. Surv., Harrisburg.

Dengler, L. and D.R. Montgomery (1989), Estimating the thickness of colluvial fill in unchanneled valleys from surface topography, *Bull. Assoc. Engr. Geol., 26*, 333–342.

De Ploey, J., and O. Cruz (1979), Landslides in the Serra Do Mar, Brazil, *Catena, 6*, 11–122.

Derbyshire, E., J. T. Wang, and I. J. Smalley (1990), Loess landslides and geotechnical classification, *Landslide News (Jpn. Landslide Soc.), 4*, 22–23.

Derbyshire, E., J. Wang, and X. Meng (2000), A treacherous terrain: background to natural hazards in northern China with special reference to the history of landslide in Gansu Province, in *Landslides in the Thick Loess Terrain of North-West China*, edited by E. Derbyshire, X. Meng, and T. A. Dijkstra, pp. 1–19, John Wiley & Sons.

DeRose, R. C. (1996), Relationships between slope morphology, regolith depth, and the incidence of shallow landslides in eastern Taranaki hill country, *Z. Geomorph. N. F., 105*, 49–60.

DeRose, R. C., N. A. Trustrum, and P. M. Blaschke (1993), Post-deforestation soil loss from steepland hillslopes in Taranaki, New Zealand, *Earth Surf. Process. Landforms, 18*(2), 131–144.

Dhakal, A. S., and R. C. Sidle (2003), Long-term modeling of landslides for different forest management practices, *Earth Surf. Process. Landforms 28*, 853–868.

Dhakal, A. S., and R. C. Sidle (2004a), Distributed simulations of landslides for different rainfall conditions, *Hydrol. Process., 18*, 757–776.

Dhakal, A. S., and R. C. Sidle (2004b), Pore water pressure assessment in a forest watershed: simulations and distributed field measurements related to forest practices, *Water Resour. Res., 40*, W02405, doi:1029/ 2003WR002017.

Dhakal, A. S., T. Amada, and M. Aniya (1999), Landslide hazard mapping and the application of GIS in the Kulekhani watershed Nepal, *Mountain Res. Develop., 19*, 3–16.

Dhakal, A. S., T. Amada, and M. Aniya (2000), Landslide hazard mapping and its evaluation using GIS: an investigation of sampling scheme for grid-cell based quantitative method, *Photogram. Eng. Rem. Sens., 66*, 981–989.

Dhakal, A. S., T. Amada, M. Aniya, and R. R. Sharma (2002), Detection of areas associated with flood and erosion caused by a heavy rainfall using multitemporal landsat TM data, *Photogram. Eng. Rem. Sens., 68*(3), 233–239.

Diemont, W. H., A. C. Smiet, and Nurdin (1991), Rethinking erosion on Java, *Netherlands J. Agric. Sci., 39*, 213–224.

Dietrich, W. E., and T. Dunne (1978), Sediment budget for a small catchment in mountainous terrain, *Z. Geomorph. N. F., Suppl. Bd. 29*, 191–206.

Dietrich, W. E., and D. R. Montgomery (1998), SHALSTAB: a digital terrain model for mapping shallow landslide potential, *National Council of the Paper Industry for Air and Stream Improvement Technical Report*, February 1998, 29 p.

Dietrich, W. E., D. Bullugi, and R. Real de Asua (2001), Validation of the shallow landslide model, SHALSTAB, for forest management, in *Land Use and Watersheds, Human Influences on Hydrology and Geomorphology in Urban and Forest Areas*, edited by M. S. Wigmosta and S. J. Burges, pp. 195–227, Water Science and Application 2, Am. Geophys. Union, Washington, DC.

Dietrich, W. E., T. Dunne, N. F. Humphrey, and L. M. Reid (1982), Construction of sediment budgets for drainage basins, in Sediment Budgets and Routing in Forested Drainage Basins, *Gen. Tech. Rep. PNW-141*, pp. 5–23, For. Serv., U.S. Dep. of Agric., Portland, OR.

Dikau, R., D. Brunsden, M. L. Ibsen, and L. Schroott (1996) (Eds.), *Landslide Recognition: Identification, Movement and Causes*, 251 p., John Wiley & Sons, Chichester, UK.

Dingman, S. L. (1994), *Physical Hydrology*, 575 p., Macmillan, New York.

Dobrovolny, E. (1971), Landslide susceptibility in and near Anchorage as interpreted from topographic and geologic maps, in *The Great Alaska Earthquake of 1964: Biology*, pp. 735–745, National Academy of Science, Washington, DC.

Donnelly, L. J., H. De La Cruz, I. Asmar, O. Zapata, and J. D. Perez (2001), The monitoring and prediction of mining subsidence in the Amaga, Angelopos, Venicia, and Bolombolo Regions, Antioquia, Colombia, *Eng., Geol., 59*, 103–114.

Donoghue, D. (1986), Review and analysis of slow mass movement mechanisms with reference to a Weardale catchment, N. England, *Z. Geomorph. N. F., 60*, 41–54.

Döös, B. R. (2000), Increasing food production at the expense of tropical forests, *Integrated Assessment, 1*, 189–202.

Douglas, I., K. Bidin, G. Balamurugan, N. A. Chappell, R. P. D. Walsh, T. Greer, and W. Sinun, (1999), The role of extreme events in the impacts of selective tropical forestry on erosion during harvesting and recovery phases at Danum Valley, Sabah, *Phil. Trans. Royal Soc. London, 354*, 1749–1761.

Duman, T. Y., T. Çan, Ö. Emre, M. Keçer, A. Doğan, Ş. Ateş, and S. Durmaz (2005), Landslide inventory of northwestern Anatolia, Turkey, *Eng. Geol., 77*, 99–114.

Duncan, J. M. (1996), Soil slope stability analysis, in *Landslides—Investigation and Mitigation*, edited by A. K. Turner and R. L. Schuster, pp. 337–371, Special Report 247, Transport. Res. Board, National Res. Council, National Academic Press, Washington, DC.

Dunne, T. (1991), Stochastic aspects of the relations between climate, hydrology, and landform evolution. *Trans. Jpn. Geomorph. Union, 12*, 1–24.

Dunne, T. (1998), Critical data requirements for prediction of erosion and sedimentation in mountain drainage basins, *J. Am. Water Resour. Assoc., 34*, 795–808.

Dunne, T., and R. D. Black (1970), Partial area contributions to storm runoff in a small New England watershed, *Water Resour. Res., 6*, 1296–1311.

Durgin, P. B. (1977), Landslides and weathering of granitic rocks, in *Reviews in Engineering Geology*, vol. 3, Landslides, pp. 127–131, Geol. Soc. Am., Boulder, CO.

Duzgoren-Aydin, N. S., A. Aydin, and J. Malpas (2002), Distribution of clay minerals along a weathered pyroclastic profile, Hong Kong, *Catena, 50*, 17–40.

Dyrness, C. T. (1965), Soil surface condition following tractor and high-lead logging in the Oregon Cascades, *J. Forestry 63*, 272–275.

Dyrness, C. T. (1967), Mass soil movements in the H.J. Andrews Experimental Forest, *Res. Pap. PNW-42*, 12 p., For. Serv., U.S. Dep. of Agric., Portland, OR.

Easterbrook, D. J. (1999), *Surface Processes and Landforms*, 2nd ed., 546 p., Prentice Hall, New Jersey.

Eden, W. J., and R. J. Mitchell (1970), The mechanics of landslides in Leda clay, *Can. Geotech. J., 7*, 285–296.

Edil, T. B., and L. E. Vallejo (1976), Shoreline erosion and landslides in the Great Lakes, Sea Grant Advisory Board, *Report No. 15*, 7 p., Univ. of Wisconsin, Madison.

Edwards, W. M., M. J., Shipitalo, L. B. Owens, and L. D. Norton (1988), Contribution of macroporosity to infiltration into a continuous corn no-tilled watershed: implications for contaminant movement, *J. Contam. Hydrol., 3*, 193–205.

Egashira, K., and S. Gibo (1988), Colloid-chemical and mineralogical differences of smectites taken from argillized layers, both from within and outside the slipsurfaces in the Kamenose landslide, *Appl. Clay Sci., 3*, 253–262.

Eis, S. (1974), Root system morphology of western hemlock, western red cedar, and Douglas-fir, *Can. J. For. Res.*, 4, 28–38.

Eisbacher, G. H. (1982), Slope stability and land use in mountain valleys, *Geoscience Canada, 9*, 14–27.

Eisbacher, G. H., and J. J. Clague (1984), Destructive mass movements in high mountains: hazard and management, *Pap. 84-16*, 230 p., Geol. Surv. of Canada.

Ekanayake, J. C., and C. J. Phillips (1999a), A model for determining thresholds for initiation of shallow landslides under near-saturated conditions in the East Coast region, New Zealand, *J. Hydrol. N. Z., 38*, 1–28.

Ekanayake, J. C., and C. J. Phillips (1999b), A method for stability analysis of vegetated hillslopes: an energy approach, *Can. Geotech. J., 36*, 1172–1184.

Ekanayake, J. C., and C. J. Phillips (2002), Slope stability thresholds for vegetated hillslopes: a composite model, *Can. Geotech. J., 39*, 849–862.

Ekanayake, J. C., M. Marden, A. Watson, and D. Rowan (1997), Tree roots and slope stability: a comparison between radiata pine and kānuka, *N. Z. J. For. Sci., 27*(2): 216–233.

El Khattabi, J., and E. Carlier (2004), Tectonic and hydrodynamic control of landslides in the northern area of the Central Rif, Morocco, *Eng. Geol., 71*(3–4), 255–264.

Ellen, S. D., and R. W. Fleming (1987), Mobilization of debris flows from soil slips, San Francisco Bay region, California, in *Debris Flows/Avalanches: Process, Recognition and Mitigation*, edited by J. E. Costa and G. F. Wieczorek, pp. 31–40, Reviews in Engineering Geology, 7, Geological Society of America.

Elmhirst, R. (1999), Space, identity politics and resource control in Indonesia's transmigration programme, *Polit. Geogr., 18*, 813–835.

Elsenbeer, H., and R. A. Vertessy (2000), Stormflow generation and flowpath characteristics in an Amazonian rainforest catchment, *Hydrol. Process., 14*, 2367–2381.

Elsenbeer, H., A. Lack, and K. Cassel (1995), Chemical fingerprints of hydrological compartments of flow paths at La Cuence, western Amazonia, *Water Resour. Res., 31*, 3051–3058.

Endo, T. (1969), Probable distribution of the amount of rainfall causing landslides, *Annual Report 1968*, pp. 122–136, Hokkaido Branch, For. Exp. Stn., Sapporo, Japan.

Endo, T., and T. Tsuruta (1969), The effect of the tree's roots on the shear strength of soil, *Annual Report, 1968*, pp.167–182, Hokkaido Branch, For. Exp. Stn., Sapporo, Japan.

Enoki, A. (1993), Comparison between graphical and other methods for investigating slope stability, *Symp. on Landslide Mechanisms and Measures*, Kansai Branch of the Jpn. Geotech. Soc. (in Japanese).

EPOCH (1993) (European Community Programme), Temporal occurrences and forecasting of landslides in the European community, edited by J.-C. Flageollet, 3 vols., Contract no. 90 0025.

Erley, D., and W. J. Kockelman (1981), Reducing landslide hazards: A guide for planners, *Plan. Advisory Serv. Rep., 359*, 29 p., Am. Plan. Assoc., Chicago, Ill.

Eslinger, E., P. Highsmith, D. Albers, and B. Demayo (1979), Role of iron reduction in the conversion of smectite to illite in bentonites in the disturbed belt, Montana, *Clay and Clay Minerals, 27*(5), 327–338.

Evans, S. G., and J. J. Clague (1994), Recent climate change and catastrophic geomorphic processes in mountain environments, *Geomorphology, 10*, 107–128.

Evans, S. G., R. Couture, and E. L. Raymond (2002), Catastrophic landslides and related processes in the southeastern Cordillera: analysis of impact on lifelines and communities, 66 p., Geol. Surv. Canada, Natural Resources Canada.

Eyles, R. J. (1971), Mass movement in Tangoio conservation reserve, northern Hawkes Bay, *Earth Sci. J., 5*(2), 79–91.

Eyles, R. J. (1979), Slip-triggering rainfalls in Wellington City, New Zealand, *N.Z. J. Sci., 22*, 117–121.

Eyles, R. J., M. J. Crozier, and R. H. Wheeler (1978), Landslips in Wellington City, *N.Z. Geogr., 34*(2), 58–74.

Fagents, S. A., and S. M., Bologa (2005), Calculation of lahar transit times using digital elevation data, *J. Volcanol. Geotherm. Res., 139*(1–2), 135–146.

Fairbairn, W. A. (1967), Erosion in the River Findhorn Valley, *Scot. Geogr. Mag., 83*(1), 46–52.

Fannin, R. J., and T. P. Rollerson (1993), Debris flows: some physical characteristics and behaviour, *Can. Geotech. J. 30*, 71–81.

Fannin, R. J., J. Jaakkola, J. M. T. Wilkinson, and E. D. Heatherington (2000), The hydrologic response of soils to precipitation at Carnation Creek, British Columbia, *Water Resour. Res., 36*, 1481–1494.

Farmer, E. E. (1980), Phosphate mine dump hydrology, in *Symposium on Watershed Management*, Vol. II, pp. 846–853, Am. Soc. Civil Eng., New York.

Farmer, E. E., and J. G. Peterson (1985), Engineering and hydrology practices in the southeastern Idaho phosphate field, Proc. of Workshop on Engineering and Hydrology Research Needs for Phosphate Mined Lands in Idaho, *Tech. Rep. INT-192*, pp. 1–11. For. Serv., U.S. Dep. of Agric., Ogden, UT.

Fell, R. (1994), Landslide risk assessment and acceptable risk, *Can. Geotech. J., 31*, 261–272.

Fernandes, N. F., R. F. Guimarães, R. A. T. Gomes, B. C. Vieira, D. R. Montgomery, and H. Greenberg (2004), Topographic controls of landslides in Rio de Janeiro: field evidence and modeling, *Catena, 55*, 163–181.

Fernandes, N. F., A. L. C. Netto, and W. A. Lacerda (1994), Subsurface hydrology of layered colluvium mantles in unchanneled valleys—south-eastern Brazil, *Earth Surf. Process. Landforms, 19*, 609–626.

Fieldes, M. (1955), Clay mineralogy of New Zealand soils, 2. Allophane and related mineral colloids, *N. Z. J. Sci. Technol., B37*, 336–350, 1955.

Fiksdal, A. J. (1974), A landslide survey of the Stequaleho Creek watershed, Supplement to final rep. *FRI-UW-7404*, 7 p., Fisheries Res. Inst., Univ. of Washington, Seattle.

Finlay P. J., R. Fell, and P. K. Meguire (1997), The relationship between the probability of landslide occurrence and rainfall, *Can. Geotech. J., 34*, 811–824.

Finlayson, B. (1981), Field measurements of soil creep, *Earth Surf. Process. Landforms, 6*, 35–48.

Fiorillo, F., and R. C. Wilson (2004), Rainfall induced debris flows in pyroclastic deposits, Campania (southern Italy), *Eng. Geol., 75*, 263–289.

Fischer, A., and L. Vasseur (2000), The crisis in shifting cultivation practices and the promise of agroforestry: a review of the Panamanian experience, *Biodiversity and Conserv., 9*, 739–756.

Flaccus, E. (1959), Revegetation of landslides in the White Mountains of New Hampshire, *Ecology, 40*, 692–703.

Fleming, R. W., and A. M. Johnson (1975), Rates of seasonal creep of silty clay soil, *Q. J. Eng. Geol., 8*, 1–29.

Fleming, R. W., and F. A. Taylor (1980), Estimating the cost of landslide damage in the United States, *Circ. 832*, 21 p., U.S. Geol. Surv.

Fleming, R. W., R. B. Johnson, and R. L. Schuster (1988), The reactivation of the Manti Landslide, Utah, in The Manti, Utah, Landslide, *Prof. Pap. 1311*, pp. 1–21, U.S. Geol. Surv.

Fletcher, J. E., K. Harris, H. B., Peterson, and V. N. Chandler (1954), Piping, *Trans. Am. Geophys. Union, 35*(2), 258–263.

Fletcher, L., O. Hungr, and S. G. Evans (2002), Contrasting failure behaviour of two large landslides in clay and silt, *Can. Geotech. J. 39*(1), 46–62.

Florsheim, J.L., E.A. Keller, and D.W. Best (1991), Fluvial sediment transport in response to moderate storm flows following chaparral wildfire, Ventura County, southern California, *Geol. Soc. Am. Bull. 103*, 504–511.

Fox, J., D. M. Truong, A. T. Rambo, N. P. Tuyen, L. T. Cuc, and S. Leisz (2000), Shifting cultivation: a new old paradigm for managing tropical forests, *BioScience, 50*, 521–528.

Fransen, P., and R. Brownlie (1995), Historical slip erosion in catchments under pasture and radiate pine forest, Hawke's Bay hill country, *N.Z. Forestry, 40*, 29–33.

Fransen, P. J. B., C. J. Phillips, and B. D. Fahey (2001), Forest road erosion in New Zealand: overview, *Earth Surf. Process. Landforms, 26*(2), 165–174.

Fredlund, D. G., and A. E. Dahlman (1972), *Statistical Geotechnical Properties of Glacial Lake Edmonton sediments*, Statistics and Probability in Civil Engineering, Oxford University Press, UK.

Fredlund, D. G., and H. Rahardjo (1993), *Soil Mechanics for Unsaturated Soils*, 517 p., John Wiley & Sons, Inc., New York.

Fredlund, D. G., N. R. Morgenstern, and R. A. Widger (1978), The shear strength of unsaturated soils, *Can. Geotech. J., 15*, 313–321.

Freeze, R. A. (1972), Role of subsurface flow in generating surface runoff, 2. Upstream source areas, *Water Resour. Res., 8*(5), 1272–1283.

Freeze, R. A. (1974), Streamflow generation, *Rev. Geophys. Space Phys., 12*(4), 627–647.

Freeze, R. A. (1978), Mathematical models of hillslope hydrology, in *Hillslope Hydrology*, edited by M. J. Kirkby, pp. 177–225, John Wiley, New York.

Froehlich, H. A. (1973), Natural and man-caused slash in headwater streams, *Loggers Handbook, 33*, Pacific Logging Congress, Portland, OR.

Froehlich, H. A. (1978), The influence of clearcutting and road building activities on landscape stability in western United States, *Proc. 5th North American Forest Soils Conf.*, pp. 165–173, Colorado State Univ., Fort Collins.

Froehlich, W., and L. Starkel (1993), The effects of deforestation on slope and channel evolution in the tectonically active Darjeeling Himalaya, *Earth Surf. Process. Landforms, 18*, 285–290.

Froehlich, W., E. Gil, I. Kasza, and L. Starkel (1990), Thresholds in the transformation of slopes and river channels in the Darjeeling Himalaya, India, *Mountain Res. and Development, 10*(4), 301–312.

Fuchu, D., C. F. Lee, and W. Sijing (1999), Analysis of rainstorm-induced slide-debris flows on the natural terrain of Lantau Island, Hong Kong, *Eng. Geol., 51*, 279–290.

Fujiwara, K. (1970), A study on the landslides by aerial photographs, *Res. Bull. Exp. Forest Hokkaido Univ., 27*(2), 297–345.

Fukuda, F., and H. Ochiai (1993), Landslides caused by the 1993, Hokkaido Nansei-oki earthquake, *Shin Sabo J. (J. Jpn. Soc. Erosion Contr. Eng.), 46*, 62–63 (in Japanese).

Fukuzono, T. (1994), Measurement of ground motion on a partly corrupted slope due to liquefaction, *Proc. 33rd Ann. Mtg. Japan Landslide Soc.*, pp. 335–336 (in Japanese).

Fukuzono, T., H. Moriwaki, T. Inokuchi, T. Maki, K. Iwanami, R. Misumi, T. Morohoshi, S. Takami, and T. Shikoku (2003), Study on landslide disaster prediction support system, *Proc. 22nd Annual Mtg. Japan Soc. for Natural Disaster Sci.*, pp. 215–216 (in Japanese).

Furbish, D. J., and S. Fagherazzi (2001), Stability of creeping soil and implications for hillslope evaluation, *Water Resour. Res., 37*(10), 2607–2618.

Furuya, G., K. Sassa, H. Hiura, and H. Fukuoka (1999), Mechanism of creep movement caused by landslide activity and underground erosion in crystalline schist, Shikoku Island, southwestern Japan, *Eng. Geol., 53*, 311–325.

Furuya, G., A. Suemine, N. Osanai, H. Hiura, H. Marui, and O. Sato (2002), Estimation of veins of groundwater related to slope failures by the investigation of soil and groundwater temperature in the landslide mass at Zentoku, Shikoku Island, southern Japan, *Interprevent Congress Publ., vol. 2*, pp. 515–523.

Gabet, E.J., and T. Dunne (2002), Landslides on coastal sage-scrub and grassland hillslopes in a severe El Niño winter: The effects of vegetation conversion on sediment delivery. *Geol. Soc. Am. Bull., 114*, 983–990.

Gage, M., and R. D. Black (1979), Slope stability and geological investigations at Mangatu State Forest, *Tech. Pap. 66*, 37 p., N. Z. For. Serv., For. Res. Inst., Wellington, NZ.

Gao, J. (1993), Identification of topographic settings conductive to landsliding from DEM in Nelson County, *Earth Surf. Process. Landforms, 18*, 579–591.

Gardner, R. B. (1979), Some environmental and economic effects of alternative forest road designs. *Trans. Am. Soc. Agric. Eng., 22*, 63–68.

Gardner, R. B., W. S. Hartsog, and K. B. Dye (1978), Road design guidelines for the Idaho Batholith based on the China Glenn Road study, *Res. Pap. INT-204*, 20 p., For. Serv., U.S. Dep. of Agric., Ogden, UT.

Garrett, J. (1980), Catchment authority work in the Rangitikei area, *Aokautere Sci. Ctr. Int. Rep. 21*, 23–26, N. Z. Ministry of Works and Develop., NZ.

Gautam, M. R., K. Watanabe, and H. Saegusa (2000), Runoff analysis in humid forest catchments with artificial network, *J. Hydrol., 235*, 117–136.

Gedney, D. S., and W. G. Weber (1978), Design and construction of soil slopes, in Landslides: Analysis and Control, *Spec. Rep. 176*, pp. 172–192, Transport. Res. Board, National Academy of Sciences, Nat. Res. Council, Washington, DC.

Geertsema, M., and J. W. Schwab (1996), A photographic overview and record of the Mink Creek earthflow, Terrace, British Columbia, *Research Report 08*, 21 p., Ministry of Forests, British Columbia, Canada.

Germann F., and L. Di Pietro (1999), Scales and dimensions of momentum dissipation during preferential flow in soils, *Water Resour. Res, 35*(5), 1443–1454.

Gerrard, A. J., and R. A. M. Gardner (2000), The role of landsliding in shaping the landscape of the Middle Hills, Nepal, *Z. Geomorph. N. F., 122*, 47–62.

Ghose, M. K. (2003), Indian small-scale mining with special emphasis on environmental management, *J. Clean. Prod., 11*, 159–165.

Giraud, A., L. Rochet, and P. Antoine (1990), Processes of slope failure in crystallophyllian formations, *Eng. Geol., 29*, 241–253.

Glade, T. (1998), Establishing the frequency and magnitude of landslide-triggering rainstorm events in New Zealand, *Environ. Geol., 35*(2–3), 160–174.

Glade, T. (2000), Modelling landslide-triggering rainfalls in different regions of New Zealand—the soil water status model, *Z. Geomorph. N. F., 122*, 63–84.

Glade, T., M. Crozier, and P. Smith (2000), Applying probability determination to refine landslide-triggering rainfall thresholds using and empirical "antecedent daily rainfall model", *Pure and Applied Geophysics, 157* (6–8), 1059–1079.

Glassey, P., D. Barrell, J. Forsyth, and R. Macleod (2003), The geology of Dunedin, New Zealand, and the management of geological hazards, *Quaternary Int., 103*, 23–40.

Golding, D. L., and R. H. Swanson (1986), Snow distribution patterns in clearings and adjacent forest. *Water Resour. Res. 22*(13): 1931–1940.

Gomi, T., and R. C. Sidle (2003), Bed load transport in managed steep-gradient headwater streams of southeastern Alaska, *Water Resour. Res., 39*(12), 1336, doi: 10.1029/2003/WR002440.

Gomi, T., R. C. Sidle, M. D. Bryant, and R. D. Woodsmith (2001), The characteristics of woody debris and sediment distribution in headwater streams, southeastern Alaska, *Can. J. For. Res., 31*, 1386–1399.

Gomi, T., R. C. Sidle, and J. S. Richardson (2002), Understanding processes and downstream linkages of headwater systems, *BioScience, 52*, 905–916.

Gomi, T., R.C. Sidle, R.D. Woodsmith, and M.D. Bryant (2003), Characteristics of channel steps and reach morphology in headwater streams, southeast Alaska, *Geomorphology, 51*(1–3), 225–242.

Gomi, T., R. C. Sidle, and D. N. Swanston (2004), Hydrogeomorphic linkages of sediment transport in headwater streams, Maybeso Experimental Forest, southeast Alaska, *Hydrol. Process., 18*, 667–683.

Gonsior, M. J., and R. B. Gardner (1971), Investigation of slope failures in the Idaho Batholith, *Res. Pap. INT-97*, 34 p., For. Serv., U.S. Dep. of Agric., Ogden, Utah.

Gostelow, T. P., M. Del Prete, and A. Simoni (1997), Slope instability in historic hilltop towns of Basilicata, southern Italy, *Q. J. Eng. Geol., 30*, 3–26.

Gokceoglu, C., H. Sonmez, H. A. Nefeslioglu, T. Y. Duman, and T. Can (2005), The 17 March 2005 Kuzulu landslide (Sivas, Turkey) and landslide-susceptibility map of its near vicinity, *Eng. Geol., 81*, 65–83.

Govi, M. (1999) The 1987 landslide on Mount Zandila in the Valtellina, northern Italy, in *Landslides of the World*, edited by K. Sassa, pp. 47–50, Kyoto University Press, Japan.

Grange, L. I., and H. S. Gibbs (1947), Soil erosion in New Zealand, 1. Southern half of North Island, N. Z., *Soil Bureau Bull. No. 1*, 22 p.

Grantz, A., G. Plafker, and R. Kachadoorian (1964), Alaska's Good Friday earthquake, March 27, 1964: a preliminary geologic evaluation, *U.S. Geol. Surv. Circ., 491*, 35 p.

Gray, D. H. (1978), Role of woody vegetation in reinforcing soils and stabilizing slopes, *Proc. Soil Reinforce. Stab. Tech. Eng. Pract.*, pp. 253–306, NSW Inst. Tech. Sydney, Australia.

Gray, D. H., and A. T. Leiser (1982), *Biotechnical Slope Protection and Erosion Control*, 271 p., Van Nostrand Reinhold, New York.

Gray, D. H., and W. F. Megahan (1981), Forest vegetation removal and slope stability in the Idaho Batholith, *Res. Pap. INT-271*, 23 p., For. Serv., U.S. Dep. of Agric., Ogden, Utah.

Gray, D. H., and H. Ohashi (1983), Mechanics of fiber reinforcement in sand, *J. Geotech. Eng. Am. Soc. Civil Eng., 109*(3), 335–353.

Green, R. D., and G. P. Askew (1965), Observation of biological development of macropores in soils of Romney Marsh, *J. Soil Sci., 16*(2), 342.

Green, W. H., and G. A. Ampt (1911), Studies in soil physics I. The flow of air and water through soils, *J. Agric. Sci., 4*, 1–24.

Greenway, D. R. (1987), Vegetation and slope stability, in *Slope Stability, Geotechnical Engineering and Geomorphology*, edited by M. G. Anderson and K. S. Richards, pp. 187–230, John Wiley & Sons, Chichester, UK.

Gregory, K. J., and D. E. Walling (1973), *Drainage Basin Form and Processes*, 456 p., John Wiley, New York.

Gresswell, S., D. Heller, and D. N. Swanston (1979), Mass movement response to forest management in the central Oregon Coast Ranges, *Resour. Bull. PNW-84*, 26 p., For. Serv., U.S. Dep. of Agric., Portland, OR.

Griffith, D. W., and L. R. West (1979), The effect of Appalachian Mountain topography on seismic waves, *Bull. Seismol. Soc. Am., 69*(1), 1018–1105.

Grim, R. E. (1962), *Applied Clay Mineralogy*, 422 p., McGraw-Hill, New York.

Grisak, G. E., and J. A. Cherry (1975), Hydrologic characteristics and response of fractured till and clay confining a shallow aquifer, *Can. Geotech., J., 12*, 23–43.

Grizzel, J. D., and N. Wolff (1998), Occurrence of windthrow in forest buffer strips and its effect on small streams in Northwest Washington, *Northwest Sci., 72*, 214–223.

Gross, M., C. Ioannou, and D. R. Ralston (1979), Inventory and classification of abandoned mine tailings, *Proc. 17th Ann. Eng. Geol. and Soils Eng. Symp.*, pp. 303–316, Univ. of Idaho, Moscow.

Guillande, R., P. Gelugne, J.-M. Bardintzeff, R. Brousse, J. Chorowicz, B. Deffontaines, and J.-F. Parrott (1995), Automated mapping of the landslide hazard on the island of Tahiti based on digital satellite data, *Map. Sci. Remote Sens., 32*(1), 59–70.

Gupta, R. P., and B. C. Joshi (1990), Landslide hazard zoning using the GIS approach—a case study from the Ramganga Catchment, Himalayas, *Eng. Geol., 28*, 119–131.

Gupta, V., N. S. Virdi (2000), On the connection between landslides and nickpoints along the Satluj River course, Higher Himalaya, India. *Z. Geomorph. N.F., 122*, 141–148.

Gutenberg, B., and C. F. Richter (1954), *Seismicity of the Earth and Associated Phenomena*, 2nd ed., 273 p., Princeton University Press, Princeton, N.J.

Guthrie, R. H. (2002), The effects of logging on frequency and distribution of landslides in three watersheds on Vancouver Island, British Columbia, *Geomorphology, 43*(3/4), 273–292.

Gysi, H. (1998), Orographic influence of the distribution of accumulated rainfall with different wind directions, *Atmosph. Res., 47–48* (1), 615–633.

Hack, J. T., and J. C. Goodlett (1960), Geomorphology and forest ecology of a mountain region in the central Appalachians, *U.S. Geol. Surv. Prof. Pap., 374*, 66 p.

Haeberli, W., and C. Burn (2002), Natural hazards in forests: glacier and permafrost effects as related to climate change, *Environmental Change and Geomorphic Hazards in Forests*, edited by R. C. Sidle, pp. 167–202., IUFRO Research Series, No. 9, CABI Publishing, Wallingford, Oxon, UK.

Hagans, D. K., W. E. Weaver, and M. A. Madej (1986), Long term on-site and off-site effects of logging and erosion in the Redwood Creek basin, Northern California, Paper presented at the Am. Geophys. Union Mtg. on Cumulative Effects, *Tech. Bull. No. 0490*, National Council for Air and Stream Improvement, Research Triangle Park, NC.

Haigh, M. J. (1984), Landslide prediction and highway maintenance in the Lesser Himalayas, India, *Z. Geomorph. N.F., 51,* 17–37.

Haigh, M. J. (1985), Geomorphic evolution of Oklahoma roadcuts, *Z. Geomorph. N.F., 29*(4), 439–452.

Haigh, M. J., L. Jansky, and J. Hellin (2004), Headwater deforestation: a challenge for environmental management, *Global Environ. Change, 14* (Suppl. 1), 51–61.

Hairiah, K. (1999). Decomposition processes and activity of soil engineers when forests are converted into agricultural use, Proc. Management of Agrobiodiversity for Sustainable Land Use and Global Environmental Benefits, *ASB-Indonesia Rep. No. 9*, pp. 53–61, Agency for Agric. Res. & Develop., Bogar, Indonesia.

Haldemann, E. G. (1956), Recent landslide phenomena in the Rungwe volcanic area, Tanganyika, *Tanganyika Notes Rec., 45,* 3–14.

Hallett, P. D., A. R. Dexter, and J. P. K. Seville (1995), The application of fracture mechanics to crack propagation in dry soil, *Euro. J. Soil Sci., 46,* 591–599.

Hammond, C., D. E. Hall, S. Miller, and P. Swetik (1992), Level I Stability Analysis (LISA), Documentation for version 2.0, *Gen. Tech. Rep. INT-285*, 190 p., For. Serv., U.S. Dep. of Agric., Ogden, UT,

Hancox, G. T., N. D. Perrin, and G. D. Dellow (2002), Recent studies of historical earthquake-induced landsliding, ground damage and MM intensity in New Zealand, *Bull. New Zealand Soc. Earthquake Eng., 35*(2), 59–94.

Haneberg, W. C. (1991a), Pore pressure diffusion and the hydrologic response of nearly-saturated, thin landslide deposits to rainfall, *J. Geol., 99,* 886–892.

Haneberg, W. C. (1991b), Observation and analysis of pore pressure fluctuations in a thin colluvium landslide complex near Cincinnati, Ohio, *Eng. Geol., 31,* 159–184.

Haneberg, W. C. (2004), A rational probabilistic method for spatially distributed landslide hazard assessment, *Envir. & Eng. Geoscience, 10,* 27–43.

Haneberg, W. C., and A. Ö. Gökce (1994), Rapid water-level fluctuations in a thin colluvium landslide west of Cincinnati, Ohio, U.S., *Bulletin 2059-C,* 16 p., U.S. Geol. Surv.

Hansen, A. (1984), Landslide hazard analysis, in *Slope Instability*, edited by D. Brunsden and D. P. Prior, pp. 523–592, John Wiley & Sons, Chichester, UK.

Hansen, W. R. (1966), Effects of the earthquake of March 27, 1964, at Anchorage, Alaska. *Prof. Pap. 542-A*, 68 p., U.S. Geol. Surv.

Harden, D. R., H. M. Kelsey, S. D. Morrison, and T. A. Stephens (1981), Geologic map of the Redwood Creek drainage basin, Humboldt County, California, scale 1:62,500, *Open-File Report 81–496*, U.S. Geol. Surv.

Hardenbicker, U., and J. Grunert (2001), Temporal occurrence of mass movements in the Bonn area, *Z. Geomorph. N. F., 125*, 13–24.

Harestad, A. S., and F. L. Bunnell (1981), Prediction of snow-water equivalents in coniferous forests, *Can. J. For. Res., 11*, 854–857.

Harmon, M. E., J. F. Franklin, F. J. Swanson, P. Sollins, S. V. Gregory, J. D. Lattin, N. H. Anderson, S. P. Cline, N. G. Aumen, J. R. Sedell, G. W. Lienkaemper, K. Cromack, and K. W. Cummins (1986), Ecology of coarse woody debris in temperate ecosystems, *Adv. Ecol. Res., 15*, 133–302.

Harp, E. L., and R.W. Jibson (1996), Landslides triggered by the 1994 Northridge, California, earthquake, *Bull. Seismol. Soc. Am., 86*, S319–S332.

Harp, E. L., W. G. Wells II, and J. G. Sarmiento (1990), Pore pressure response during failure in soils, *Geol. Soc. Am. Bull., 102*(4), 428–438.

Harp, E. L., R. C. Wilson, and G. F. Wieczorek (1981), Landslides from the February 4, 1976, Guatemala earthquake, *Prof. Pap. 1204-A*, 35 p., U.S. Geol. Surv.

Harper, S. B. (1993), Use of approximate mobility index to identify areas susceptible to landsliding by rapid mobilization to debris flows in southern Thailand, *J. Southeast Asian Earth Sci., 8*(1–4), 587–596.

Harr, R. D. (1977), Water flux in soil and subsoil on a steep forested slope, *J. Hydrol., 33*, 37–58.

Harris, C., and A. G. Lewkowicz (1993), Form and internal structure of active-layer detachment slides, Fosheim Peninsula, Ellesmere Island, Northwest Territories, Canada, *Can. J. Earth Sci., 30*, 1708–1714.

Harris, C., M. C. R. Davies, and J.-P. Coutard (1997), Rates and processes of periglacial solifluction: an experimental approach, *Earth Surf. Process. Landforms, 22*, 849–868.

Haruyama, M. (1974), Features of slope-movements due to heavy rainfalls in the Shirasu region of southern Kyushu, *Memoirs of Faculty of Agriculture (Kagoshima Univ.), 10*, pp. 151–163, Kagoshima, Japan.

Harvey, A. M. (1991), The influence of sediment supply on channel morphology of upland streams, *Earth Surf. Process. Landforms, 16*, 675–884.

Harwood, R. R. (1996), Development pathways toward sustainable systems following slash-and-burn, *Agriculture, Ecosystems & Environment 58*(1), 75–86.

Hashino, M., and F. Sasaki (1990), A stochastic forecasting model of landslides due to heavy storms and confidence intervals of forecasts, *J. Water Eng., 34*, 385–390.

Haugerud, R. A., D. J. Harding, S. T. Johnson, J. L. Harless, C. S. Weaver, and B. L. Sherrod (2003), High-resolution lidar topography of the Puget Lowland, Washington—a bonanza for earth science, *GSA Today, 13*(6), 4–10.

Havenith, H.–B., D. Jongmans, K. Abdrakhmatov, P. Trefois, D. Delvaux, and I. A. Torgoev (2000), Geophysical investigations of seismically induced surface effects: case study of a landslide in the Suusamyr valley, Kyrgyzstan, *Surv. Geophys., 21*, 349–369.

Hawley, J. G. (1980), Introduction, in Workshop on the influence of soil slip erosion on hill country pastoral productivity, *Intern. Rep. 21*, pp. 4–6, Aokautere Sci. Ctr., N. Z. Ministry of Works and Dev.

Hawley, J. G. (1981), Summary comments, *J. Hydrol. N. Z., 20*(1), 118–120.

Hawley, J. G., and J. R. Dymond (1988), How much do trees reduce landsliding?, *J. Soil Water Conserv., 43*, 495–498.

Heatherington, E. D. (1987), The importance of forests in hydrological regime, in *Canadian Aquatic Resources*, edited by M. C. Healey and R. R. Wallace, pp. 179–209, Can. Bull. Fish. Aquatic Sci., 215, Dept. Fish. and Oceans, Ottawa.

Heede, B. H., M. D. Harvey, and J. R. Laird (1988), Sediment delivery linkages in a chaparral watershed following a wildfire, *Environ. Manage., 12*, 349–358.

Heim, A. (1932), Bergsturz and Menschenleben, *Naturf. Gesell., 77*, 218 p.

Heimsath, A. M., W. E. Dietrich, K. Nishiizumi, and R. C. Finkel (1997), The soil production function and landscape equilibrium, *Nature, 338*, 358–361.

Heimsath, A. M., W. E. Dietrich, K. Nishiizumi, and R. C. Finkel (2001), Stochastic processes of soil production and transport: erosion rates, topographic variation and cosmogenic nuclides in the Oregon coast range, *Earth Surf. Process. Landforms, 26*, 531–552.

Helvey, J. D. (1980), Effects of a north central Washington wildfire on runoff and sediment production, *Water Resour. Bull., 16*(4), 627–634.

Herbert, D. W. M., and J. C. Merkens (1961), The effect of suspended mineral solids on the survival of trout, *Int. J. Air Water Pollut., 5*, 46–55.

Hewitt, K. (1998), Catastrophic landslides and their effects on the Upper Indus streams, Karakoram Himalaya, northern Pakistan, *Geomorphology 26*(1–3), 47–80.

Hewlett, J. D., and A. R. Hibbert (1963), Moisture and energy conditions within a sloping soil mass during drainage, *J. Geophys. Res., 68*, 1081–1087.

Hicks, B. G., and R. D. Smith (1981), Management of steeplands impacts by landslide hazard zonation and risk evaluation, *J. Hydrol. N. Z., 20*, 63–70.

Hicks, D. L. (1991), Erosion under pasture, pine plantations, scrub and indigenous forest: a comparison from Cyclone Bola, *N.Z. Forestry, 36*(3), 21–22.

Hicock, S. R., and A. Dreimanis (1985), Glaciotectonic structures as useful ice-movement indicators in glacial deposits: four Canadian case studies, *Can. J. Earth Sci., 22*, 339–345.

Highland, L. M. (2003), An account of preliminary landslide damage and losses resulting from the February 24, 2001, Nisqually, Washington, earthquake, *Open-file report 03-211*, U.S. Geol. Surv.

Hinch, L. W., (1986), The location and geotechnical design of roads in mountainous terrain, *Proc. Sino-British Highways and Urban Traffic Conf.*, pp. 87–96, Beijing, Nov. 17–22, 1986.

Hinton, J. J., M. M. Veiga, and A. T. C. Veiga (2003), Clean artisanal gold mining: a utopian approach?, *J. Clean. Prod., 11*, 99–105.

Hoffman, G. J., G. O. Schwab, and R. B. Curry (1964), Slope stability of coal strip mine spoil banks, Final Rep. on State Proj. 231, *A. E. Dep. Ser. 8*, 24 p., Ohio Agric. Exp. Stn., Wooster, Ohio.

Hogan, D. L. (1987), The influence of large organic debris on channel recovery on the Queen Charlotte Islands, British Columbia, Canada, *IAHS Publ. 165*, 343–353.

Hogan, D. L., S. A., Bird, and M. A. Hassan (1995), Spatial and temporal evolution of small coastal gravel-bed streams: the influence of forest management on channel morphology and fish habitat, *Proc. 4th Int. Workshop on Gravel Bed Rivers*, August 20–26, Gold Bar, Washington.

Holden, J., T. P. Burt, and M. Vilas (2002), Application of ground-penetrating radar to the identification of subsurface piping in blanket peat, *Earth Surf. Process. Landforms, 23*, 235–249.

Hollenbeck, K. J., and K. H. Jensen (1998), Experimental evidence of randomness and nonuniqueness in unsaturated outflow experiments designed for hydraulic parameter estimation, *Water Resour. Res, 34*, 595–602.

Holtz, R. D., and R. L. Schuster (1996), Stabilization of soil slopes, in *Landslides – Investigation and Mitigation*, edited by A. K. Turner and R. L. Schuster, pp. 439–473, Special Report 247, Transport. Res. Board, National Res. Council, National Academic Press, Washington, DC.

Horton, R. E. (1933), The role of infiltration in the hydrological cycle, *Trans Am. Geophys. Union, 14*, 446–460.

Hough, B. K. (1951), An analysis of the effect of particle interlocking on the strength of cohesionless soils, *Am. Soc. Testing and Materials Bull. 176*.

Howes, D. E., and E. Kenk (1988), Terrain classification system for British Columbia (revised ed.), *Ministry of Environment Manual 10*, Ministry of Crown Lands, Victoria, BC, Canada.

Howes, D. E., and D. N. Swanston (1994), A technique for stability hazard assessment, in *A Guide for Management of Landslide-prone Terrain in the Pacific Northwest*, edited by S. C., Chatwin,

D. E. Howes, J. W. Schwab, and D. N. Swanston, pp. 19–89, Land Management Handbook No. 18, B.C Ministry of Forests, Crown Publications Inc., Victoria, BC, Canada.

Howorth, R., and N. Penn (1979), Survey of Viti Levu storm of 5 May 1979, *Univ. S. Pacific Inf. Bull., 12*, 3–10.

Hsu, T. L. (1954), On the geomorphic features and the recent uplifting movement of the coastal range, eastern Taiwan, *Taiwan Geol. Surv. Bull., 7*, 51–57.

Hungr, O., F. M. Salgado, and P. M. Byrne (1989), Evaluation of a three-dimensional method of slope stability analysis, *Can. Geotech. J., 26*, 679–686.

Hürlimann, M., J. Martí, and A. Ledesma (2004), Morphological and geological aspects related to large slope failures on oceanic islands. The huge La Orotava landslides on Tenerife, Canary Islands, *Geomorphology, 62*, 143–458.

Hürlimann, M., E. Turon, and J. Marti (1999), Large landslides triggered by caldera collapse events in Tenerife, Canary Islands, *Phys. Chem. Earth(A), 24*(10), 921–924.

Hutchinson, J. N. (1977), Assessment of the effectiveness of corrective measures in relation to geological conditions and types of slope movement, *Bull. Int. Assoc. Eng. Geol., 16*, 131–155.

Hutchinson, J. N. (1986), A sliding consolidation model for flow slides, *Can. Geotech. J., 23*, 115–126.

Hutchinson, J. N. (1988), Morphological and geotechnical parameters of landslides related to geology and hydrology, general report, in *Landslides*, edited by C. Bonnard, Vol.1, pp. 3–35, Proc. 5th Int. Symp. on Landslides, Lausanne, Switzerland.

Hylland, M. D., and M. Lowe (1997), Regional landslide-hazard evaluation using landslide slopes, western Wasatch Country, Utah, *Env. & Eng. Geoscience, 3*(1), 31–43.

Ibetsberger, H. J. (1996), The Tsergo Ri landslide: an uncommon area of high morphological activity in the Langthang valley, Nepal, *Tectonophysics, 260*, 85–93.

Ibsen, M.-L., D. Brunsden, E. Broomhead, and A. Collison (1996), Block Slide, in *Landslide Recognition—Identification, Movement, and Courses*, edited by R. Dikau, D. Brunsden, L. Schrott, and M.-L. Ibsen, pp. 64–77, John Wiley & Sons Ltd., Chichester, UK.

Iida, T. (1993), A probability model of the slope failure and the hillslope development, *Trans. Jpn. Geomorph. Union, 14*, 17–31.

Iida, T. (2004), Theoretical research on the relationship between return period of rainfall and shallow landslides, *Hydrol. Process., 18*, 739–756.

Iiritano, G., P. Versace, and B. Sirangelo (1998), Real-time estimation of hazard for landslides triggered by rainfall, *Environ. Geol., 35*(2–3), 175–183.

Imaizumi, F., and R. C. Sidle (2005), Relationship between sediment supply and transport processes in Miyagawa Dam catchment, *Ann. Disaster Prevention Res. Inst., Kyoto Univ. 48C*, 209–217.

Imaizumi, F., S. Tsuchiya, and O. Ohsaka (2005), Behaviour of debris flows located in a mountainous torrent on the Ohya landslide, Japan, *Can. Geotech. J., 42*, 919–931.

Innes, J. L. (1983), Debris flows, *Prog. Phys. Geogr., 7*, 469–501.

Innes, J. L. (1985), Lichenometric dating of debris-flow deposits on alpine colluvial fans in southwest Norway, *Earth. Surface. Process, 10*, 519–524.

Innes, J. L. (1997), Historical debris-flow activity and climate in Scotland, in *Rapid Mass Movement as a Source of Climatic Evidence for the Holocene*, edited by F. Burkhard et al., pp. 233–240, Paleoklimaforschung Bd. 19, European Sci. Foundation, Straasbourg, France.

Ishihara, K. (1985), Lecture: Important Notice of Design and Construction, 6. *Soil Dynamics, Soils and Foundations, 33*(6), pp. 61–66 (in Japanese).

Istanbulluoglu, E., D. G. Tarboton, R. T. Pack, and C. H. Luce (2004), Modeling of the interaction between forest vegetation, disturbances, and sediment yields, *J. Geophys. Res., 109*, F01009.

Istok, J. D., and M. E. Harward (1982), Clay mineralogy in relation to landscape instability in the Coast Range of Oregon, *Soil Sci. Soc. Am. J., 46*(6), 1326–1331.

Iverson, R. M. (1995), Can magma-injection and groundwater forces cause massive landslides on Hawaiian volcanoes?, *J. Volcanol. Geotherm. Res., 66*, 295–308.

Iverson, R. M. (2000), Landslide triggering by rain infiltration, *Water Resour. Res, 36*(7), 1897–1910.

Iverson, R. M., and R. G. LaHusen (1989), Dynamic pore-pressure fluctuations in rapid shearing granular materials, *Science, 246*, 796–798.

Iverson, R. M., and J. J. Major (1986), Groundwater seepage vectors and the potential for hillslope failure and debris flow mobilization, *Water Resour. Res, 22*(11), 1543–1548.

Iverson, R. M., and J. J. Major (1987), Rainfall, ground-water flow, and seasonal movement at Minor Creek landslide, northwestern California: physical interpretation of empirical relations, *Geol. Surv. Am. Bull., 99*, 579–594.

Iverson, R. M., and M. E. Reid (1992), Gravity-driven groundwater flow and slope failure potential. 1. Elastic effective-stress model, *Water Resour. Res, 28*(3), 925–938.

Iverson, R. M., M. E. Reid, and R. G. LaHusen (1997), Debris-flow mobilization from landslides, *Annu. Rev. Earth Planet. Sci. 25*, 85–138.

Iverson, R. M., M. E. Reid, N. R., Iverson, R. G., LaHusen, M. Logan, J. E. Mann, and D. L. Brien (2000), Acute sensitivity of landslide rates to initial soil porosity, *Science, 290*, 513–516.

Iverson, R. M., S. P. Schilling, and J. W. Vallence (1998), Objective delineation of lahar-inundation hazard zones, *Geol. Surv. Am. Bull., 100*(8), 972–984.

Ives, J. D., and B. Messerli (1981), Mountain hazards mapping in Nepal. Introduction to an applied mountain research project, *Mountain Res. Develop., 1*, 223–230.

Iwamatsu, A., A. Fukushige, and S. Koriayma (1989), Applied geology of so-called "Shirasu", non-welded ignimbrite, *J. Geography (Tokyo Geogr. Soc.), 98*(4), 379–400 (in Japanese with English abstract).

Jackson, C. R., and C. A. Sturm (2002), Woody debris and channel morphology in first and second order forested channels in Washington's coast ranges, *Water Resour. Res 38*(9), 16-1–16-14.

Jacobsen, N. K. (1987), Studies on soils and potential for soil erosion in the sheep farming area of south Greenland, *Arct. Alp. Res., 19*(4), 498–507.

Jahn, A. (1967), Some features of mass movement on Spitsbergen slopes, *Geograf. Ann., 49A*(2-4), 213–224.

Jahn, A. (1989), Soil creep on slopes in different altitudinal and ecological zones of Sudetes Mountains, *Geograf. Ann., 71A*(3-4), 161–170.

Jahns, R. H. (1958), Residential ills of the Heartbreak Hills of southern California, *Caltech Alumni Mag., 22*, 13–20.

Jakob, M. (2000), The impacts of logging on landslide activity at Clayoquot Sound, British Columbia, *Catena, 38*(4), 279–300.

Jakob, M., and O. Hungr (2005) (eds.), *Debris-Flow Hazards and Related Phenomena*, 700 p., Springer Verlag.

Jakob, M., and H. Weatherly (2003), A hydroclimatic threshold for landslide initiation on the North Shore Mountains of Vancouver, British Columbia, *Geomorphology, 54*, 137–156.

Jeppson, R. W., and E. E. Farmer (1980), Composite water movement in a sloping phosphate mine dump, *Proc. Symp. on Watershed Management*, vol. II. pp. 855–866, Am. Soc. Civil Eng., New York.

Jeyapalan, J. K., J. M. Duncan, and H. B. Seed (1981), Summary of research on analyses of flow failures of mine tailings impoundments, *Rep. BMIC 8857*, pp. 54–61, Bur. of Mines, Denver, CO.

Jibson, R. W. (1992), The Mameyes, Puerto Rico, landslide disaster of October 7, 1985, in *Landslides/Landslide Mitigation* edited by J. A. Johnson and J. E. Slosson, vol. 9, pp. 37–54, Rev. Eng. Geol., Geol. Soc. Am., Boulder, CO.

Jibson, R. W. (2005), Landslide hazards at La Conchita, California, *Open-file rep. 2005–1067*, 11 p. U.S. Geol. Surv.

Jibson, R. W., and D. K. Keefer (1989), Statistical analysis of factors affecting landslide distribution in the New Madrid seismic zone Tennessee and Kentucky, *Eng. Geol, 27*, 509–542.

Jibson, R. W., E. L. Harp, and J. A. Michael (2000), A method for producing digital probabilistic seismic landslide hazard maps, *Eng. Geol., 58*, 271–289.

Jibson, R. W., C. S. Prentice, B. A. Borissoff, E. A. Rogozhin, and C. J. Langer (1994), Some observations of landslides triggered by the 29 April 1991 Racha Earthquake, Republic of Georgia, *Bull. Seismol. Soc. Am., 4*, 963–973.

Johnson, A. C., D. N. Swanston, and K. E. McGee (2000), Landslide initiation, runout, and deposition within clearcuts and old-growth forests of Alaska, *J. Am. Water Res. Assoc., 36*, 1–13.

Johnson, A. M. (1984), Debris flow, in *Slope Instability*, edited by D. Brundsen and D. B. Prior, pp. 257–361, Wiley & Son, Chichester, UK.

Johnson, K., E. A. Olson, and S. Manandhar (1982), Environmental knowledge and response to natural hazards in mountainous Nepal, *Mountain Res. Develop., 2*, 175–188.

Johnson, K. A., and N. Sitar (1990), Hydrologic conditions leading to debris-flow initiation, *Can. Geotech. J., 27*, 789–801.

Jonasson, C. (1988), Slope processes in periglacial environments of northern Scandinavia, *Geograf. Ann., 70A* (3), 247–253.

Jones, A. (1971), Soil piping and stream channel initiation, *Water Resour. Res., 7*(3), 602–610,

Jones, D. K. C., and E. M. Lee (1989), The Catak landslide disaster, Turkey, *Q. J. Eng. Geol., 22*, 93–95.

Jones, F. O. (1973), Landslides of Rio de Janeiro and the Serra des Araras escarpment, Brazil, *Prof. Pap. 697*, 42 p., U.S. Geol. Surv.

Jones, F. O., D. R. Embody, and W. L. Petersen (1961), Landslides along the Columbia River Valley northeastern Washington, *Prof. Pap. 367*, 98 p., U.S. Geol. Surv.

Jones, J. A. A. (1981), The natures of soil piping: a review of research, *Research Monograph 3*, 301 p., British Geomorph. Res. Group, GeoBooks, Norwich, UK.

Jones, J. A. A. (1990), Piping effects in humid lands, in *Groundwater Geomorphology: the Role of Subsurface Water in Earth Surface Processes and Landforms*, edited by C. G. Higgins and D. R. Coates, pp. 111–138, Geol. Soc. Am. Spec. Pap. 252, Boulder, CO.

Jones, J. A. A. (1997), *Global Hydrology: Processes, Resources and Environmental Management*, 399 p., Longman, Essex, UK.

Jordan, R. P. (1994), Debris flows in the southern Coast Mountains, British Columbia: dynamic behaviour and physical properties, Ph.D. thesis, Univ. of British Columbia, Dept. of Geography, Vancouver, BC.

Julian, M., and E. Anthony (1996), Aspects of landslide activity in the Mercantour Massif and the French Riviera, southeastern France, *Geomorphology, 15*(3/4), 275–289.

Jungerius, P. D., J. A. M. van den Ancker, and H. J. M. van Zon (1989), Long term measurements of forest soil exposure and creep in Luxembourg, *Catena, 16*(4–5), 437–447.

Kaliser, B. N. and R. W. Fleming (1986), The 1983 landslide dam at Thistle, Utah, in *Landslide Dams: Processes, Risk, and Mitigation*, edited by R. L. Schuster, Geotech. Special Publ. No. 3, pp. 59–83, Am. Soc. Civ. Eng., New York.

Kamai, T. (1989), Movement patterns of Ogawamura-Sodechi landslide, in Nagano, central Japan—dermination and application of the surface strain-rate sensor, *J. Jpn. Landslide Soc., 26*(2), 1–8.

Kamai, T. (1998), Monitoring the process of ground failure in repeated landslides and associated stability assessments, *Eng. Geol., 50*, 71–84.

Kamai, T., and H. Shuzui (2002), *Landslides in Urban Region*, 200 p., Rikou-tosho, Tokyo (in Japanese).

Kamai, T., H. Shuzui, R. Kasahara, and Y. Kobayashi (2004), Earthquake risk assessments of large residential fill-slope in urban areas, *J. Jpn. Landslide Soc., 40*(5), 389–399.

Karambiri, H., O. Ribolzi, J. P. Delhoume, J. Ducloux, A. Coudrain-Ribstein, and A. Casenave (2003), Importance of soil surface characteristics on water erosion in a small grazed Sahelian catchment, *Hydrol. Proc. 17*, 1495–1507.

Kardos, L. T., P. I. Vlasoff, and S. N. Twiss (1944), Factors contributing to landslides in the Palouse region, *Soil Sci. Soc. Am. Proc., 8*, 437–440.

Kazi, A., and J. Moum (1973), Effect of leaching on the fabric of normally consolidated marine clays, *Proc. Int. Symp. on Soil Structure*, pp. 137–152, Gothenburg, Sweden, Swedish Geotech. Soc.

Keaton, J. R., and J. V. DeGraff (1996), Surface observation and geologic mapping, in *Landslides – Investigation and Mitigation*, edited by A. K. Turner and R. L. Schuster, pp. 178–230, Special Report 247, Transport. Res. Board, National Res. Council, National Academic Press, Washington, DC.

Keefer, D. K. (1984a), Landslides caused by earthquakes, *Geol. Soc. Am. Bull., 95*, 406–421.

Keefer, D. K. (1984b), Rock avalanches caused by earthquake: source characteristics, *Science, 223*, 1288–1290.

Keefer, D. K. (2000), Statistical analysis of an earthquake-induced landslide distribution—the 1989 Loma Prieta, California, event, *Eng. Geol., 58*, 213–249.

Keefer, D. K. (2002), Investigating landslides caused by earthquakes—a historic review, *Surv. of Geophys., 23*, 473–510.

Keefer, D. K., and A. Johnson (1983), Earth flows: morphological, mobilization, and movement, *Prof. Pap. 1264*, U.S. Geol. Surv., 56 p.

Keefer, D. K., and R. C. Wilson (1989), Predicting earthquake-induced landslides, with emphasis on arid and semi-arid environments, in *Landslides in Semi-arid Environments*, edited by P. M. Sadler and D. M. Morton, vol. 2, part 1, pp. 118–149, Inland Geol. Soc. S. Calif. Publ., Riverside, CA.

Keefer, D. K., R. C. Wilson, R. K. Mark, E. E. Brabb, W. M. Brown, S. D. Ellen, E. L. Harp, G. F. Wieczorek, C. S. Alger, and R. S. Zatkin (1987), Real-time landslide warning during heavy rainfall, *Science, 238*, 921–925.

Keilis-Borok, V., A. Ismail-Zadeh, V. Kossobokov, and P. Shebalin (2001), Non-linear dynamics of the lithosphere and intermediate-term earthquake prediction, *Tectonophysics, 338*, 247–260.

Keim, R. F., and A. E. Skaugset (2003), Modelling effects of forest canopies on slope stability, *Hydrol. Process., 17*(7), 1457–1467.

Keller, E. A,, and F. J. Swanson (1979), Effects of large organic material on channel form and fluvial processes, *Earth Surface Processes*, 4, 361–380.

Kellerhals, R., and M. Church (1990), Hazard management on fans, with examples from British Columbia, in *Alluvial Fans: a Field Approach*, edited by A.H. Rachocki and M. Church, pp. 335–354, John Wiley & Sons, Chichester, UK.

Kelsey, H. M. (1978), Earthflows in Franciscan mélange, Van Duzen River basin, California, *Geology, 6*, 361–364.

Kelsey, H. M. (1982), Hillslope evolution and sediment movement in a forested headwater basin, Van Duzen River, north coastal California, in Sediment Budgets and Routing in Forested Drainage Basins, edited by F. J. Swanson, R. J. Janda, T. Dunne, and D. N. Swanston, *Gen. Tech. Rep. PNW-141*, pp. 86–96, For. Serv., U.S. Dept. of Agric., Portland, OR.

Kelsey, H. M., M. A. Madej, J. Pitlick, P. Stroud, and M. Coghlan (1981), Major sediment sources and limits to the effectiveness of erosion control techniques in the highly erosive watersheds of north coastal California, *IAHS Publ., 132*, 493–509.

Ker, D. S. (1970), Renewed movement on a slump at Utiku, *N. Z. J. Geol. Geophys., 13*(4), 996–1017.

Kerr, P. F. (1963), Quick clay, *Sci. Am., 209*(5), 132–142.

Kerr, P. F. (1979), Quick clays and other slide-forming clays, *Eng. Geol., 14*, 173–181.

Ketcheson, G. L., and H. A. Froehlich (1978), Hydrology factors and environmental impacts of mass soil movements in the Oregon Coast Range, *Report*, 94 p., Water Resour. Res. Inst., Oregon State Univ., Corvallis.

Khazai, B., and N. Sitar (2003), Evaluation of factors controlling earthquake-induced landslides caused by Chi-Chi earthquake and comparison with the Northridge and Loma Prieta events, *Eng. Geol., 71*, 79–95.

Kienholz, H., G. Schneider, M. Bichsel, M. Grunder, and P. Mool (1984), Mapping of mountain hazards and slope stability, *Mountain Res. Develop. 4*(3), 247–266.

Kiersch, G. A. (1964), Vaiont reservoir disaster, *Civ. Eng., 34*(3), 32–39.

Kiersch, G. A. (2001), Development of engineering geology in the western United States, *Eng. Geol., 59*, 1–49.

Kim, H.-J., R. C. Sidle, R. D. Moore, and R. Hudson (2004), Throughflow variability during snowmelt in a forested mountain catchment, coastal British Columbia, Canada, *Hydrol. Process., 18*, 1219–1236.

King, J., I. Loveday, and R. L. Schuster (1989), The 1985 Bairaman landslide dam and resulting debris flow, Papua New Guinea, *Q. J. Eng. Geol., 22*, 257–270.

Kirkby, M. J., and I. Statham (1975), Surface stone movement and scree formation, *J. Geol., 83*, 349–62.

Kitahara, H. (1989), Characteristics of pipe flow in a subsurface soil layer on a gentle slope (II) Hydraulic properties of pipes, *J. Jpn. For. Soc., 71*, 317–322 (in Japanese with English abstract).

Kitahara, H. (1994), A study on the characteristics of soil pipes influencing water movement in forested slopes, *Forestry and Forest Products Res. Inst. Bull., No. 337*, pp. 63–115, Ibaraki, Japan (in Japanese).

Kitahara, H., A. Shimizu, and Y. Mashima (1988), Characteristics of pipe flow in a subsurface soil layer of a gentle hillside, *J. Jpn. For. Soc., 70*, 318–323 (in Japanese with English abstract).

Kitahara, H., Y. Terajima, and Y. Nakai (1994), Ratio of pipe flow to throughflow discharge, *J. Jpn. For. Soc., 76*(1), 10–17 (in Japanese with English abstract).

Kitamura, T., and S. Namba (1981), The function of tree roots upon landslide prevention presumed through the uprooting test, *Forestry and Forest Products Res. Inst. Bull., No. 313*, pp. 175–208, Ibaraki, Japan (in Japanese).

Klages, M. G., and Y. P. Hsieh (1975), Suspended solids carried by the Gallatin River of southwestern Montana, 2, Using mineralogy for inferring sources, *J. Environ. Qual., 4*(1), 68–73.

Klazura, G. E., and D. A. Imy (1993), A description of the initial set of analysis products available from the NXRAD WSR-88D system, *Bull. Am. Meteorol. Soc., 74*(7), 1293–1311.

Klock, G. O., and J. D. Helvey (1976), Debris flows following wildfire in north central Washington, *Proc. 3rd Inter-Agency Sedimentation Conf.*, pp. 1-91–1-98, Denver, Colo., Water Resour. Council, Washington, DC.

Klukanova, A., and S. Rapant (1999), Impact of mining activities upon environment of the Slovak Republic: two case studies, *J. Geochem. Expl., 66*, 299–306.

Knapen, A., M. G. Kitutu, J. Poessen, W. Breugelmans, J. Deckers, and A. Muwanga (2006), Landslides in a densely populated country at the footslopes of Mount Elgon (Uganda): Characteristics and causal factors, *Geomorphology, 73*, 149–165.

Koaze, T. (1983), Slow mass movement in periglacial regions, *Trans. Jpn. Geomorph. Union, 4*, 189–203.

Kobashi, S., and K. Sassa (1990), *To Prevent Landslides and Slope Disasters*, 165 p., Sankaido Publ. Co., Ltd., Japan (in Japanese).

Kockelman, W. J. (1986), Some techniques for reducing landslide hazards, *Bull. Assoc. Eng. Geol., 23*, 29–52.

Koga, Y. (1989), *A Guide to Slope Stability Analysis 2.3.2 Methods of Slope Stability Calculation (During Earthquakes)*, Jpn. Soc. Soil Mechanics and Foundation Eng., 188 p. (in Japanese).

Koide, H. (1955), *Landslides*, 205 p., Kokon-Syoin, Japan (in Japanese)

Komatsu, Y., and Y. Onda (1996), Spatial variation in specific discharge of base flow in small catchments, Oe-yama Region, western Japan, *J. Jpn. Soc. Hydrol. Water. Resour., 9*, 489–497.

Koons, P. O. (1989), The topographic evolution of collisional mountain belts: a numerical look at the Southern Alps, New Zealand, *Am. J. Sci., 289*, 1041–1069.

Kosugi, K., T. Mizuyama, and M. Fujita (2002), Accuracy of shallow-landslide prediction model to estimate groundwater table, *Shin Sabo (J. Jpn. Soc. Erosion Contr. Eng.), 55*(3), 21–32 (in Japanese with English abstract).

Kotarba, A. (1976), Studies of mass movements in Poland (1970–1975), *Geograf. Ann., 58A*(3), 173–178.

Krammes, J. S. (1965), Seasonal debris movement from steep mountainside slopes in southern California, Proc. of the Federal Inter-agency Sedimentation Conference, *Misc. Publ. 970*, pp. 85–88, Jackson, Miss., U. S. Dep. of Agric.

Krauter, E., and K. Steingötter (1983), Die Hangstabilitätskarte des linkrheinnischen Mainzer Beckens, *Geologisches Jahrbuch, C34*, 3–31 (in German with English abstract).

Krejčí, O., I. Baroň, M. Bíl, F. Hubatka, Z. Jurová, and K. Kirchner (2002), Slope movements in the Flysch Carpathians of Eastern Czech Republic triggered by extreme rainfalls in 1997: a case study, *Phys. Chem. Earth, 27*, 1567–1576.

Krohn, J. P. (1992), Landslide mitigation using horizontal drains, Pacific Palisades area, Los Angeles, California, in *Landslides/Landslide Mitigation*, edited by J. E. Slosson, A. G. Keens and J. A. Johnson, pp. 63–68, Rev. of Eng. Geol., vol. 9, Geol. Soc. Am., Boulder, CO.

Kuenzi, W. D., O. H. Horst, and R. V. McGehee (1979), Effect of volcanic activity on fluvial-deltaic sedimentation in a modern arc-trench gap, southwestern Guatemala, *Geol. Soc. Am. Bull., 90*, 827–838.

Kumar, K., and G. S. Satyal (1999), Cost analysis of losses caused by the Malpa landslide in Kumaun Himalaya—a basic framework for risk assessment, *Curr. Sci., 77*(8), 1023–1028.

Kuruppuarachchi, T., and K. H. Wyrwoll (1992), The role of vegetation clearing in the mass failure of hillslopes: Moresby Ranges, Western Australia, *Catena, 19*, 193–208.

Kwong, A. K. L., M. Wang, C. F. Lee, and K. T. Law (2004), A review of landslide problems and mitigation in Chongqing and Hong Kong: similarities and differences, *Eng. Geol., 76*, 27–39.

Lafleur, J., and G. Lefebvre (1980), Groundwater regime associated with slope stability in Champlain clay deposits, *Can. Geotech., J., 17*(1),44–53.

Lambe, T. W., and R. V. Whitman (1969), *Soil Mechanics*, 553 p., John Wiley, New York.

Lambert, M. G. (1980), Pastoral production on eroded and uneroded slopes in the dry Wararapa hill country: An interim (12 month) report on the joint Grasslands Division and Aokautere Science Centre Project, *Aokautere Sci. Ctr. Intern. Rep. 21*, pp. 7–12, N. Z. Ministry of Works and Dev.

Lamberti, G. A., S. V. Gregory, L. R. Ashkenas, R. C. Wildman, and K. M. Moore (1991), Stream ecosystem recovery following a catastrophic debris flow, *Can. J. Fish. Aquatic Sci., 48*, 196–208.

Lan, H. X., C. H. Zhou, L. J. Wang, H. Y. Zhang, and R. H. Li (2004), Landslide hazard spatial analysis and prediction using GIS in the Xiaojiang watershed, Yunnan, China, *Eng. Geol., 76*, 109–128.

Lancaster, S. T., S. Hayes, and G. E. Grant (2003), Effect of wood on debris flow runout in small mountain watersheds, *Water Resour. Res., 39*(6), 1168, doi: 1029/2001WR001227.

Lanly, J. P. (1969), Regression de la Foret Dense en Cote d' Ivoire, *Bois For. Tropiques, 127*, 45–59.

Larsen, M. C., and J. E. Parks (1997), How wide is a road? The association of roads and mass-wasting in a forested mountain environment, *Earth Surf. Process. Landforms, 22*(9), 835–848.

Larsen, M. C., and A. Simon (1993), A rainfall intensity-duration threshold for landslides in a humid-tropical environment, Puerto Rico, *Geograf. Ann. 75A*, 13–23.

Laudon, H., and O. Slaymaker (1997), Hydrographic separation using stable isotopes, silica and electrical conductivity: an alpine example, *J. Hydrol., 201*, 82-101.

Lavigne, F., and J. C. Thouret (2003), Sediment transportation and deposition by rain-triggered lahars at Merapi Volcano, central Java, Indonesia, *Geomorphology, 49*(1-2), 45–69.

Lavigne, F., J. C. Thouret, B. Voight, H. Suwa, and A. Sumaryono (2000), Lahars at Merapi volcano, central Java: an overview, *J. Volcanol. Geotherm. Res., 100*(1-4), 423–456.

LeBaron, A., L. K. Bond, P. Aitken, and L. Michaelsen (1979), An explanation of the Bolivian highlands grazing → erosion syndrome, *J. Range Manage., 32*(3), 201–208.

Lee, S., U. Chwae, and K. Min (2002), Landslide susceptibility mapping by correlation between topography and geological structure: the Janghung area, Korea, *Geomorphology, 46*, 149–162.

Lee, S., J.-H. Ryu, J.-S. Won, and H.-J. Park (2004), Determination and application of the weights for landslide susceptibility mapping using an artificial neural network, *Eng. Geol., 71*, 289–302.

Lefebvre, G. (1996), Soft sensitive clays, in *Landslides—Investigation and Mitigation*, edited by A. K. Turner and R. L. Schuster, pp. 607–619, Special Report 247, Transport. Res. Board, National Res. Council, National Academic Press, Washington, DC.

Lehre, A. K. (1981), Sediment budget of a small California Coast Range drainage basin, *IAHS Publ., 132*, 123–139.

Lehre, A. K. (1982), Sediment budget of a small Coast Range drainage basin in north-central California, in Sediment Budgets and Routing in Forested Drainage Basins, *Gen. Tech. Rep. PNW-141*, pp. 67–77, For. Serv., U.S. Dep. of Agric., Portland, OR.

Leighton, F. B. (1966), Landslides and hillside development, in *Engineering Geology in Southern California*, Spec. Publ., pp. 149–200, Assoc. of Eng. Geologists, Los Angeles, CA.

Leighton, F. B. (1972), Origin and control of landslides in the urban environment of California, *Proc. 24th Int. Geol. Congr.*, Sec. 13, pp. 89–96, Montreal, Canada.

Leighton, F. B. (1976), Urban landslides: Targets for land-use planning in California, *Spec. Pap. 174*, pp. 37–60. Geol. Soc. Am., Boulder, Colo.

Leroi, E., O. Rouzeau, J.-Y. Scanvic, C. C. Weber, and C. G. Varges (1992), Remote sensing and GIS technology in landslide hazard mapping in the Columbian Andes, *Episodes, 15*, 32–35.

Lessing, P., and R. B. Erwin (1977), Landslides in West Virginia, in *Rev. Eng. Geo., 3, Landslides*, pp. 245–254, Geol. Soc. Am., Boulder, CO.

Lewis, J. (2002), Quantifying recent erosion and sediment delivery using probability sampling: a case study, *Earth Surf. Process. Landforms, 27*, 559–572,

Lewkowicz, A. G., and J. Hartshorn (1998), Terrestrial records of rapid mass movements in the Sawtooth Range, Ellesmere Island, Northwest Territories, Canada, *Can. J. Earth Sci., 35*, 55–64.

Li, T. (1989), Landslides: extent and economic significance in China, in *Landslides: Extent and Economic Significance*, edited by E. E. Brabb and B. L. Harrod, pp. 271–287, A.A. Balkema, Rotterdam.

Li, T., and M. Li (1985), A preliminary study on landslide triggered by heavy rainfall, *Int. Symp. on Erosion, Debris Flow, and Disaster Prevention*, pp. 317–320, Sept. 3–5, Tsukuba, Japan.

Lin, J.-C., D. Petley, C.-H. Jen, A. Koh, and M.-L. Hsu (2006), Slope movements in a dynamic environment—A case study of Tachia River, Central Taiwan, *Quaternary Int., 147*, 103–112.

Lin, S.-H. (1995), Study on the growth and root strength of Roxburgh sumac and dense-flowered false-nettle in limestone mining area, *Journal of Soil and Water Conservation (Taiwan), 27*(1), 90–99 (in Chinese).

Lineback Gritzner, M., W. A. Marcus, R. Aspinall, and S. G. Custer (2001), Assessing landslide potential using GIS, soil wetness modeling and topographic attributes, Payette River, Idaho, *Geomorphology, 37*, 149–165.

Lisle, T. E. (1989), Sediment transport and resulting deposition in spawning gravels, north coastal California, *Water Resour. Res., 25*(6), 1303–1319.

Lisle, T. E., and M. A. Madej (1992), Spatial variation in armoring in a channel with high sediment supply, in *Dynamics of Gravel-bed Rivers*, edited by P. Billi, R. D. Hey, C. R. Thorne, and P. Tacconi, pp. 277–296, John Wiley & Sons, London.

Loaiciga, H. A., J. B. Valdes, R. Vogel, J. Garvey, and H. Schwarz (1996), Global warming and the hydrologic cycle, *J. Hydrol., 174*, 83–127.

Lohnes, R. A., and R. L. Handy (1968), Slope angles in friable loess, *J. Geol., 76*(3), 247–258.

Loope, D. B., J. A. Mason, and L. Dingus (1999), Geological notes. Lethal sandslides from eolian dunes, *J. Geol., 107*(6), 707–713.

Lorenzini, G., and N. Mazza (2003), *Debris Flow: Phenomenology Rheological Modelling*, p. 216, Wit Press, Billerica, MA.

Louisville Herald (1923), Outlet barred as avalanche – hits tunnel, *Louisville Herald*, Dec. 5, 1923.

Lu, S. Y., J. D. Cheng, and K. N. Brooks (2001), Managing forests for watershed protection in Taiwan, *For. Ecol. Manage., 143*, 77–85.

Lu, X. X., and D. L. Higgitt (1998), Recent changes of sediment yield in the Upper Yangtze, China, *Environ. Manage., 22*(5), 697–709.

Luckman, P. G., R. D. Gibson, and R. C. De Rose (1999), Landslide erosion risk to New Zealand pastoral steeplands productivity, *Land Degrad. Dev., 10*, 49–65.

Luckman, P. G., R. L. Hathaway, and W. R. N. Edwards (1982), Root systems, root strength and slope stability, *Internal Rep. 34.*, 31 p., Natl. Water and Soil Conserv. Org., Min. of Works and Dev., Wellington, N.Z.

Lumb, P. (1962), The properties of decomposed granite, *Geotechnique, 12*, 226–243.

Lumb, P. (1965), The residual soils of Hong Kong, *Geotechnique, 15*, 180–194.

Lumb, P. (1975), Slope failures in Hong Kong, *Q. J. Eng. Geol., 8*, 31–65.

Luxmoore, R. J., and L. A. Ferrand (1993), Towards pore-scale analysis of preferential flow and chemical transport, in *Water Flow and Solute Transport in Soils*, edited by D. Russo and D. Dagan, pp. 45–60, Springer-Verlag, Berlin.

Luxmoore, R. J., P. M. Jardine, G. V. Wilson, J. R. Jones, and L. W. Zelazny (1990), Physical and chemical controls of preferred path flow through a forested hillslope, *Geoderma, 46*, 139–154.

Luxmoore, R. J., B. P. Spalding, and L. M. Munro (1981), Areal variation and chemical modification of weathered shale infiltration characteristics, *Soil Sci. Soc. Am. J., 45*(4), 687–691.

Luzi, L., and F. Pergalani (1999), Slope instability in the static and dynamic conditions for urban planning: the 'Oltre Po Pavese' Case History (Regione Lombardia – Italy), *Nat. Hazards, 20*, 57–82.

MacFarlane, W. A., and E. Wohl (2003), Influence of step composition on step geometry and flow resistance in step-pool streams of the Washington Cascades, *Water Resour. Res., 39*, 1037, doi: 10.1029/2001 WR001238.

Madej, M. A. (2001), Erosion and sediment delivery following removal of forest roads, *Earth Surf. Process. Landforms, 26*(2), 175–190.

Maharaj, R. J. (1993), Landslide processes and landslide susceptibility analysis from an upland watershed: A case study from St. Andrew, Jamaica, West Indies, *Eng. Geol., 34*, 53–79.

Maharaj, R. J. (1995), Engineering-geological mapping of tropical soils for land-use planning and geotechnical purposes: a case study from Jamaica, West Indies, *Eng. Geol., 40*, 243–286.

Major, J. J., S. P. Schilling, C. R. Pullinger, C. D. Escobar, and M. M. Howell (2001), Volcano-hazard zonation for San Vicente Volcano, El Salvador, *Open-File Report 01-367*, 21 p., U.S. Geol. Surv.

Malet, J.-P., O. Maquaire, and E. Calais (2002), The use of Global Positioning System techniques for the continuous monitoring of landslides: application to the Super-Sauze earthflow (Alpes-de-Haute-Provence, France), *Geomorphology, 43*(1/2), 33–54.

Malik, M. H., and S. Farooq (1996), Landslide Hazard Management and Control in Pakistan—a review, *International Centre for Integrated Mountain Development* (ICIMOD), 68 p.

Mallants, D., B. P. Mohanty, A. Vervoort, and J. Feyen (1997), Spatial analysis of saturated hydraulic conductivity in a soil with macropores, *Soil Technol., 10*, 115–131.

Mann, D. H., D. M. Peseet, R. E. Reanier, and M. L. Kunz (2002), Response of an arctic landscape to Lateglacial and early Holocene climatic changes: the importance of moisture, *Quaternary Sci. Rev., 21*, 997–1021.

Manville, V., E. H. Newton, and J. D. L. White (2005), Fluvial response to volcanism: resedimentation of the 1800a Taupo ignimbrite eruption in the Rangitaiki River catchment, North Island, New Zealand, *Geomorphology, 65*, 49–70.

Marden, M., and D. Rowan (1993), Protective value of vegetation on Tertiary terrain before and during Cyclone Bola, East Coast, North Island, New Zealand. *N.Z. J. For. Sci. 23*(3), 255–263.

Marion, D. A. (1981) Landslide occurrence in the Blue River drainage, Oregon, M.S. thesis, 114 p., Oregon State University, Corvallis, OR.

Mark, A. F., G. A. M. Scott, F. R. Sanderson, and P. W. James (1964), Forest succession on landslides above Lake Thomson, Fjordland, New Zealand, *N.Z. J. Botany, 2*, 60–89.

Marron, D. C. (1985), Colluvium in bedrock hollows on steep slopes, Redwood Creek drainage basin, northwestern California, *Catena, Suppl. 6*, 59–68.

Marshall, C. E. (1977), *The Physical Chemistry and Mineralogy of Soils. II. Soils in Place*, 313 p. Wiley, New York.

Marston, R. A., M. M. Miller, and L. P. Devkota (1998), Geoecology and mass movement in the Manaslu-Ganesh and Langtang-Jugal Himals, Nepal, *Geomorphology, 26*, 139–150.

Martin, D. J., and L. E. Benda (2001), Patterns of instream wood recruitment and transport at the watershed scale, *Trans. Am. Fish. Soc., 130*, 940–958.

Martin, E. F. (1989), "A tale of two certificates" the California forests practice program, 1976 through 1988, 299 p., State of California, Resources Agency, Dept. of Forestry and Fire Protection, Sacramento.

Marui, H. (1981), Study on shallow landslides in natural slopes, Ph.D. thesis, Kyoto Univ., Japan (in Japanese).

Massari, R., and P. M. Atkinson (1999), Modelling susceptibility to landsliding: an approach bases on individual landslide type, *Trans. Jpn. Geomorph. Union, 20*, 151–168.

Mather, J. D., D. A. Gray, and D. G. Jenkins (1969), The use of tracers to investigate the relationship between mining subsidence and groundwater occurrence at Aberfan, South Wales, *J. Hydrol., 9*, 136–154.

Mathews, W. H. (1979), Landslides of central Vancouver Island and the 1946 earthquake, *Bull. Seismol. Soc. Am., 69*(2), 445–450.

Mathewson, C. C., and J. H. Clary (1977), Engineering geology of multiple landsliding along I-45 road cut near Centerville, Texas, *Rev. in Eng. Geol.*, vol. 3, Landslides, pp. 213–223, Geol. Soc. Am., Boulder, CO.

Matsuoka, N. (1994), Continuous recording of frost heave and creep on a Japanese alpine slope, *Arct. Alp. Res., 26*, 245–254.

Matsuoka, N. (2001), Solifluction rates, processes and landforms: a global review, *Earth Sci. Rev., 55*, 107–134.

Matsuura, S. (1985), Characteristics of mass-wasting in the midland area of Nepal Himalayas, *Proc. 4th Int. Conf. and Field Workshop on Landslides*, pp. 463–471, Tokyo, Japan.

Mavko, G. M., and E. L. Harp (1984), Analysis of wave-induced pore pressure changes, *Bull. Seismol. Soc. Am., 74*, 139–1407.

May, C. L., and R. E. Gresswell (2004), Spatial and temporal patterns of debris-flow deposition in the Oregon Coast Range, USA, *Geomorphology, 57*, 135–149.

McConchie, J. A. (1980), Implication of landslide activity for urban drainage, *J. Hydrol. N. Z.*, *19*, 27–34.

McDonnell, J. J. (1990), The influence of macropores on debris flow initiation, *Q. J. Eng. Geol.*, *23*, 325–331.

McDonnell, J. J., J. Freer, R. Hopper, C. Kendall, D. Burns, K. Beven, and N. Peters (1996), New method developed for studying flow on hillslopes, *Eos Trans. AGU*, *77*, 465 and 472.

McEven, L. J., and A. Werrity (1988), The hydrology and long-term geomorphic significance of flash flood in the Cairngorm Mountains, Scotland, *Catena*, *15*, 361–377.

McKay, L. D., and J. Fredericia (1995), Distribution, origin, and hydraulic influence of fractures in a clay-rich glacial deposit, *Can. Geotech. J.*, *32*, 957–975.

McKean, J., and R. Roering (2004), Objective landslide detection and surface morphology mapping using high-resolution airborne laser altimetry, *Geomorphology*, *57*(3–4), 331–351.

McKean, J., S. Buechel, and L. Gaydos (1991), Remote sensing and landslide hazard assessment, *Photogram. Eng. Remote Sens.*, *57*, 1185–1193.

McNaughton, K. G., and P. G. Jarvis (1983), Predicting effects of vegetation changes on transpiration and evaporation, in *Water Deficits and Plant Growth*, vol. 7, edited by T. T. Kozlowski, pp. 1–47, Academic, New York.

McSaveney, M. J., T. C. Chinn, and G. T. Hancox (1999), Mount Cook rock avalanche of 14 December 1991, New Zealand, in *Landslides of the World*, edited by K. Sassa, pp. 56–59, Kyoto University Press, Japan.

Meehan, W. R. (1974), Fish habitat and timber harvest in southeast Alaska, *Naturalist*, *25*, 28–31.

Megahan, W. F. (1972), Subsurface flow interception by a logging road in mountains of central Idaho, *Proc. National Symp. on Watersheds in Transition*, pp. 350–356, Am. Water Resour. Assoc., Fort Collins, CO.

Megahan, W. F. (1976), Tables of geometry for low-standard roads for watershed management considerations, slope staking, and end areas, *Gen. Tech. Rep. INT-32*, 104 p., For. Serv., U.S. Dep. of Agric., Ogden, Utah.

Megahan, W. F. (1977), Reducing erosional impacts of roads, in *Guidelines for Watershed Management*, FAO Conservation Guide, pp. 237–251, U.N. Food and Agric. Organ., Rome.

Megahan, W. F. (1978), Erosion Processes on steep granitic road fills in central Idaho, *Soil Sci. Soc. Am. J.*, *42*(2), 350–357.

Megahan, W. F. (1982), Channel sediment storage behind obstructions in forested drainage basins draining the granitic bedrock of the Idaho batholith, in Sediment Budgets and Routing in Forested Drainage Basins, edited by F. J. Swanson, R. J. Janda, T. Dunne, and D. N. Swanston, *Gen. Tech. Rep. PNW-141*, pp. 114–121, For. Serv., U.S. Dep. of Agric., Portland, OR.

Megahan, W. F. (1983), Hydrologic effects of clearcutting and wildlife on steep granitic slopes in Idaho, *Water Resour. Res.*, *19*(3), 811–819.

Megahan, W. F. (1984), Snowmelt and logging influence on piezometric levels in steep forested watersheds in Idaho, *Transport. Res. Record*, *965*, 1–8, National Research Council, Wash., DC.

Megahan, W. F. (1986), Roads and forest site productivity, in *Degradation of Forest Land: Forest Soils at Risk* edited by J. D. Lousier and D. W. Stills, Land Mgmt. Rep. 56, pp. 54–65, B.C. Min. of Forests, Victoria.

Megahan, W. F. (1987), Effects of forest roads on watershed function in mountainous areas, in *Environmental Geotechnics and Problematic Soils and Rocks*, edited by A. S. Balasubramaniam et al., pp. 335–348, A.A. Balkema, Rotterdam, The Netherlands.

Megahan, W. F., and J. L. Clayton (1986), Saturated hydraulic conductivities of granitic material of the Idaho Batholith, *J. Hydrol. 84*, 169–180.

Megahan, W. F., and W. J. Kidd (1972), Effect of logging roads on sediment production rates in the Idaho Batholith, *Res. Paper INT-123*, For. Serv., U.S. Dep. of Agric, Ogden, UT, 15 p.

Megahan, W. F., and J. Schweithelm (1983), Guidelines for reducing negative impacts of logging, in *Tropical Forested Watersheds—Hydrologic and Soils Response to Major Uses or Conversions*, edited by L. S. Hamilton and P. H. King, pp. 143–154, Westview Press, Boulder, CO.

Megahan, W. F., N. F. Day, and T. M. Bliss (1978), Landslide occurrence in the western and central Northern Rocky Mountain physiographic province in Idaho, *Proc. 5th North American Forest Soils Conf.*, pp. 116–139, Colo. State Univ., Fort Collins, CO.

Megahan, W. F., S. B. Monsen, and M. D. Wilson (1991), Probability of sediment yields form surface erosion on granitic roadfills in Idaho, *J. Environ. Qual.*, *20*(1), 53–60.

Megahan, W. F., K. A. Seyedbagheri, and P. C. Dodson (1983), Long term erosion on granitic roadcuts based on exposed tree roots, *Earth Surf. Process. Landforms, 8*, 19–28.

Megahan, W. F., M. Wilson, and S. B. Monsen (2001), Sediment production from granitic cutslopes on forest roads in Idaho, USA, *Earth Surf. Process. Landforms, 26*, 153–163

Mehrotra, G. S., S. Sarkar, D. P. Kanungo, and K. Mahadevaiah (1996), Terrain analysis and spatial assessment of landslide hazards in parts of Sikim Himalaya, *J. Geol. Soc. India, 47*, 491–498.

Merifield, P. M. (1992), Surficial slope failures: the role of vegetation and other lessons from the 1978 and 1980 rainstorms, in *Engineering Geology Practice in Southern California*, edited by B. W. Pipkin and R. J. Proctor, pp. 613–637, Special Publ., No. 4, S. Calif. Sect., Assoc. of Eng. Geol., Star Publ. Co., Belmont, CA.

Mersereau, R. C., and C. T. Dyrness (1972), Accelerated mass wasting after logging and slash burning in western Oregon, *J. Soil Water Conserv.*, *27*(3), 112-1 14.

Meyer, G. A., and S. G. Wells (1997), Fire-related sedimentation events on alluvial fans, Yellowstone National Park, USA, *J. Sediment. Process., A37*, 776–791.

Meyer, G. A., J. L. Pierce, S. H. Woods, and A. J. Tull (2001), Fire, storms and erosional events in the Idaho batholith, *Hydrol. Process., 15*, 3025–3038.

Meyer, G. A., S. G. Wells, R. C. Balling, and A. J. T. Jull (1992), Response of alluvial systems to fire and climate change in Yellowstone National Park, *Nature, 357*, 147–149.

Mileti, D. S., P. A. Bolton, G. Fernandez, and R. G. Updike (1991), The eruption of Nevado del Ruiz volcano, Colombia, South America, November 13, 1985, *Natural Disaster Studies*, vol. 4, 109 p., Committee on Natural Disasters, National Res. Council, National Academy Press, Washington, DC.

Millard, T. (2000), Channel disturbances and logging slash in S5 and S6 streams: an examination of streams in the Nitinat Lake area, southwest Vancouver Island, *Forest Res. Tech. Rep. TR-005*, B.C. Forest Serv., Nanaimo, Canada.

Millard, T. (2003), Schmidt Creek sediment sources and the Johnstone Strait Killer Whale Rubbing Beach, *Forest Res. Tech. Rep. TR-025*, B.C. Forest Serv., Nanaimo, Canada.

Miller, D., C. Luce, and L. Benda (2003), Time, space and episodicity of physical disturbance in streams, *For. Ecol. Manage., 178*, 121–140.

Miller, D. C., P. W. Birkelund, and D. T. Rodbell (1993), Evidence of Holocene stability of steep slopes, northern Peruvian Andes, based on soils and radiocarbon dating, *Catena, 20*, 1–12.

Miller, D. J. (1995), Coupling GIS with physical models to assess deep-seated landslide hazards, *Environ. Eng. Geosci., 1*, 263–276.

Mills, H. H. (1991), Temporal variation of mass-wasting activity in Mount St. Helens crater, Washington, as indicated by seismic activity, *Arct. Alp. Res., 23*, 417–423.

Mills, K. (1997), Forest roads, drainage, and sediment delivery in the Kilchis River watershed, Oregon Dept. of Forestry, Salem, 21 p.

Ministry of Works and Development (1970), Wise land use and community development, Report of Tech. Comm. of Inquiry into the Problems of the Poverty Bay–East Cape Dist. of New Zealand, 119 p., Wellington, New Zealand.

Mitchell, J. K. (1993), *Fundamentals of Soil Behavior* (2nd ed.), John Wiley & Sons, Inc., New York, 437 p.

Miyabuchi, Y. (1993), Debris movement by heavy rainfall during snowmelt season in gullies of Tarumae volcano, Hokkaido, *Shin Sabo J. (J. Jpn. Soc. Erosion Contr. Eng.), 45*(5), 11–16.

Miyabuchi, Y. (1999), Deposits associated with the 1990–1995 eruption of Unzen volcano, Japan, *J. Volcanol. Geotherm. Res., 89*(1/4), 139–158.

Miyazaki, Y., M. Chigira, and U. Kurokawa (2005), Geological and geomorphological factors of slope failures caused by the 2000 Niijima and Kozushima earthquake and a subsequent rainstorm: a case study of slope failures of rhyolitic lava and pyroclastics, *Trans. Jpn. Geomorph. Union, 26*, 205–224 (in Japanese with English abstract).

Mizukoshi, H., and M. Aniya (2002), Use of contour-based DEMs for deriving and mapping topographic attributes, *Photogram. Eng. Remote Sens., 68*(1), 83–93.

Mockler, S., and J. Croke (1999), Prescriptive measures for the prevention of road to stream linkage, *Proc. Second Australian Stream Management Conf.*, pp. 451–455, Adelaide, South Australia.

Moeyersons, J. (1989), A possible causal relationship between creep and sliding on the Rwaza Hill, southern Rwanda, *Earth Surf. Process. Landforms, 14*, 597–614.

Monteith, J. L. (1976), *Vegetation and the Atmosphere*, Vol. 2, 439 p., Academic, New York.

Montgomery, D. R. (1994), Road surface drainage, channel initiation, and slope instability, *Water Resour. Res., 30*(6), 1925–1932.

Montgomery, D. R., and M. T. Brandon (2002), Topographic controls on erosion rates in tectonically active mountain ranges, *Earth Planet. Sci. Let., 201*, 481–489.

Montgomery, D. R., and J. M. Buffington (1997), Channel reach morphology in mountain drainage basins, *Geol. Soc. Am. Bull., 109*, 596–611.

Montgomery, D. R., and W. E. Dietrich (1994), A physically based model for the topographic control on shallow landsliding, *Water Resour. Res., 30*(4), 1153–1171.

Montgomery, D. R., T. B. Abbe, J. M. Buffington, N. P. Peterson, K. M. Schmidt, and J. D. Stock (1996), Distribution of bedrock and alluvial channels in forested mountain drainage basins, *Nature, 381*, 587–588.

Montgomery, D. R., W. E. Dietrich, R. Torres, S. P. Anderson, J. T. Heffner, and K. Loague (1997), Hydrologic response of a steep, unchanneled valley to natural and applied rainfall, *Water Resour. Res., 33*, 91–109.

Montgomery, D. R., H. M. Greenberg, W. T. Laprade, and W. D. Nashem (2001), Sliding in Seattle: test of a model of shallow landsliding potential in an urban environment, in *Land Use and Watersheds, Human Influences on Hydrology and Geomorphology in Urban and Forest Areas*, edited by M. S. Wigmosta and S. J. Burgess, pp. 59–73, Water Science and Application 2, Am. Geophys. Union, Washington, DC.

Montgomery, D. R., K. Sullivan, and H. M. Greenberg (1998), Regional test of a model for shallow landsliding, *Hydrol. Process., 12*, 943–955.

Moore, I. D., E. M. O'Loughlin, and G. J. Burch (1988), A contour based topographic model and its hydrologic and ecological applications, *Earth Surf. Process. Landforms, 13*, 305–320.

Moreiras, S. M. (2004), Landslide susceptibility zonation in the Rio Mendoza Valley, Argentina, *Geomorphology, 66*(1–4), 345–357.

Morgenstern, N. R., and D. A. Sangrey (1978), Methods of stability analysis, in Landslides: Analysis and Control, *Spec. Rep. 176*, pp. 155–171, Transport. Res. Board, National Academy of Science, National Res. Council, Washington, DC.

Mori, Y., K. Iwama, T. Maruyama, and T. Mitsuno (1999), Discriminating the influence of soil texture and management-induced changes in macropore flow using soft X-rays, *Soil Sci., 164*(7), 467–682.

Moroto, K. (1925), Earthquakes, mountains, and landslides, *Shinsai Yobo Chosakai*, Vol. 100, 79 p., Otsu, Japan (in Japanese).

Morris, S. E., and T. A. Moses (1987), Forest fire and the natural soil erosion regime in the Colorado Front Range, *Ann. Assoc. Am. Geogr. 77*, 245–254.

Morrison, P. H. (1975), Ecological and geomorphological consequences of mass movements in the Alder Creek watershed and implications for forest land management, B.A. thesis, 102 p., Univ. of Oregon, Eugene.

Mosley, M. P. (1979), Streamflow generation in a forested watershed, New Zealand, *Water Resour. Res.*, *15*, 795–806.

Mosley, M. P. (1980), The impact of forest road erosion in the Dart Valley, Nelson, *N.Z. J. Forestry*, *25*(2), 184–198.

Mulder, H. F. H. M., and Th. W. J. van Asch (1988), Quantitative approaches in landslide hazard analysis, *Travaux de l'Institut de Géographie de Reims*, *69–72*, 43–53.

Murphy, W. (1995), The geomorphological controls on seismically triggered landslides during the 1908 Straits of Messina earthquake, Southern Italy, *Q. J. Eng. Geol.*, *28*, 61–74.

Murray, J. B., B. Voight, and J.-P. Glot (1994), Slope movement crisis on the east flank of Mt. Etna volcano: models for eruption triggering and forecasting, *Eng. Geol.*, *38*, 245–259.

Nakamura, F. (1986), Analysis of storage and transport processes based on age distribution of sediment, *Trans. Jpn. Geomorph. Union*, *7*, 165–184.

Nakamura, F., H. Maita, and T. Araya (1995), Sediment routing analyses based on chronological changes in hillslope and riverbed morphologies, *Earth Surf. Process. Landforms*, *20*, 333–336.

Nakamura, F., F. J. Swanson, and S. M. Wondzell (2000), Disturbance regimes of stream and riparian systems—a disturbance-cascade perspective, *Hydrol. Process.*, *14*, 2849–2860.

Nakamura, H. (1999), Landslide prevention law and law concerning prevention of failure of steep slopes in Japan, in *Landslides of the World*, edited by K. Sassa, pp. 205–208, Kyoto Univ. Press, Japan.

Nakamura, S. (1996) (editor), *Landslides in Japan*, Natl. Conf. on Landslide Control, 57 p., Jpn. Landslide Soc.

Nakano, H. (1971), Soil and water conservation functions of forest on mountainous land, *Report*, 66 p., Gov. Forest Exp. Stn. (Japan), Forest Influences Div.

Natarajan, T. K., and S. C. Gupta (1980), Techniques of erosion control for surficial landslides, *Proc. of Int. Symp. Landslides*, vol. 1, pp. 413–417, New Delhi, India,

National Research Center for Disaster Prevention (1980), Background and general view of research activities on landslides in Japan, pp. 1–10, Tsukuba, Japan.

New Zealand Commission of Inquiry (1980), Report of the Commission of Inquiry into the Abbotsford Landslip Disaster, 196 p., Govt. Printer, Wellington, New Zealand.

Newland, D. H. (1916), Landslides in unconsolidated sediments with a description of some occurrences in the Hudson Valley, *N. Y. State Museum Bull.*, *187*, 79–105.

Newman, E. B., A. R. Paradis, and E. E. Brabb (1978), Feasibility and cost of using a computer to prepare landslide susceptibility maps of the San Francisco Bay Region, California, *Bull. 1443*, U.S. Geol. Surv., 27 p.

Newmark, N. M. (1965), Effects of earthquakes on dams and embankments, *Geotechnique*, *15*(2), 139–159.

Ngecu, W. M., C. M. Nyamai, and G. Erima (2004), The extent and significance of mass-movement in Eastern Africa: case studies of some major landslides in Uganda and Kenya, *Environ. Geol.*, *46*(8), 1123–1133.

Nieber, J. L., and M. F. Walter (1981), Two-dimensional soil moisture flow in a sloping rectangular region: experimental and numerical studies, *Water Resour. Res.*, *17*(6), 1722–1730.

Nilsen, T. H., and E. E. Brabb (1977), Slope-stability studies in the San Francisco Bay region, California, in *Rev. in Eng. Geology*, vol. 3, Landslides, pp. 235–243, Geol. Soc. Am., Boulder, CO.

Nilsen, T. H., and B. L. Turner (1975), The influence of rainfall and ancient landslide deposits on recent landslides (1950–1971) in urban areas of Contra Costa County, California, *U.S. Geol. Surv. Bull. 1388*, 18 p.

Nilsen, T. H., R. H. Wright, T. C. Vlasic, and W. E. Spangle (1979), Relative slope stability and land-use planning in the San Francisco Bay region, California, *Prof. Pap. 944*, 96 p., U.S. Geol. Surv.

Nishida, A., Kobashi, S. and T. Mizuyama (1996), Analysis of mountain slope failure distribution during the Hyogo-ken Nanbu earthquake using a database of soil disasters, *Shin Sabo J. (J. Jpn. Soc. Erosion Control Eng.)*, *49*, 19–24 (in Japanese).

Nishimura, K. and Morii (1983), An observed effect of topography on seismic ground motions, *Jisin (J. Seismol. Soc. Jpn.)*, 36, 383–392 (in Japanese).

Nishiyama, K., and M. Chigira (2002), Geological features of slope failures due to the 1982 heavy rainfall disaster in Nagasaki, Japan, *Ann. Disaster Prevent. Res. Inst. Kyoto Univ., 45B*, 47–59.

Niu, F., G. Cheng, W. Ni, and D. Jin (2005), Engineering-related slope failure in permafrost regions of the Qinghai-Tibet Plateau, *Cold Reg. Sci. & Tech., 42*, 215–225.

Niyazou, R. A. (1982) The formation of large landslides in Middle Asia. Fan Publ., Uzbek Republic, 156 p. (in Russian).

Noguchi, S., Y. Tsuboyama, R. C. Sidle, and I. Hosoda (1999), Morphological characteristics of macropores and distribution of preferential flow pathways in a forested slope segment, *Soil Sci. Soc. Am. J., 63*, 1413–1423.

Noguchi, S., Y. Tsuboyama, R. C., Sidle, and I. Hosoda (2001), Subsurface runoff characteristics from a forest hillslope soil profile including macropores, Hitachi Ohta, Japan, *Hydrol. Process., 15*, 2131–2149.

Nolan, M. F. (1984), Vegetation of U.S. Army Corps of Engineers project levees on the Sacramento/San Joaquin Valley, California, in *California Riparian Systems: Ecology, Conservation and Productive Management*, edited by R.E. Warner and K.M. Hendrix, pp. 738–747, University of California Press, Berkeley.

Nunamoto, S., M. Suzuki, and T. Ohta (1999), Decreasing trend of deaths and missings caused by sediment disasters in the last fifty years in Japan, *Shin Sabo J. (J. Jpn. Soc. Erosion Contr. Eng.) 51*, 3–12.

Nyberg, L., A. Rodhe, and K. Bishop (1999), Water transit times and flow paths from two line injections of ^{3}H and ^{36}Cl in a microcatchment at Gårdsjön, Sweden, *Hydrol. Process., 13*, 1557–1575.

Nyberg, R. (1989), Observations of slushflows and their geomorphological effects in the Swedish mountain area, *Geograf. Ann., 71A*(3–4), 185–198.

Nyssen, J., J. Moeyersons, J. Poesen, and M. Haile (2002), The environmental significance of the remobilisation of ancient mass movements in the Atbara-Tekeze headwaters, northern Ethiopia, *Geomorphology, 49*, 303–322.

Ocakoglu, F., C. Gokceoglu, and M. Ercanoglu (2002), Dynamics of a complex mass movement triggered by heavy rainfall: a case study from NW Turkey, *Geomorphology, 42*(3/4), 329–341.

Ochiai, H. (1997), Mechanisms of slope movement during an earthquake, slope disasters during an earthquake, Hokkaido Branch, Japan Landslide Society, pp. 3–12 (in Japanese).

Ochiai, H., S. Matsuura, and H. Yanase (1987), Pore pressure change in soil layer of a landslide area during earthquakes, *Proc. 5th Int. Conf. and Field Workshop on Landslides*, pp. 205–210, Tokyo, Japan.

Ochiai, H., H. Kitahara, T. Sammori, and K. Abe (1995), Earthquake-induced landslides and earthquake response analysis, Research Report of Landslides and Slope Failures triggered the Hyogoken-Nanbu Earthquake, Japan Landslide Soc., 119–132 (in Japanese).

Ochiai, H., H. Kitahara, T. Sammori, and K. Abe (1996), Landslides triggered by the 1995 Hyogo-Ken Nanbu Earthquake in the Rokko Mountains, in *Landslide/Glissements de Terrain*, edited by K. Senneset, A.A. Balkema, Rotterdam.

Ochiai, H., Y. Okada, G. Furuya, Y. Okura, T. Matsui, T. Sammori, T. Terajima, T., and K. Sassa (2004), A fluidized landslide on a natural slope by artificial rainfall, *Landslides, 1*, 211–219.

Ochiai, H., H. Yanase, and S. Matsuura (1985), Measurements of earthquake motion and pore water pressure at the Yui landslide area, *Proc. 4th Int.. Conf. and Field Workshop on Landslides*, pp. 203–208, Tokyo, Japan.

Oguchi, C. T., and Y. Matsukura (1999), Effect of porosity on the increase in weathering-rind thicknesses of andesite gravel, *Eng. Geol., 55,* 77–89.

Ohta, T., Y. Tsukamoto, and H. Noguchi (1981), An analysis of pipeflow and landslide, *Proc. Ann. Mtg. Jpn. Soc. Erosion Control Eng.*, pp. 92–93 (in Japanese).

Okamoto, T., J. O. Larsen, S. Matsuura, S. Asano, Y. Takeuchi, and L. Grande (2004), Displacement properties of landslide masses at the initiation of failure in quick clay deposits and the effects of meteorological and hydrological factors, *Eng. Geol., 72,* 233–251.

Okimura, T. (1982), Situation of surficial slope failure based on the distribution of soil layer, *Shin Sabo J. (J. Jpn. Soc. Erosion Contr. Eng.)* 35, 9–18 (in Japanese with English abstract).

Okimura, T., and Y. Tanaka (1999), Geographical and geo-technical considerations of the damage caused by the Kobe earthquake, 1995—researches for the distribution of damaged housing in urban area, *Trans. Jpn. Geomorph. Union, 20,* 373–382.

Okuda, S., K. Ahida, Y. Gocho, K. Okunishi, T. Sawada, and K. Yokoyama (1979), Characteristics of heavy rainfall and debris hazard, *J. Nat. Disaster Sci., 1*(2), 41–55.

Okunishi, K. (1982), Kinematics of large-scale landslides—a case study in Fukuchi, Hyogo Prefecture, Japan, *Trans. Jpn. Geomorph. Union, 3,* 41–56.

Okunishi, K., and T. Iida (1981), Evolution of hillslopes including landslides, *Trans. Jpn. Geomorph. Union, 2,* 291–300.

Okunishi, K., M. Sonoda, and K. Yokoyama (1999), Geomorphic and environmental controls of earthquake-induced landslides, *Trans. Jpn. Geomorph. Union, 20,* 351–368.

Okura, Y., H. Kitahara, H. Ochiai, T. Sammori, and A. Kawanami (2002), Landslide fluidization process by flume experiments, *Eng. Geol., 66,* 65–78.

O'Loughlin, C. L. (1972), The stability of steepland forest soils in the Coast Mountains, southwest British Columbia. Ph.D. thesis, 147 p., Univ. of British Columbia, Vancouver, Canada.

O'Loughlin, C. L. (1974a), The effect of timber removal on the stability of forest soils, *J. Hydrol. N. Z., 13,* 121–123.

O'Loughlin, C. L. (1974b), A study of tree root strength deterioration following clearfelling, *Can. J. For. Res., 4*(1), 107–113.

O'Loughlin, C. L., and M. Gage (1975), A report on the status of slope erosion on selected steep areas, West Coast beech project area, New Zealand, 86 p.

O'Loughlin, C. L., and A. J. Pearce (1976), Influence of Cenozoic geology on mass movement and sediment yield response to forest removal, North Westland, New Zealand. *Bull. Int. Assoc. Eng. Geol. 14,* 41–46.

O'Loughlin, C. L., and A. J. Watson (1979), Root wood strength deterioration in radiata pine after clearfelling, *N. Z. J. For. Sci., 9*(3), 284–293.

O'Loughlin, C. L., and A. J. Watson (1981), Root wood strength deterioration in beech (*Nothofagus fusca* and *M. truncata*) after clearfelling, *N. Z. J. For. Sci., 11*(2), 183–185.

O'Loughlin, C. L., and R. R. Ziemer (1982), The importance of root strength and deterioration rates upon edaphic stability in steepland forests, in *Carbon Uptake and Allocation in Subalpine Ecosystems as a Key to Management*, pp. 70–78, Proc. of an IUFRO Workshop, Aug. 2–3, 1982, Oregon State Univ., Corvallis.

O'Loughlin, C. L., L. K. Rowe, and A. J. Pearce (1982), Exceptional storm influences on slope erosion and sediment yield in small forest catchments, North Westland, New Zealand, *The First National Symp. on Forest Hydrology* edited by E. M. O'Loughlin and E. J. Bren, pp. 84–91, Inst. of Eng., Melbourne, Australia.

O'Loughlin, E. M. (1986), Prediction of surface saturation zones in natural catchments by topographic analysis, *Water Resour. Res., 22,* 794–804.

Olshansky, R. B. (1986), Geologic hazard abatement districts, *Calif. Geol., 39*(7), 158–159.

Olshansky, R. B. (1989), Landslide hazard reduction: a need for greater government involvement, *Zoning & Planning Law Rep., 12*(3), 106–112.

Olshansky, R. B. (1996), Financing hazard mitigation in the United States, *J. Environ. Plan. Manage., 39*, 371–385.

Olshansky, R. B. (1998), Regulation of hillside development in the United States, *Environ. Manage., 22*, 383–392.

Oltchev, A., J. Constantin, G. Gravenhorst, A. Ibrom, J. Schmidt, M. Falk, K. Morgenstern, I. Richter, and N. Vygodskaya (1996), Application of a six-layer SVAT model for simulation of evapotranspiration and water uptake in a spruce forest, *Phys. Chem. Earth, 21*(3), 195–199.

Omura, H., and F. Nakamura, (1983), Some features of landslide caused by Typhoon 18, 1982 at different land use in Fujieda, *Proc. Ann. Congress of Erosion Control Eng.*, pp. 20–21, Jpn. Soc. Erosion Control Eng., Tokyo.

Onda, Y. (1992), Influence of water storage capacity in the regolith zone on hydrologic characteristics, slope processes, and slope form, *Z. Geomorph. N.F., 36*, 165–178.

Onda, Y. (1994), Seepage erosion and its implication to the formation of the amphitheatre valley heads: a case study at Obara, Japan, *Earth Surf. Process. Landforms, 19*, 627–640.

Onda, Y., and N. Itakura (1997), An experimental study on the burrowing activity of river crabs on subsurface water movement and piping erosion, *Geomorphology, 20*, 279–288.

Onda, Y., and Y. Matsukura (1997), Mechanism for the instability of slopes composed of granular materials. *Earth Surf. Process. Landforms, 22*, 401–411.

Onda, Y., Y. Komatsu, M. Tsujimura, and J. Fujihara (2001), The role of subsurface runoff through bedrock on storm flow generation, *Hydrol. Process., 15*, 1693–1706.

Onda, Y., A. Mori, and S. Shindo (1992), Effect of topographic convergence and location of past landslides on subsurface water movement on granitic hillslope, *J. Nat. Disaster Sci., 14*, 45–58.

Onda, Y., M. Tsujimura, and H. Tabuchi (2004), The role of subsurface water flow paths on hillslope hydrological processes, landslides and landform development in steep mountains of Japan, *Hydrol. Process., 18*, 637–650.

O'Neill, M. P., and D. M. Mark (1987), On the frequency distribution of land slope, *Earth Surf. Process. Landforms, 12*, 127–136.

Oostwoud Wijdenes, D. J., and P. Ergenzinger (1998), Erosion and sediment transport on steep marly hillslopes, Draix, Haute-Provence, France: an experimental field study, *Catena, 33*(3/4), 179–200.

Oregon Department of Forestry (2003a), High landslide hazard locations, shallow, rapidly moving landslides and public safety: screening and practices, *Forest Practices Tech. Note No. 2*, vers., 2.0., 12 p., Oregon Depart. of Forestry, Salem, OR.

Oregon Department of Forestry (2003b), Determination of rapidly moving landslide impact rating, *Forest Practices Tech. Note No. 6*, vers., 1.0, 11 p., Oregon Dept. of Forestry, Salem, OR.

O'Rouke, T. D., and C. J. F. P. Jones (1990), Overview of earth retention systems: 1970–1990, in *Design and Performance of Earth Retaining Structures*, edited by P. C. Lambe and L. A. Hansen, pp. 22–51, Proc. Specialty Conf., Ithaca, New York, Geotech. Special Publ. 25, Am. Soc. Civ. Engr., New York.

Osei, W. Y. (1993), Woodfuel and deforestation—answers for a sustainable environment, *J. Environ. Manage., 37*, 51–62.

Owens, I. F. (1969), Causes and rates of soil creep in the Chilton Valley, Cass, New Zealand, *Arct. Alp. Res., 1*(3), 213–220.

Oyagi, N. (1977) (editor), *Guide-book for excursions of landslides in central Japan*, Jpn. Soc. of Landslide, Tokyo, Japan, 29 p.

Oyagi, N. (1992), Sediment Disasters, in *Encyclopedia of Disasters*, edited by Y. Hagiwara, pp. 179–252, Asakura Publ Co., Japan (in Japanese).

Pachauri, A.K., and M. Pant (1992), Landslide hazard mapping based on geological attributes, *Eng. Geol. 32*, 81–100.

Pachauri, A. K., P. V. Gupta, and R. Chander (1998), Landslide zoning in a part of the Garhwal Himalayas, *Environ. Geol., 36*(3–4), 325–334.

Pack, R. T. (1997), New developments in terrain stability mapping in B.C., *Proc. 11th Vancouver Geotech. Soc. Symp.*, May 30, 1997, Canada.

Packer, P. E. (1967), Criteria for designing and locating roads to control sediment, *Forest Sci., 13*(1): 1–18.

Paeth, R. C., M. E. Harward, E. G. Knox, and C. T. Dyrness (1971), Factors affecting mass movement of four soils in the western Cascades of Oregon, *Soil Sci. Soc. Am. Proc., 35*(6), 943–947.

Page, M. J., and N. A. Trustrum (1997), A late Holocene lake sediment record of the erosion response to land use change in a steepland catchment, New Zealand, *Z. Geomorph. N.F., 41*(3), 369–392.

Painter, D. J. (1981), Steeplands erosion processes and their assessment, *IAHS Publ. 132*, 2–20.

Palacios, D., R. García, V. Rubio, and R. Vigil (2003), Debris flows in a weathered granitic massif: Sierra de Gredos, Spain, *Catena, 51*, 115–140.

Palacios, D., J. Zamorano, and A. Gómez (2001), The impact of present lahars on the geomorphologic evolution of proglacial gorges: Popocatepetl, Mexico, *Geomorphology, 37*, 15–42.

Palkovics, W. E., and G. W. Petersen (1977), Contribution of lateral soil water movement above a fragipan to streamflow, *Soil Sci. Soc. Am. J., 41*, 394–400.

Palladino, D. J., and R. B. Peck (1972), Slope failures in an overconsolidated clay, Seattle, Washington, *Geotechnique, 22*(4), 563–595.

Palmer, L. (1977), Large landslides of the Columbia River Gorge, Oregon and Washington, in *Reviews in Eng. Geol.*, vol. 3, Landslides, pp. 69–83, Geol. Soc. Am., Boulder, CO.

Papadopoulos, G. A., and A. Plessa (2000), Magnitude-distance relations for earthquake-induced landslides in Greece, *Eng. Geol., 58*(3–4), 377–386.

Parise, M. (2001), Landslide mapping technique and their use in the assessment of the landslide hazard, *Phys. Chem. Earth, 26*(9), 697–703.

Parry, S., S. D. C. Campbell, and C. J. Churchman (2000), Kaolin-rich zones in Hong Kong saprolites—their interpretation and engineering significance, *Proc., GeoEng 2000*, Nov. 19–24, Melbourne, Australia.

Pasuto, A., and S. Silvano (1998), Rainfall as a trigger of shallow mass movements. A case study in the Dolomites, Italy, *Environ. Geol., 35*(2–3), 184–189.

Pearce, A. J. (1982), Complex mass movement terrain in the eastern Raukumara Peninsula, New Zealand: Lithologic and structural-tectonic influences and the effect of recent deforestation and reafforestation, *Proc. Int. Seminar on Landslides and Mudflows*, p. 235–249, Alma-Ata, USSR, UNESCO, Paris.

Pearce, A. J., and J. A. Elson (1973), Postglacial rates of denudation by soil movement, free face retreat, and fluvial erosion, Mont. St. Hilaire, Quebec, *Can. J. Earth Sci., 10*, 91–101.

Pearce, A. J., and A. J. Watson (1983), Medium-term effects of two landsliding episodes on channel storage of sediment, *Earth Surf. Process. Landforms, 8*, 29–39.

Pearce, A. J., R. D. Black, and C. S. Nelson (1981), Lithologic and weathering influences on slope form and process, eastern Raukumara Range, New Zealand, *IAHS Publ., 132*, 95–122.

Pearce, A. J., C. L. O'Loughlin, and A. J. Watson (1985), Medium-term effects of landsliding and related sedimentation evaluated fifty years after an M7.7 earthquake, *Int. Symp. on Erosion, Debris Flow and Disaster Prevention*, Sept. 3–5, 1985, Tsukuba, Japan.

Pearce, A. J., M. K. Stewart, and M. K. Sklash (1986), Storm runoff generation in humid headwater catchments, 1. Where does the water come from?, *Water Resour. Res. 22*, 1263–1272.

Peart, M. R., K. Y. Ng, and D. D. Zhang (2005), Landslides and sediment delivery to a drainage system: some observations from Hong Kong, *J. Asian Earth Sci., 25*(5), 821–836.

Peck, D. L., A. B. Griggs, H. G. Schlicker, F. G. Wells, and H. M. Dole (1964), Geology of the central and northern parts of the western Cascade Range in Oregon, *Prof. Pap. 449*, 56 p., U.S. Geol. Surv.

Pellegrini, G. B. (1979), Geomorphological field experiments in the Italian Alps, *Studia Geomorphol. Carpatho-Balcanica, 13*, 81–95.

Pender, M. J. (1976), Probabilistic assessment of the stability of a cut slope, *N. Z. Eng., 15*, 239–241.

People's Daily Online (2002), Landslide suspends traffic on Beijing–Kowloon Railway, *People's Daily Online* August 20, 2002.

Peresan, A., V. Kossobokov, L. Romahkova, and G. F. Panza (2005), Intermediate-term middle-range earthquake predictions in Italy: a review, *Earth Sci. Rev., 69*, 97–132.

Perotto-Baldiviezo, H. L., T. L. Thurow, C. T. Smith, R. F. Fisher, and X. B. Wu (2004), GIS-based spatial analysis and modeling for landslide hazard assessment in steeplands, southern Honduras, *Agric. Ecosyst. Environ., 103*(1), 165–176.

Peters, P. A., and C. J. Biller (1986), Preliminary evaluation of the effect of vertical angle of pull on stump uprooting failure, *Proc. Improving Productivity through Forest Engineering*, pp. 90–93, Council of Forest Engineering, Mobile, Alabama.

Pettapiece, W. W., and S. Pawluk (1972), Clay mineralogy of soils developed partially from volcanic ash, *Soil Sci. Soc. Am. Proc., 36*, 515–519.

Phien-Wej, N., P. Nutalaya, Z. Aung, and T. Zhibin (1993), Catastrophic landslides and debris flows in Thailand, *Bull. Int. Assoc. Eng. Geol., 48*, 93–100.

Philip, J. R. (1957), The theory of infiltration, 4, Sorptivity and algebraic infiltration equations, *Soil Sci., 84*, 257–264.

Phillips, C. J. (1988), Geomorphic effects of two storms on the upper Waitahaia River catchment, Raukumara Peninsula, New Zealand, *J. Hydrol. N.Z., 27*, 99–112.

Phillips, C. J., and A. J. Watson (1994), Structural tree root research in New Zealand: a review, *Landcare Res. Sci. Series No. 7*, 71 p., Lincoln, New Zealand.

Phillips, C. J., M. Marden, and A. J. Pearce (1990), Effectiveness of reforestation in prevention and control of landsliding during large cyclonic storms, *Proc. 19th IUFRO World Congress*, Div. 1, vol. 1., pp. 340–350, Montreal, Canada.

Phillips, R. W. (1971), Effects of sediment on the gravel environment and fish production, in *Forest Land Use and Stream Environment*, pp. 64–74, Oregon State Univ., Corvallis.

Piehl, B. T., R. L. Beschta, and M. R. Pyles (1988), Ditch-relief culverts and low-volume forest roads in the Oregon Coast Range, *Northwest Sci. 62*(3), 91–98.

Pierson, T. C. (1977), Factors controlling debris flow initiation on forested hill slopes in the Oregon Coast Range, Ph.D. Thesis, 167 p., Dept. of Geological Sciences, Univ. of Washington, Seattle, WA.

Pierson, T. C. (1980a), Erosion and deposition by debris flows at Mt. Thomas, North Canterbury, New Zealand, *Earth Surf. Process., 5*, 227–247.

Pierson, T. C. (1980b), Piezometric response to rainstorms in forested hillslope drainage depressions, *J. Hydrol. N. Z., 19*, 1–10.

Pierson, T. C. (1983), Soil pipes and slope stability, *Q. J. Eng. Geol., 16*, 1–11.

Pierson, T. C., and K. M. Scott (1985), Downstream dilution of a lahar: Transition from debris flow to hyperconcentrated streamflow, *Water Resour. Res., 21*(10), 1511–1524.

Pierson, T. C., R. J. Janda, J. C. Thouret, and C. A. Borrero (1990), Perturbation and melting of snow and ice by the 13 November 1985 eruption of Nevado del Ruiz, Columbia, and consequent mobilization, flow, and deposition of lahars, *J. Volcanol. Geotherm. Res., 41*, 17–66.

Pitlick, J. (1995), Sediment routing in tributaries of the Redwood Creek basin, northwestern California, edited by K. M. Nolan, H. M. Kelsey, and D. C. Marron, Geomorphic processes and aquatic habitat in the Redwood Creek basin, northwestern California, *Prof. Pap. 1454*, 10 p., U.S. Geol. Surv..

Plafker, G., and R. Kachadoorian (1966), Geologic effects of the March, 1964, earthquake and associated seismic sea waves on Kodiak and nearby islands, Alaska, *Prof Pap. 543-D*, 46 p., U.S. Geol. Surv.

Plafker, G., G. E. Ericksen, and J. F. Concha (1971), Geological aspects of the May 31, 1970, Peru earth quake, *Bull. Seismol. Soc. Am., 61*(3), 543–578.

Platts, W. S., and W. F. Megahan (1975), Time trends in riverbed sediment composition in salmon and steelhead spawning areas: South Fork Salmon River, Idaho, *Trans. N. Am. Wildlife and Nat. Resour. Conf., 40*, pp. 229–239.

Pole, M. W., and D. R. Satterlund (1978), Plant indicators of slope instability, *J. Soil Water Conserv., 33*, 230–232.

Pomeroy, J. S. (1982), Mass movement in two selected areas of western Washington County, Pennsylvania, *Prof. Pap. 1170-B*, 17 p., U.S. Geol. Surv.

Popescu, M. E., A. Trandafir, A. Fedrrico, and V. Simeone (1998), Probabilistic risk assessment of landslide related geohazards, in *Geotechnical Hazards*, edited by B. Mari , Z. Lisac, and A. Szavits-Nossan, pp. 863–870, Balkema, Rotterdam.

Post, J. L., and L. Borer (2002), Physical properties of selected illites, beidellites and mixed-layer illite-beidellites from southwestern Idaho, and their infrared spectra, *Appl. Clay Sci., 22*(3), 77–91.

Potts, D. F., and B. K. M. Anderson (1990), Organic debris and the management of small channels, *Western J. App. For., 5*, 25–28.

Preston, N. J., and M. J. Crozier (1999), Resistance to shallow landslide failure through root-derived cohesion in east coast country soils, North Island, New Zealand, *Earth Surf. Process. Landforms, 24*, 665–675.

Prior, D. B., and C. Ho (1972), Coastal and mountain slope instability on the islands of St. Lucia and Barbados, *Eng. Geol., 6*, 1–18.

Pulinets, S. A. (2006), Space technologies for short-term earthquake warning, *Adv. Space Res.*, doi:10.1016/ j.asr.2004.12.074.

Pyles, M. R., and J. Stoupa (1987), Load-carrying capacity of second-growth Douglas-fir stump anchors, *Western J. App. For., 2*, 77–80.

Quinton, W. L., and P. Marsh (1999), A conceptual framework for runoff generation in a permafrost environment, *Hydrol. Process., 13*, 2563–2581.

Rahn, P. H. (1969), The relationship between natural forested slopes and angles of repose for sand and gravel, *Geol. Soc. Am. Bull., 80*, 2123–2128.

Rao, S. M. (1995), Mechanistic approach to the shear strength behaviour of allophonic soils, *Eng. Geol., 40*, 215–211.

Rapp, A. (1960), Recent development of mountain slopes in Karkevagge and surroundings, northern Scandinavia, *Geograf. Ann., 42*(2–3), 71–200.

Rapp, A., and H. J. Åkerman (1993), Slope processes and climate in the Abisko Mountains, northern Sweden, *Paläoklimaforschung, 11*, 163–177.

Rapp, A., and R. Nyberg (1981), Alpine debris flows in northern Scandinavia: morphology and dating by lichenometry, *Geograf. Ann., 63*, 183–196.

Reddi, L. N., and T. H. Wu (1991), Probabilistic analysis of ground-water levels in hillside slopes, *J. Geotech. Eng, 117*(6), 872–890.

Reeves, G. H., L. E. Benda, K. M. Burnett, P. A. Bisson, and J. R. Sedell (1995), A disturbance-based eco-system approach to maintaining and restoring freshwater habitats of evolutionarily significant units of anadromous salmonids in the Pacific Northwest, in *Evolution and the Aquatic Ecosystem: Defining Unique Units in Population Conservation*, edited by J. Nielsen, pp. 334–349, Am. Fish. Soc. Mono. Ser. No. 17, Bethesda, MD.

Refice, A., and D. Capolongo (2002), Probabilistic modeling of uncertainties in earthquake-induced landslide hazard assessment, *Comp. Geosci., 28,* 735–749.

Regalado, C. M. (2005), On the distribution of scaling hydraulic parameters in a spatially aniso-tropic banana field, *J. Hydrol., 307,* 112–125.

Reid, M. E. (1994), A pore-pressure diffusion model for estimating landslide-inducing rainfall, *J. Geol., 102,* 709–717.

Reid, M. E. (1997), Slope instability caused by small variations in hydraulic conductivity, *J. Geotech. Geoenviron. Eng., 8,* 717–725.

Reid, M. E., and R. M. Iverson (1992), Gravity-driven groundwater flow and slope failure potential. 2. Effects of slope morphology, material properties, and hydraulic heterogeneity, *Water Resour. Res., 28,* 939–950.

Reid, M. E., S. B. Christian, and D. L. Brien (2000), Gravitational stability of three-dimensional stratovolcano edifices, *J. Geophys. Res., 105*(B3), 6043–6056.

Reid, M. E., R. G. LaHusen, and R. M. Iverson (1997), Debris-flow initiation experiments using diverse hydrologic triggers, *Proc. 5th Int. Conf. on Debris-flow hazards and mitigation: mechanics, prediction and assessment*, pp. 1–11, Water Resour. Eng. Div., San Francisco, CA.

Reid, M. E., R. G. LaHusen, and W. L. Ellis (1999), Real-time monitoring of active landslides, *Fact Sheet 091–99*, 4 p., U.S. Geol. Surv.

Reid, M. E., T. W. Sisson, and D. L. Brien (2001), Volcano collapse promoted by hydrothermal alteration and edifice shape, Mount Rainier, Washington, *Geology, 29*(9), 779 – 782.

Reifsnyder, W. E., and H. W. Lull (1965), Radiant energy in relation to forests, *Tech. Bull. No. 1344*, For. Serv. U.S. Dept. of Agric., Washington, DC.

Reneau, S. L., and W. E. Dietrich (1991), Erosion rates in the southern Oregon Coast Range: evidence for an equilibrium between hillslope erosion and sediment yield, *Earth Surf. Process. Landforms, 16,* 307–322.

Reneau, S. L., W. E. Dietrich, D. J. Donahue, A. J. T. Juli, and M. Rubin (1990) Late Quaternary history of colluvial deposition and erosion in hollows, central California Coast Ranges, *Geol. Soc. Am. Bull., 102,* 969–982.

Reneau, S. L., W. E. Dietrich, R. I. Dorn, C. R. Berger, and M. Rubin (1986), Geomorphic and paleoclimatic implications of latest Pleistocene radiocarbon dates from colluvium-mantled hollows, California, *Geology, 14,* 655–658.

Reneau, S. L., W. E. Dietrich, M. Rubin, D. J. Donahue, and A. J. T. Juli (1989), Analysis of hillslope erosion rates using dated colluvial deposits, *J. Geol., 97,* 45–63.

Rice, R. M. (1974), The hydrology of chaparral watersheds, *Proc. Symp. on Living with the Chaparral*, pp. 27–34, Sierra Club, San Francisco, CA.

Rice, R. M. (1977), Forest management to minimize landslide risk, in *Guidelines for Watershed Management*, edited by S. H. Kunkle and J. L. Thames, pp. 271–287, Food and Agric. Organ. of the UN, Rome, Italy.

Rice, R. M. (1982), Sedimentation in the chaparral: How do you handle unusual events? in *Sediment Budgets and Routing in Forested Drainage Basins, Gen. Tech. Rep. PNW-141*, pp. 39–49, For. Serv., U.S. Dep. of Agric., Portland, Oreg.

Rice, R. M. (1999), Erosion on logging roads in Redwood Creek, northwestern California, *J. Am. Water Resour. Assoc., 35*(5), 1171–1180.

Rice, R. M., and J. Lewis (1991), Estimating erosion risks associated with logging and forests roads in northwestern California, *Water Resour. Bull., 27*(5), 809–818.

Rice, R. M., E. S. Corbett, and R. G. Bailey (1969), Soil slips related to vegetation, topography, and soil in southern California, *Water Resour. Res., 5*(3), 647–659.

Rice, R. M., N. H. Pillsbury, and K. W. Schmidt (1985), A risk analysis approach for using discriminant functions to manage logging-related landslides on granitic terrain, *Forest Sci., 31,* 772–784.

Rice, R. M., R. R. Ziemer, and S. C. Hankin (1982), Slope stability effects of fuel management strategies: Inferences from Monte Carlo simulations, *Gen. Tech. Rep. PSW-58,* pp. 365–371, For Serv., U.S. Dep. of Agric., Berkeley, CA.

Rice, S., and M. Church (1996), Bed material texture in low order streams on the Queen Charlotte Islands, British Columbia, *Earth Surf. Process. Landforms, 21,* 1–18.

Richards, L. A. (1931), Capillary conduction of liquids through porous mediums, *Physics, 1,* 318–333.

Rickenmann, D., and C.-L. Chen (2003) (eds.), Debris-flow hazards mitigation: mechanics, prediction, and assessment, *Proc. 3rd Int. Conf. on Debris-flow Hazards Mitigation,* Millpress, Rotterdam, The
Netherlands, 1392 p.

Riestenberg, M. M., and S. Sovonick-Dunford (1983), The role of woody vegetation in stabilizing slopes in the Cincinnati area, Ohio, *Geol. Soc. Am. Bull., 94,* 506–518.

Roberge, J., and A. P. Plamondon (1987), Snowmelt runoff pathways in a boreal forest hillslope, the role of pipe throughflow, *J. Hydrol., 95,* 39–54.

Roberts, B., B. Ward, and T. Rollerson (2004), A comparison of landslide rates following helicopter and conventional cable-based clear-cut logging operations in the Southwest Coast Mountains of British Columbia , *Geomorphology, 61*(3/4), 337–346.

Robison, E. G., K. A. Mills, J. Paul, L. Dent, and A. Skaugset (1999), Storm impacts and landslide of 1996: final report, *Forest Practices Tech. Rep. No. 4.,* 145 p., Oregon Dept. of Forestry, Salem.

Rockey, R. E., and K. E. Bradshaw (1962), Assessing soil stability for road location on the Six Rivers National Forest, California, Report, 39 p., For. Serv., U.S. Dep. of Agric., Reg. 5, Div. of Watershed Mgmt. and Div. of Eng., San Francisco, CA.

Rodrigues, D., and F. J. Ayala-Carcedo (2003), Rain-induced landslides and debris flows on Madeira Island, Portugal, *Landslide News (Jpn. Landslide Soc.), 14/15,* 43–45.

Roering, J., K. M. Schmidt, J. D. Stock, W. E. Dietrich, and D. R. Montgomery (2003), Shallow landsliding, root reinforcement, and the spatial distribution of tress in the Oregon Coast range, *Can. Geotech. J., 40,* 237–253.

Roering, J. J., J. W. Kirchner, L. S. Sklar, and W. E. Dietrich (2001), Hillslope evolution by nonlinear creep and landsliding; an experimental study, *Geology, 29,* 143–146.

Rogers, H. H., G. B. Runion, and S. V. Krupa (1994), Plant responses to atmospheric CO_2 enrichment with emphasis on roots and the rhizosphere, *Environ. Pollut., 83,* 155–189.

Rogers, J. D. (1992), Recent developments in landslide mitigation techniques, in *Landslides/Landslide Mitigation,* edited by J. E. Slosson, A. G. Keens and J. A. Johnson, vol. 9, pp. 95–118, Rev. Eng. Geol., Geol. Soc. Am., Boulder, Colo.

Rogers, N. W., and M. J. Selby (1980), Mechanisms of shallow translational landsliding during summer rainstorms: North Island, New Zealand, *Geograf. Ann., 62A,* 11–21.

Rogers, W. P., E. English, R. L. Schuster, and R. M. Kirkham (1999), Large rock slide/debris avalanche in the San Juan Mountains, southwestern Colorado, USA, July 1991, in *Landslides of the World,* edited by K. Sassa, pp. 51–53, Kyoto University Press, Japan.

Roghair, C. N., C. A. Dolloff, and M. K. Underwood (2002), Response of a brook trout population and instream habitat to a catastrophic flood and debris flow, *Trans. Am. Fish. Soc., 131,* 718–730.

Rohdenburg, H. (1989), *Landscape Ecology—Geomorphology,* Catena Paperback, Reiskirchen, Germany, 177 p.

Rohn, J., T. Fernandez-Steeger, D. Sidow, and K. Czurda (2003), Rock fall triggers earthflow by undrained loading at Steinbach, Austria, April 1995, *Landslide News (Jpn. Landslide Soc.),14/15*, 33–35.

Rollerson, T. P., and M. W. Sondheim (1985), Predicting post-logging terrain stability: a statistical-geographical approach, in *Improving Mountain Logging Planning, Techniques and Hardware*, Proc. Joint Symp. IUFRO Mountain Logging Sect. and the 6th Pacific Northwest Skyline Logging Symp., Forest Eng. Res. Inst. of Canada.

Rollerson, T. P., D. E. Howes, and M. W. Sondheim (1986), An approach to predicting post-logging slope stability for coastal British Columbia, NCASI West Coast Regional Meeting, Portland, Oregon.

Rollerson, T. P., B. Thomson, and T. H. Millard (1997), Identification of coastal British Columbia terrain susceptible to debris flows, in *Debris-Flow Hazards Mitigation: Mechanics, Prediction and Assessment*, Am. Soc. Civ. Engr., San Francisco, CA.

Romme, W. H., and D. G. Despain (1989), The Yellowstone fires, *Scientific Am. 261*(5): 37–46.

Rooyani, F. (1985), A note on soil properties influencing piping at the contact zone between albic and argillic horizons of certain duplex soils (aqualfs) in Lesotho, southern Africa, *Soil Sci., 139*(6), 517–522.

Rosenfeld, C. L. (1999), Forest engineering implications of storm-induced mass wasting in the Oregon Coast Range, USA, *Geomorphology, 31*, 217–228.

Rosenquist, I. T. (1953), Sensitivity of Norwegian quick clays, *Geotechnique*, 3, 195–200.

Rost, K. T. (1999), Observations on deforestation and alpine turf destruction in the Central Wutai Mountains, Shanxi Province, China, *Mountain Res. Develop., 19*(1), 31–40.

Rozier, I. T., and M. J. Reeves (1979), Ground movement at Runswick Bay, North Yorkshire, *Earth Surf. Process. Landforms, 4*, 275–280.

Rupke, J., E. Cammeraat, A. C. Seijmonsbergen, and C. J. van Westen (1988), Engineering geomorphology of the Widentobel catchment, Appenzell and Sankt Gallen, Switzerland. A geomorphological inventory system applied to geotechnical appraisal of slope stability, *Eng. Geol., 26*, 33–68.

Russell, M. J. (1979), Spoil-heap revegetation at open-cut coal mines in the Bowen Basin of Queensland, Australia, in *Ecology and Coal Resource Development, Vol. 1*, edited by M. K. Wali, pp. 516–523, Pergamon, New York.

Sah, M. P., and R. K. Mazari (1998), Anthropogenically accelerated mass movement, Kulu Valley, Himachal Pradesh, India, *Geomorphology, 26*, 123–138.

Saito, M., T. Araya, and F. Nakamura (1995), Sediment production and storage processes associated with earthquakes-induced landslides in Okushiri Island, 1993, *Shin Sabo J. (J. Jpn. Soc. Erosion Contr. Eng.), 47*(6), 28–33 (in Japanese with English abstract).

Sakals, M. E., and R. C. Sidle (2004), A spatial and temporal model of root strength in forest soils, *Can. J. For. Res., 34*, 950–958.

Sakellariou, M. G., and M. D. Ferentinou (2001), GIS-based estimation of slope stability, *Nat. Hazards Rev.*, 12–21.

Samarakoon, L., S. Ogawa, N. Ebisu, R. Lapitan, and Z. Kohri (1993), Inference of landslide susceptibility areas by Landsat Thematic Mapper data, Proc. of the Yokohama Symp., *IAHS Publ. 217*, 83–90.

Sammori, T. (1995), A numerical method of quasi three dimensional analysis of rain-induced landslide by axisymmetric coordinates, *Shin Sabo J. (J. Jpn. Soc. Erosion Contr. Eng.), 48*(2) (in Japanese with English abstract).

Sammori, T., and Y. Tsuboyama (1990), Study on method of slope stability considering infiltration phenomenon, *Shin Sabo J. (J. Jpn. Soc. Erosion Contr. Eng.), 43*(4), 14–21 (in Japanese with English abstract).

Sammori, T., and Y. Tsuboyama (1992), Parametric simulation of shallow-sheeted landslides with consideration of seepage process, *Proc. of Japan-US Workshop on Snow Avalanche, Landslide, Debris Flow Prediction and Control*, pp. 297–304.

Sammori, T., Y. Okura, and Y. Horie (1993), Numerical experimental investigation of the effects of parameters on shallow landslides, *Shin Sabo J. (J. Jpn. Soc. Erosion Contr. Eng.), 46*(1), pp. 3–12 (in Japanese with English abstract).

Sammori, T., Y. Okura, H. Ochiai, and Y. Kitahara (1995), Effects of soil thickness on slope failures during rain, *Shin Sabo J. (J. Jpn. Soc. Erosion Contr. Eng.), 48*(1), pp. 12–23 (in Japanese with English abstract).

Santoso, D., Landslide in fill material controlled by old morphology and ground-water condition (1990), *Landslide News (Jpn. Landslide Soc.), 4*, 18–20.

Sasaki, Y., A. Fujii, and K. Asai (2000), Soil creep process and its role in debris slide generation—field measurements on the north side of Tsukuba Mountain in Japan, *Eng. Geol., 56*, 163–183.

Sassa, K. (1984), The mechanism starting liquefied landslides and debris flows, *Proc. 4th Int. Symp. on Landslides*, vol. 2, pp. 349–354.

Sassa, K. (1985), The geotechnical classification of landslides, *Proc. 4th Int. Conf. and Field Workshop on Landslides*, pp. 31–45.

Sassa, K. (1989), Geotechnical classification of landslides, *Landslide News (Jpn. Landslide Soc.), 3*, 21–24.

Sassa, K. (1996), Prediction of earthquake induced landslides, in *Landslides/Glissements de Terrain*, edited by K. Senneset, vol. 1, pp. 115–132, A.A. Balkema, Rotterdam.

Sassa, K., and H. Fukuoka (1996), The Nikawa landslide and assessment of earthquake induced landslide, Res. Rep. of Landslides and Slope Failures triggered the Hyogoken–Nanbu Earthquake, Jpn. Landslide Soc., 145–170 (in Japanese).

Sassa, K., H. Fukuoka, G. Scarascia-Mugnozza, and S. Evans (1996), Earthquake-induced landslides: distribution, motion and mechanisms, Special Issue on Geotechnical Aspects of the January 17, 1995 Hyogoken-Nambu earthquake, *Soils and Foundations (Jpn. Geotech. Soc.)*, 53–64.

Sassa, K., G. Wang, H. Fukuoka, F. Wang, H. Ochiai, M. Sugiyama, and T. Sekiguchi (2004), Landslide risk evaluation and hazard zoning for rapid and long travel landslides in urban development areas, *Landslides, 1*, 221–235.

Satterlund, D. R. (1972), *Wildland Watershed Management*, 370 p., Ronald Press, New York.

Satterlund, D. R., and H. F. Haupt (1970), Disposition of snow caught by conifer crowns. *Water Resour. Res. 6*, 649–652.

Savage, C. N. (1950), Earthflow associated with strip mining, *Min. Eng., 187*, 337–339.

Savage, W. Z., and R. W. Fleming (1996), Slumgullion landslide fault creep studies, in The Slumgullion Earth Flow: a Large-scale Natural Laboratory, edited by D. J. Varnes and W. Z. Savage, *Bull. 2130*, U.S. Geol. Surv., pp. 73–76.

Schmidt, K. M., and D. R. Montgomery (1995), Limits to relief, *Science, 270*, 617–620.

Schmidt, K. M., J. J. Roering, J. D. Stock, W. E. Dietrich, D. R. Montgomery, and T. Schaub (2001), The variability of root cohesion as an influence on shallow landslide susceptibility in the Oregon Coast Range, *Can. Geotech. J., 38*, 995–1024.

Schroeder, W. L. (1983), Geotechnical properties of southeast Alaskan forest soils, Oregon State University, Civil Eng. Dept., Corvallis, 97 p.

Schroeder, W. L. (1985), The engineering approach to landslide risk analysis, in Proc. of a Workshop on Slope Stability: Problems and Solutions in Forest Management, *Gen. Tech. Rep. PNW-180*, pp. 43–50, For. Serv., U.S. Dept. of Agric., Portland, OR.

Schroeder, W. L., and G. W. Brown (1984), Debris torrents, precipitation, and roads in two coastal Oregon watersheds, *Proc. of Symp. on Effects of Forest Land Use on Erosion and Slope Stability*, edited by C. L. O'Loughlin and A. J. Pearce, pp. 117–122, Univ. of Hawaii, Honolulu.

Schroeder, W. L., and D. N. Swanston (1987), Application of geotechnical data to resource planning in southeast Alaska, *Gen. Tech. Rep. PNW-198*, 22 p., For. Serv., U.S. Dept. of Agric., Portland, OR.

Schultze, E. (1972), Frequency distributions and correlations of soil properties, *Statistics and Probability in Civil Engineering*, Oxford University Press, UK.

Schuster, R. L. (1996), Socioeconomic significance of landslides, in *Landslides—Investigation and Mitigation*, edited by A. K. Turner and R. L. Schuster, pp. 12–35, Special Report 247, Trans. Res. Board, National Res. Council, National Academic Press, Washington, DC.

Schuster, R. L., and L. M. Highland (2001), Socioeconomic and environmental impacts of landslides in the Western Hemisphere, *Open-file report 01-0276*, U.S. Geol. Surv.

Schuster, R. L., and W. K. Kockelman (1996), Principles of landslide hazard reduction, in *Landslides—Investigation and Mitigation*, edited by A. K. Turner and R. L. Schuster, pp. 91–105, Special Report 247, Trans. Res. Board, National Res. Council, National Academic Press, Washington, DC.

Schwab, J. W. (1983), Mass wasting: October–November 1978 storm, Rennel Sound, Queen Charlotte Islands, British Columbia, *Publ. 91*, 23 p. Ministry of Forests, Victoria, BC, Canada.

Schwab, J. W. (1988), Mass wasting impacts to forest land: forest management implications, Queen Charlotte timber supply area, in *Degradation of Forested Land: Forest Soils and Risk*, edited by J. D. Louisier and G. W. Still, pp. 104–115, B.C. For. Serv., Canada.

Schwab, J. W. (1994), Erosion control: planning, forest road deactivation and hillslope revegetation, in *A Guide for Management of Landslide-prone Terrain in the Pacific Northwest*, edited by S. C. Chatwin, D. E. Howes, J. W. Schwab, and D. N. Swanston, pp. 173–213, Land Management Handbook No. 18, B.C. Ministry of Forests, Crown Publications Inc., Victoria, BC, Canada.

Schwarzhoff, J. C. (1975), Retaining wall practice and the selection of low volume forest roads, unpublished rep., 39 p., Region 6, For. Serv., U.S. Dep. of Agric, Portland, OR.

Scott, D. F., and D. B. Van Wyk (1990), The effects of wildfire on soil wettability and hydrological behaviour of an afforested catchment, *J. Hydrol., 121,* 239–256.

Scott, K. M. (1989), Magnitude and frequency of lahars and lahar-runout flows in the Toutle-Cowlitz River system, *Prof. Pap. 1447-B,* 33 p., U.S. Geol. Surv.

Scott, R. L., C. Watts, J. G. Payan, E. Edwards, D. C. Goodrich, D. Williams, and W. J. Shuttleworth (2003), The understory and overstory partitioning of energy and water fluxes in an open canopy, semiarid woodland, *Agric. For. Meteorol., 114*(3/4), 127–139.

Seed, H. B., and Lee (1966), Liquefaction of saturated sands during cyclic loading, *J. Soil Mech. Found. Div. Am. Soc. Civil Eng., 92,* 105–134.

Selby, M. J. (1966), Some slumps and boulder fields near Whitehall, *J. Hydrol. N. Z., 5*(2), 35–44.

Selby, M. J. (1974), Dominant geomorphic events in landform evolution, *Bull. Int. Assoc. Eng. Geol., 9,* 85–89.

Selby, M. J. (1982), *Hillslope Materials and Processes*, 264 p., Oxford Univ. Press, Oxford, UK.

Selkregg, L. L., E. B. Crittenden, and N. Williams (1970), Urban planning in the reconstruction, in *The Great Alaska Earthquake of 1964: Human Ecology*, pp. 186–242, National Academy of Sci., Washington, DC.

Sells, D. E., R. G. Darmody, and F. W. Simmons (1992), The effects of coal mine subsidence on soil macroporosity and water flow, *Preprint 1992-2*, pp. 137–145, Illinois Mine Subsidence Res. Program, Illinois State Geol. Surv., Champaign, IL.

Shakoor, A., and A. J. Smithmyer (2005), An analysis of storm-induced landslide in the colluvial soils overlaying mudrock sequences, southeastern Ohio, USA, *Eng. Geol., 78,* 257–274.

Sharpe, C. F., and E. F. Dosch (1942), Relation of soil-creep to earthflow in the Appalachian Plateaus, *J. Geomorph., 5,* 312–324.

Sharpe, C. F. S. (1938), *Landslides and Related Phenomena, a Study of Mass-Movement of Soil and Rock*, Columbia Univ. Press, New York, 137 p., reprinted 1960 by Pageant Books Inc., Paterson, NJ.

Shea-Albin, V. R. (1992), Effects of longwall subsidence on escarpment stability, *Proc. 3rd Subsidence Workshop due to Underground Mining*, pp. 272–279, West Virginia University, Morgantown.

Sherlock, M. D., N. A. Chappell, and J. J. McDonnell (2000), Effects of experimental uncertainty on the calculation of hillslope flow paths, *Hydrol. Process., 14*, 2457–2471.

Shimokawa, E. (1980), Creep deformation of cohesive soils and its relationship to landslide, *Mem. Fac. Agr. Kagoshima Univ.*, 16, 129–156.

Shimokawa, E. (1984), A natural recovery process of vegetation on landslide scars and landslide periodicity in forested drainage basins, *Proc. of Symp. on Effects of Forest Land Use on Erosion and Slope Stability*, pp. 99–107, Honolulu, Hawaii.

Shimokawa, E., T. Jitousono, and S. Takano (1989), Periodicity of shallow landslides in Shirasu (Ito pyroclastic flow deposits) steep slopes and prediction of potential landslide sites, *Trans. Jpn. Geomorph. Union, 10*, 267–284 (in Japanese with English abstract).

Shoaei, Z., and J. Ghayoumian (2000), Seimareh Landslide, western Iran, one of the worlds largest complex landslides, *Landslide News (Jpn. Landslide Soc.), 13*, 23–26.

Shrestha, D. P., J. A. Zinck, and E. Van Ranst (2004), Modelling land degradation in the Nepalese Himalaya, *Catena, 57*, 135–156.

Shroder, J. F. Jr., and M. P. Bishop (1998), Mass movement in the Himalaya: new insights and research directions, *Geomorphology, 26*, 13–35.

Shuzui, H. (2001), Process of slip-surface development and formation of slip-surface clay in landslides in Tertiary volcanic rocks, Japan, *Eng. Geol., 61*, 199–220.

Sidle, R. C. (1980), Slope stability on forest land, *Pac. NW Ext. Publ. PNW209*, 23 p. Oregon State Univ., Corvallis.

Sidle, R. C. (1984a), Shallow groundwater fluctuations in unstable hillslopes of coastal Alaska, *Z. für Gletscherkunde und Glazialgeol., 20*, 79–95.

Sidle, R. C. (1984b), Relative importance of factors influencing landsliding in coastal Alaska, *Proc. 21st Ann. Eng. Geol. and Soils Eng. Symp.*, pp. 311–325, Univ. of Idaho, Moscow.

Sidle, R. C. (1986), Groundwater accretion in unstable hillslopes of coastal Alaska, *IAHS Publ. 156*, 335–343.

Sidle, R. C. (1987), A dynamic model of slope stability in zero-order basins, *IASH Publ. 165*, 101–110.

Sidle, R. C. (1988), Evaluating long-term stability of previously failed hillslope depressions, *Proc. 24th Ann. Eng. Geol. and Soils Eng. Symp.*, pp. 279–297, Idaho Dept. of Transport., Boise, Idaho.

Sidle, R. C. (1991), A conceptual model of changes in root cohesion in response to vegetation management, *J. Environ. Quality, 20*(1), 43–52.

Sidle, R. C. (1992), A theoretical model of the effects of timber harvesting on slope stability, *Water Resour. Res., 28*(7), 1897–1910.

Sidle, R. C. (2005), Influence of forest harvesting activities on debris avalanches and flows, in *Debris Flow Hazards and Related Phenomena*, edited by M. Jakob and O. Hungr, pp. 345–367, Springer-Praxis, Heidelberg.

Sidle, R. C. (2006), Field observations and process understanding in hydrology: essential components in scaling, *Hydrol. Processes, 20*, 1439–1445.

Sidle, R. C., and R. W. Brown (1992), Decision model for successful reclamation of disturbed lands, in *Land Reclamation: Advances in Research and Technology*, edited by T. Younos et al., pp. 14–23, Am. Soc. Agric. Engr., St. Joseph, Michigan.

Sidle, R. C., and M. Chigira (2004), Landslides and debris flows strike Kyushu, Japan, *Eos, Trans. AGU*, *85*(15): 145–151.

Sidle, R. C., and A. S. Dhakal (2002), Potential effects of environmental change on landslide hazards in forest environments, in *Environmental Change and Geomorphic Hazards in Forests*, edited by R. C. Sidle, pp. 123–165, IUFRO Research Series, No. 9, CABI Publishing, Wallingford, Oxen, UK.

Sidle, R. C., and A. S. Dhakal (2003), Recent advances in the spatial and temporal modeling of shallow landslides, in *Integrated Modelling of Biophysical, Social and Economic Systems for Resource Management Solutions*, Proc. of MODSIM 2003, vol. 2, pp. 602–607, Modelling and Simulation Society of Australia and New Zealand, Inc., Canberra, Australia.

Sidle, R. C., and D. M. Drlica (1981), Soil compaction from logging with a low-ground pressure skidder in the Oregon Coast Ranges, *Soil Sci. Soc. Am. J.*, *45*, 1219–1224.

Sidle, R. C., and J. W. Hornbeck (1991), Cumulative effects: a broader approach to water quality research, *J. Soil & Water Conserv.*, *46*, 268–271.

Sidle, R. C., and A. Sharma (1996), Stream channel changes associated with mining and grazing in the Great Basin, *J. Environ. Quality*, *25*(5), 1111–1121.

Sidle, R. C., and D. N. Swanston (1982), Analysis of a small debris slide in coastal Alaska, *Can. Geotech. J.*, *19*, 167–174.

Sidle, R. C., and Y. Tsuboyama (1992), A comparison of piezometric response in unchanneled hillslope hollows: coastal Alaska and Japan, *J. Jpn. Soc. Hydrol. Water. Resour.*, *5*, 3–11.

Sidle, R. C., and W. Wu (1999), Simulating effects of timber harvesting on the temporal and spatial distribution of shallow landslides, *Z. Geomorphol. N.F.*, *43*, 185–201.

Sidle, R. C., R. W. Brown, and B. D. Williams (1993), Erosion processes on arid minespoil slopes, *Soil Sci. Soc. Am. J. 57*, 1341–1347.

Sidle, R. C., E. E. Farmer, and B. D. Williams (1994), Subsidence and rock creep in a cross-valley fill, *Environ. Geol.*, *24*(3), 159–165.

Sidle, R. C., T. Kamai, and A. C. Trandafir (2005), Landslide damage during the Chuetsu earthquake, Niigata, Japan, *Eos, Trans. AGU, 86*(13), 133–140.

Sidle, R. C., I. Kamil, A. Sharma, and S. Yamashita (2000b), Stream response to subsidence from underground coal mining in central Utah, *Environ. Geol.*, *39*(3–4), 279–291.

Sidle, R. C., J. Negishi, N. Abdul Rahim, and R. Siew (2004c), Erosion processes in steep forested terrain – truths, myths, and uncertainties related to land use in Southeast Asia, in *Forests and Water in Warm, Humid Asia*, edited by R. C. Sidle, pp. 20–23, Proc. of a IUFRO Forest Hydrology Workshop, 10–12 July 2004, Kota Kinabalu, Malaysia, Disaster Prevention Research Inst., Uji, Japan.

Sidle, R. C., B. Nilsson, and M. Hansen (1998), Spatially varying hydraulic characteristics of a fractured clayey till determined by field tracer tests, Fuenen, Denmark, *Water Resour. Res.*, *34*, 2515–2527.

Sidle, R. C., S. Noguchi, Y. Tsuboyama, and K. Laursen (2001), A conceptual model of preferential flow systems in forested hillslopes: evidence of self-organization, *Hydrol. Processes, 15*(10), 1675–1692.

Sidle, R. C., A. J. Pearce, and C. L. O'Loughlin (1985), *Hillslope Stability and Land Use*, Water Resources Monograph, vol. 11, Am. Geophys. Union, Washington, DC., 140 p.

Sidle, R. C., S. Sasaki, M. Otsuki, S. Noguchi, and N. Abdul Rahim (2004b), Sediment pathways in a tropical forest: effects of logging roads and skid trails, *Hydrol. Process. 18*, 703–720.

Sidle, R. C., D. Taylor, X. X. Lu, W. N. Adger, D. J. Lowe, W. P. deLange, R. N. Newnham, and J. R., Dodson (2004a), Interactions of natural hazards and humans: evidence in historical and recent records, *Quaternary Int., 118–119*, 181–203.

Sidle, R. C., Y. Tsuboyama, S. Noguchi, I. Hosoda, M. Fujieda, and T. Shimizu (1995), Seasonal hydrologic response at various spatial scales in a small forested catchment, Hitachi Ohta, Japan, *J. Hydrol.*, *168*, 227–250.

Sidle, R. C., Y. Tsuboyama, S. Noguchi, I. Hosoda, M. Fujieda, and T. Shimizu (2000a), Stream-flow generation in steep headwaters: a linked hydrogeomorphic paradigm, *Hydrol. Process.,* *14*, 369–385.

Sidle, R. C., A. D. Ziegler, J. N. Negishi, N. Abdul Rahim, R. Siew, and F. Turkelboom (2006), Erosion processes in steep terrain—truths, myths, and uncertainties related to forest management in Southeast Asia. *For. Ecol. Manage., 224(1–2),* 199–225.

Sidorova, T. (1998), Regime of mudflows and its potential changes due to global warming, *Publication No. 23,* pp. 249–253, Norwegian Geotechnical Inst., Oslo.

Siebert, L., H. Glicken, and T. Ui (1987), Volcanic hazards from Bezymianny and Bandai-type eruptions, *Bull. Volcanol., 49,* 435–459.

Simoni, A., M. Berti, M. Generali, C. Elmi, and M. Ghirott (2004), Preliminary result from pore pressure monitoring on an unstable clay slope, *Eng. Geol., 73*(1/2), 117–128.

Singh, N. (2004), Quantitative analysis of partial risk from debris flows and debris floods: community of Swansea Point, Sicamous, British Columbia, in *Landslide Risk Case Studies in Forest Development Planning and Operations,* edited by M. Wise, G. Moore, and D. VanDine, pp. 45–52, British Columbia, Ministry of Forest, Forest Science Program, Victoria, BC, Canada.

Singh, R. B. (1998), Land use/cover changes, extreme events and ecohydrological responses in the Himalayan region, *Hydrol. Process. 12,* 2043–2055.

Sjöberg, J. (1996), Large scale slope stability open pit mining—a review, *Technical Report, ISSN 0349-3571,* pp. 177–187, Luleå University of Technology, Sweden.

Skaugset, A., and B. Wemple (1999), The response of forest roads on steep, landslide-prone terrain in western Oregon to the February 1996 storm, in *Proc. of the Int. Mountain Logging and 10th Pac. Northwest Skyline Symp.,* edited by J. Sessions and W. Chung, pp. 193–203, Dept. Forest Eng., Oregon State Univ., Corvallis.

Skaugset, A., S. Swall, and K. Martin (1996), The effect of forest road location, construction, and drainage standards on road-related landslides in western Oregon associated with the February 1996 storm, *Proc. Pacific Northwest Floods of February 1996 Water Issues Conf.,* pp. 201–206, Portland, OR.

Skempton, A. W. (1953), The colloidal "activity" of clays, *Proc. 3rd Int. Conf. Soil Mech. Found. Eng., 1,* pp. 57–61.

Skempton, A. W. (1970), First-time slides in over-consolidated clays, *Geotechnique, 20*(3), 320–324.

Skempton, A. W., and F. A. Delory (1957), Stability of natural slopes in London clay, *Proc. of 4th Int. Conf. Soil Mech. Found. Eng.,4,* 379–381.

Skempton, A. W., and J. N. Hutchinson (1969), Stability of natural slopes and embankment foundations, *Proc. 7th Int. Conf. Soil Mech. Found. Eng., 3,* 291–340, Mexico.

Slaughter, C. B., G. W. Freethey, and L. E. Spangler (1995), Hydrology of the North Fork of the Right Fork of Millar Creek, Carbon County, Utah, before, during and after underground coal mining, *Water Resources Invest. Rep. 95-4025,* U.S. Geol. Surv., Denver, CO.

Slaymaker, O. (2000), Assessment of the geomorphic impacts of forestry in British Columbia, *Ambio, 29,* 381–387.

Slaymaker, O. (2001), Why so much concern about climate change and so little attention to land use change, *The Canadian Geographer, 45,* 71–78.

Slosson, J. E. (1969), The role of engineering geology in urban planning, *Colorado Geol. Surv. Spec. Publ. 1,* 8–15.

Slosson, J. E., D. D. Yoakum, and G. Shuirman (1992), Thistle landslide: was mitigation possible? in *Landslides/Landslide Mitigation,* edited by J. E. Slossen, A. G. Keene, and J. A. Johnson, pp. 83–93, Rev. Eng. Geol., vol. 9., Geol. Soc. Am., Boulder, CO.

Smalley, I. (1976), Factors relating to the landslide process in Canadian quickclays, *Earth Surf. Processes, 1,* 163–172.

Smart, P. L., and C. M. Wilson (1984), Two methods for the tracing of pipeflow on hillslopes, *Catena, 11*, 159–168.

Smith, K. (1996), *Environmental Hazards: Assessing Risk and Reducing Disaster*, 2nd ed., 389 p. Routledge, London.

Smith, R. B., P. R. Commandeur, and M. W. Ryan (1983), Natural vegetation, soil development and forest growth on Queen Charlotte Islands, *Working Paper 7/83*, 44 p., British Columbia Min. of Forest and Min. of the Environ., Fish/Forestry Interaction Program.

Smith, R. B., P. R. Commandeur, and M. W. Ryan (1986), Soils, vegetation, and forest growth on landslides and surrounding logged and old-growth areas on the Queen Charlotte Islands, *Land Management Rep. No. 41*, British Columbia Ministry of Forests.

Smyth, C. G., and S. A. Royle (2000), Urban landslide hazards: incidence and causative factors in Niterói, Rio de Janeiro State, Brazil, *Appl. Geogr., 20*, 95–117.

So, C. L. (1971), Mass movements associated with the rainstorm of June 1966 in Hong Kong, *Trans. Inst. Brit. Geogr., 53*, 55–65.

Soeters, R., and C. J. van Westen (1996), Slope instability recognition, analysis, and zonation, in *Landslides—Investigation and Mitigation*, edited by A. K. Turner and R. L. Schuster, pp. 129–177, Special Report 247, Trans. Res. Board, National Res. Council, National Academic Press, Washington, DC.

Somerville, A. (1979), Root anchorage and root morphology of *pinus radiata* on a range of ripping treatments, *N.Z. J. For. Sci, 9*(3), 294–315.

Sonmez, H., R. Ulusay, and C. Gokceoglu (1998), A practical procedure for the back analysis of slope failures in closely jointed rock masses, *Int. J. Rock Mech. Min. Sci., 35*(2), 219–233.

Sonoda, M. (1998), A numerical simulation of displacement of weathered granite on a forest slope, *Trans. Jpn. Geomorph. Union, 19*, 135–154.

Sonoda, M., and K. Okunishi (1994), Downslope soil movement on a forested hillslope with granite bedrock, *Proc. Intern. Symp. on Forest Hydrology*, edited by T. Ohta, pp. 479–486, Univ. of Tokyo, Japan.

Sonoda, M., and K. Okunishi (1999), Measurement of soil creep on a forest slope, *Transaction Jpn. Geomorphol. Union, 20-5*, 519–540 (in Japanese).

Spangle, W., and Associates Inc. (1988), Geology and planning: the Portola Valley experience, 66 p. plus appendices, W. Spangle and Assoc. Inc., Portola, CA.

Spencer, J. (1966), Shifting cultivation in Southeast Asia, *Univ. of California Publ. in Geography*, vol. 19, Univ. of California Press, Berkeley, CA.

Spiker, E. C., and P. L. Gori (2000), National landslide hazards mitigation strategy, *Open-file report 11-450*, 49 p., U.S. Geol. Surv.

Stacey, F. D. (1969), *Physics of the Earth*, 324 p., John Wiley, New York.

Stal, T., and L. Viberg (1981), Surveying potential landslide risk zones in Sweden, *Särtryck ur Väg-och vatten-byggaren, 5–6*, 4 p.

Starkel, L. (1972a), The role of catastrophic rainfall in the shaping of the relief of the lower Himalaya (Darjeeling Hills), *Geogr. Polonica, 21*, 103–147.

Starkel, L. (1972b), The modeling of monsoon areas of India as related to catastrophic rainfall, *Geogr. Polonica, 23*, 151–173.

Starkel, L. (1976), The role of extreme (catastrophic) meteorological events in the contemporary evolution of slopes, in *Geomorphology and Climate*, edited by E. Derbyshire, pp. 203–246, John Wiley & Sons, New York.

Stead, D., and E. Eberhardt (1997), Developments in the analysis of footwall slopes in coal surface mining, *Eng., Geol., 46*, 41–61.

Steiner, F. (1988), Agroforestry's coming of age, *J. Soil & Water Conserv., 43*, 157–158.

Stephenson, G. R., and R. A. Freeze (1974), Mathematical simulation of subsurface flow contributions to snowmelt runoff, Reynolds Creek Watershed, Idaho, *Water Resour. Res.*, *10*(2),
284–294.

Stevenson, P. C. (1977), An empirical method for the evaluation of relative landslide risk, *Int. Bull.
Assoc. Eng. Geol.*, *16*, 69–72.

Stott, P. (2000), Combustion in tropical biomass fires: a critical review, *Prog. Phys. Geogr. 24*(3),
355–377.

Stout, M. L. (1969), Radiocarbon dating of landslides in southern California and engineering geology implications, in *United States Contributions to Quaternary Research*, edited
by S. A. Schumm and W. E. Bradley, Special Paper 123, pp. 167–179, Geol. Soc. Am.,
Boulder, CO.

Sukhija, B. S., M. N. Rao, D. V. Reddy, P. Nagabhushanam, S. Hussain, R. K. Chadha, and H. K.
Gupta (1999), Paleoliquefaction evidence and periodicity of large prehistoric earthquakes in
Shillong Plateau, India. *Earth Planet. Sci. Let.*, *167*, 269–282.

Sulebak, J. R., L. M. Tallaksen, and E. Erichsen (2000), Estimation of areal soil moisture by used
of terrain data, *Geograf. Ann.*, *82A*, 89–105.

Sumsion, O. B. (1983), *Thistle... Focus on disaster*, 1st ed., Art City Publishing, Springville, UT.

Suwa, H., and T. Yamakoshi (1999), Sediment discharge by storm runoff at volcanic torrents
affected by eruption, *Z. Geomorph. N.F.*, *114*, 63–88.

Swanson, F. J., and C. T. Dyrness (1975), Impact of clearcutting and road construction on soil erosion by landslides in the western Cascade Range, Oregon, *Geology*, *3*(7), 393–396.

Swanson, F. J., and R. L. Fredriksen (1982), Sediment routing and budgets: Implications for judging
impacts of forestry practices, in Sediment Budgets and Routing in Forested Drainage Basins,
Gen. Tech. Rep. PNW-141, pp. 129–137, For. Serv., U.S. Dep. of Agric., Portland, OR.

Swanson, F. J., and G. Grant (1982), Rates of soil erosion by surface and mass erosion processes
in the Willamette National Forest, Final Rep. to Willamette Natl. For., 50 p., For. Sci. Lab.,
Corvallis, OR.

Swanson, F. J., and D. N. Swanston (1977), Complex mass movement terrains in the western
Cascade Range, Oregon, in *Rev. Eng. Geology, Landslides.* vol. 3, pp. 113–124, Geol. Soc. Am.,
Boulder, CO.

Swanson, F. J., L. E. Benda, S. H. Duncan, G. E. Grant, W. F. Megahan, L. M. Reid, and R. R.
Ziemer (1987), Mass failures and other processes of sediment production in Pacific Northwest
forest landscapes, in *Streamside Management: Forestry and Fisheries Interactions*, edited by E.
O. Salo and T. W. Cundy, pp. 9–38, Inst. Forest Resour., Univ. of Washington, Seattle, WA.

Swanson, F. J., R. L. Fredriksen, and F. M. McCorison (1982b), Material transfer in a western
Oregon forested watershed, in *Analysis of Coniferous Forest Ecosystems in the Western United
States*, edited by R. L. Edmonds, pp. 233–266, Hutchison Ross, Stroudsburg, PA.

Swanson, F. J., S. V. Gregory, J. R. Sedell, and A. G. Campbell (1982a), Land-water interactions:
the riparian zone, in *Analysis of Coniferous Forest Ecosystems in the Western United States*,
edited by R.L. Edmonds, pp. 267–291, Hutchison Ross, Stroudsburg, PA.

Swanson, F. J., M. M. Swanson, and C. Woods (1977), Inventory of mass erosion in the Mapleton
Ranger District, Su,islaw National Forest, final report, 62 p., For. Sci. Lab., Corvallis, OR.

Swanson, F. J., M. M. Swanson, and C. Woods (1981), Analysis of debris-avalanche erosion in steep
forested lands; an example from Mapleton, Oregon, USA, *IAHS Publ. 132*, 67–75.

Swanston, D. N. (1967), Soil-water piezometry in a southeast Alaska landslide area, *Res. Note
PNW-68*, 17 p., For. Serv., U.S. Dep. of Agric., Portland, OR.

Swanston, D. N. (1969), Mass wasting in coastal Alaska, *Res. Paper, PNW-83*, 15 p., For. Serv.,
U.S. Dep. of Agric., Portland, OR.

Swanston, D. N. (1970), Mechanics of debris avalanching in shallow till soils of southeast Alaska,
Res. Pap. PNW-103, 17 p., For. Serv., U.S. Dep. of Agric., Portland, OR.

Swanston, D. N. (1973), Judging landslide potential in glaciated valleys in southeastern Alaska, *Explorers Journal, 51*(4) 214–217.

Swanston, D. N. (1974a), The forest ecosystem of southeast Alaska. 5. Soil mass movement, *Gen. Tech. Rep. PNW-17*, 22 p., For. Serv., U.S. Dep. of Agric., Portland, OR.

Swanston, D. N. (1974b), Slope stability problems associated with timber harvesting in mountainous regions of the western United States, *Gen. Tech. Rep. PNW-21*, 14 p., For. Serv., U.S. Dep. of Agric., Portland, OR.

Swanston, D. N. (1978), Effect of geology on soil mass movement activity in the Pacific Northwest, *Proc. 5th North American Forest Soils Conf.*, pp. 89–115. Colo. State Univ., Fort Collins, CO.

Swanston, D. N. (1981), Creep and earthflow erosion from undisturbed and management impacted slopes in the Coast and Cascade Ranges of the Pacific Northwest, U.S.A., *IAHS Publ. 132*, 76–94.

Swanston, D. N., and C. T. Dyrness (1973), Stability of steep land, *J. For., 71*(5), 264–269.

Swanston, D. N., and D. E. Howes (1994), Slope movement processes and characteristics, in *A Guide for Management of Landslide-Prone Terrain in the Pacific Northwest*, edited by S. C. Chatwin, D. E. Howes, J. W. Schwab, and D. N. Swanston, pp. 1–17, Land Management Handbook No. 18, B.C Ministry of Forests, Crown Publications Inc., Victoria, Canada.

Swanston, D. N., and D. A. Marion (1991), Landslide response to timber harvest in Southeast Alaska, *Proc. 5th Federal Interagency Sedimentation Conf.*, pp. 10–49, March 18–21; Las Vegas, NV.

Swanston, D. N., and R. L. Schuster (1989), Long-term landslide hazard mitigation programs: structure and experience from other countries, *Bull. Assoc. Eng. Geol., 26*, 109–133.

Swanston, D. N., and F. J. Swanson (1976), Timber harvesting, mass erosion, and steepland forest geomorphology in the Pacific Northwest, in *Geomorphology and Engineering*, edited by D. R. Coates, pp. 199–221, Dowden, Hutchison and Ross, Stroudsburg, PA.

Swanston, D. N., R. R. Ziemer, and R. J. Janda (1995), Rate and mechanics of progressive hillslope failure in the Redwood Creek Basin, northwestern California, Geomorphic Processes and Aquatic Habitat in the Redwood Creek Basin, northwestern California, *Prof. Pap. 1451-E, E1-E16*, U.S. Geol. Surv.

Taber, S. (1930), Mechanics of frost heaving, *J. Geol., 38*, 303–317.

Takada, Y. (1964), On the landslide mechanism of the Tertiary-type landslide in the thaw time, *Ann. Disaster Prevention Res. Inst. Kyoto Univ., 14*, 11–21, Japan.

Takahashi, T. (1991), *Debris Flow*, 165 p., IAHR Monograph, A.A. Balkema, Rotterdam.

Takahashi, T. (1994), Debris flow hazards in Japan and China, *Proc. Japan-China Joint Research on the Prevention from Debris Flow Hazards*, edited by T. Takahashi, pp. 1–4, Res. Rep. No. 0344085.

Takasao, T., and M. Shiiba (1988), Incorporation of the effect of concentration of flow into the kinematic wave equations and its applications to runoff system lumping, *J. Hydrol., 102*, 301–322.

Tamrakar, N.K., S. Yokota, and O. Osaka (2002), A toppled structure with sliding in the Siwalik Hills, Midwestern Nepal, *Eng. Geol., 64*(4), 339–350.

Tanaka, K. (1985), Features of slope failures by the Naganoken-seibu earthquake, 1984, *Soil Mech. Found. Eng., 33*(11), 5–11 (in Japanese).

Tanaka, T. (1982), The role of subsurface water exfiltration in soil erosion processes, *IAHS Publ. 137*, 73–80.

Tang, C., and J. Grunert (1999), Inventory of landslides triggered by the 1996 Lijiang Earthquake, Yunnan Province, China, *Trans. Jpn. Geomorph. Union, 20*, 335–349.

Tang, Z. (1991), A study of landslides in weathered granitic slopes in Amphoe Phi Pun, Nakhon Sin Thammarat, Thailand, M.Sc. thesis, Inst. of Technology, Bangkok, Thailand, 113 p.

Tani, M. (1997), Runoff generation processes estimated from hydrological observations on a steep forested hillslope with a thin soil layer, *J. Hydrol., 200*, 84–109.

Tarboton, D. (1997), A new method for the determination of flow directions and upslope areas in grid digital elevation models, *Water Resour. Res., 33*, 309–319.

Taskey, R. D. (1977), Relationship of clay mineralogy to landscape stability in western Oregon, Ph.D. thesis, 223 p., Oregon State Univ., Corvallis.

Taskey, R. D., M. E. Harward, and C. T. Youngberg (1978), Relationship of clay mineralogy to landscape stability, *Proc. 5th North American Forest Soils Conf.*, pp. 140–164, Colo. State Univ., Fort Collins, CO.

Taylor, D. W. (1948), *Fundamentals of Soil Mechanics*, 700 p., John Wiley, New York.

Temple, P. H., and A. Rapp (1972), Landslides in the Mgeta area, western Uluguru Mountains, Tanzania, *Geograf. Ann., 54A*, 157–193.

Templeton, S. R., and S. J. Scherr (1999), Effects of demographic and related microeconomic change on land quality in hills and mountains of developing countries, *World Dev., 27*(6), 903–918.

Terajima, T., and K. Moroto (1990), Stream flow generation in a small watershed in granitic mountain, *Trans. Jpn. Geomorph. Union, 11*, 75–96 (in Japanese with English abstract).

Terajima, T., and Y. Sakura (1993), Effects of subsurface flow on the topographic change at the valley head of granitic mountains, *Trans. Jpn. Geomorph. Union, 14*, 365–384 (in Japanese with English abstract).

Terajima, T., T. Sakamoto, and T. Shirai (2000), Morphology, structure and flow phases in soil pipes developing in forested hillslopes underlain by a Quaternary sand-gravel formation, Hokkaido, northern main island in Japan, *Hydrol. Process., 14*(4), 713–726.

Terlien, M. T. J. (1997), Hydrological landslide triggering in ash-covered slopes of Manizales (Columbia), *Geomorphology, 20*, 165–175.

Terlien, M. T. J. (1998), The determination of statistical and deterministic hydrological landslide-triggering thresholds, *Environ. Geol., 35*(2–3), 124–130.

Terlien, M. T. J., C. J. van Westen, and T. W. J. van Asch (1995), Deterministic modelling in GIS-based landslide hazard assessment, in *Geographical Information Systems in Assessing Natural Hazards.*, edited by A. Carrara and F. Guzzetti, pp. 57–78, Kluwer Academic Publ., Dordrecht.

Terwilliger, V. J., and L. J. Waldron (1990), Assessing the contribution of roots to the strength of undisturbed, slip prone soils, *Catena, 17*, 151–162.

Terwilliger, V. J., and L. J. Waldron (1991), Effects of root reinforcement on soil-slip patterns in the Transverse Ranges of southern California, *Geol. Soc. Am. Bull., 103*, 775–785.

Terzaghi, K., and R. B. Peck (1967), *Soil Mechanics in Engineering Practice.* 2nd ed., John Wiley, New York.

Thapa, G. B. (2001), Changing approaches to mountain watersheds management in mainland South and Southeast Asia, *Environ. Manage., 27*, 667–679.

Thomas, B. R. (1985), Use of soils and geomorphic information for road location and timber management in the Oregon Coast Ranges, in Proc. of a Workshop on Slope Stability: Problems and Solutions in Forest Management, *Gen. Tech. Rep. PNW-180*, pp. 68–78, For. Serv., U.S. Dep. of Agric., Portland, OR.

Thomas, R. B., and W. F. Megahan (1998), Peakflow responses to clear cutting and roads in small and large basins, western Cascades, Oregon: a second opinion. *Water Resour. Res., 34*, 3393–3403.

Thompson, J. C., and R. D. Moore (1996), Relations between topography and water table depth in a shallow forest soil, *Hydrol. Process., 10*, 1513–1525.

Thomson, S. (1971), Analysis of a failed slope, *Can. Geotech. J., 8*, 596–599.

Thomson, S., and C. E. Tiedemann (1982), A review of factors affecting landslides in urban areas, *Bull. Assoc. Eng. Geol., 19*(1), 55–65.

Thorne, C. R. (1990), Effects of vegetation on riverbank erosion and stability, in *Vegetation and Erosion*, edited by J. B. Thornes, pp. 125–144, John Wiley & Sons, Chichester, UK.

Toews, D. A. A. (1991), Climatic and hydrologic circumstances antecedent to mass wasting events in southeastern British Columbia, *Proc. 59th Western Snow Conf.*, pp. 91–102, Stowe, VT.

Toews, D. A. A., and D. R. Gluns (1986), Snow accumulation and ablation on adjacent forested and clearcut sites in southeastern British Columbia, *Proc. 54th Western Snow Conf.*, pp. 101–111, Phoenix, AZ.

Torrance, J. K. (1979), Post-depositional changes in the pore-water chemistry of the sensitive marine clays of the Ottawa area, Eastern Canada, *Eng. Geol.*, *14*, 135–147.

Torrance, J. K. (1999), Physical, chemical and mineralogical influences on the rheology of remoulded low-activity sensitive marine clay, *Appl. Clay Sci.*, *14*, 199–223.

Torres, R., W. E. Dietrich, D. R. Montgomery, S. P. Anderson, and K. Loague (1998), Unsaturated zone processes and the hydrologic response of a steep, unchanneled catchment, *Water Resour. Res.*, *34*(8), 1865–1879.

Trandafir, A. C., and K. Sassa (2005), Seismic triggering of catastrophic failures on shear surfaces in saturated cohesionless soils, *Can. Geotech. J.*, *42*(1), 229–251.

Trandafir, A. C., K. Sassa, and H. Fukuoka (2002), A computational method for residual excess pore pressure, response in sand under cyclic loading, *Ann. Disaster Prevent. Res. Inst., Kyoto University, 45B*, 61–69.

Troendle, C. A., and R. M. King (1985), The effect of timber harvest on the Fool Creek watershed, 30 years later, *Water Resour. Res.*, *21*(12): 1915–1922.

Trustrum, N. A., and R. C. DeRose (1988), Soil depth-age relationship of landslides on deforested hillslopes, Taranaki, New Zealand, *Geomorphology, 1*, 143–160.

Trustrum, N. A., and P. R. Stephens (1981), Selection of hill country pasture measurement sites by interpretation of sequential aerial photographs, *N. Z. J. Exp. Agric., 9*, 31–34.

Trustrum, N. A., M. G. Lambert, and V. J. Thomas (1983), The impact of sol slip erosion on hill country pasture production in New Zealand, *Proc. 2nd Int. Conf. on Soil Erosion and Conserv.*, Univ. of Hawaii, Honolulu.

Trustrum, N. A., V. J. Thomas, and M. G. Lambert (1984), Soil slip erosion as a constraint to hill country pasture production, *Proc. New Zealand Grasslands Assoc. Conf., 45*, pp. 66–76.

Trustrum, N. A., R. C. Wallace, and R. C. DeRose (1989), Tephrochronological dating of regolith in landslide prone steeplands, New Zealand, *Int. Symp. on Erosion and Volcanic Debris Flow Technology*, Yogyakarta, Indonesia, pp. S34-1–S34-11.

Tsuboyama, Y., I. Hosoda, S. Noguchi, and R. C. Sidle (1994a), Piezometric response in a zero-order basin, Hitachi Ohta, Japan. in *Proc. of Int. Symp. on Forest Hydrology*, edited by T. Ohta, pp. 217–224, Univ. of Tokyo, Japan.

Tsuboyama, Y., R. C. Sidle, S. Noguchi, and I. Hosoda (1994b), Flow and solute transport through the soil matrix and macropores of a hillslope segment, *Water Resour. Res. 30*, 879–890.

Tsuboyama, Y., R. C. Sidle, S. Noguchi, S. Murakami, and T. Shimizu (2000), A zero-order basin—its contribution to catchment hydrology and internal hydrological processes, *Hydrol. Processes, 14*, 387–401.

Tsukamoto, Y. (1987), Evaluation of the effect of tree roots on slope stability, *Bull. of the Exp. Forest, 23*, 65–124 (in Japanese).

Tsukamoto, Y., and O. Kusabe (1984), Vegetative influence on debris slide occurrence on steep slopes in Japan, *Proc. Symp. on Effects of Forest Land Use on Erosion and Slope Stability*, pp. 63–72, Honolulu, Hawaii.

Tsukamoto, Y., and H. Minematsu (1986), Evaluation of the effect of lateral roots on slope stability, *Proc. 18th IUFRO World Congress*, Ljubliana, Yugoslavia.

Tsukamoto, Y., and H. Minematsu (1987), Evaluation of the effect of deforestation on slope stability and its application to watershed management, *IAHS Publ. 167*, 81–189.

Tsukamoto, Y., and T. Ohta (1988), Runoff processes on a steep forested slope, *J. Hydrol., 102*, 165–178.

Tsukamoto, Y., T. Ohta, and H. Noguchi (1982), Hydrological and geomorphological studies of debris slides on forested hillslopes in Japan, *IAHS Publ. 137*, 89–98.

Tsutsumi, D., R. C. Sidle, and K. Kosugi (2005) Development of a simple lateral preferential flow model with steady state application in hillslope soils. *Water Resour. Res., 41*, W12420, doi:101029/2004WR003877.

Turkelboom, F. (1999), On-farm diagnosis of steepland erosion in northern Thailand, Ph.D. thesis, 309 p.

Faculty of Agricultural and Applied Biological Sciences, K.U. Leuven, The Netherlands.

Turner, A. K., and R. L. Schuster (editors) (1996), *Landslides – Investigation and Mitigation*, Special Report 247, Trans. Res. Board, National Res. Council, National Academic Press, Washington, DC., 673 p.

Turton, D. J., C. T. Haan, and E. L. Miller (1992), Subsurface flow response of a small forested catchment in the Ouachita mountains, *Hydrol. Processes, 6*, 111–125.

Uchida, T., Y. Asano, N. Ohte, and T. Mizuyama (2003), Seepage area and rate of bedrock groundwater discharge at a granitic unchanneled hillslope, *Water Resour. Res., 39*, 1018, doi:10.1029/2002WR001298.

Uchida, T., K. Kosugi, and T. Mizuyama (1999), Runoff characteristics of pipeflow and effects of pipeflow on rainfall-runoff phenomena in a mountainous watershed, *J. Hydrol., 222*, 18–36.

Uchida, T., K. Kosugi, and T. Mizuyama (2001), Effects of pipeflow on hydrological process and its relation to landslide: review of pipeflow studies in the forested headwater catchments, *Hydrol. Process., 15*, 2151–2174.

Uchida, T., K. Kosugi, N. Ohte, and T. Mizuyama (1996), The influence of pipe flow on slope stability, *J. Japan Soc. Hydrol & Water Resour, 9*(4), 330–338 (in Japanese with English abstract).

UNDRO (1991), Mitigating natural disasters, phenomena, effects and options, Office of the United Nations Disaster and Relief Coordinator, United Nations Publications, 164 p., New York.

Updike, R. G., J. A. Egan, Y. Morikawa, I. M. Idris, and T. L. Moses (1988), A model for earthquake-induced translatory landslides in Quaternary sediments, *Geol. Soc. Am. Bull., 100*, 783–792.

Ural, S., and F. Yuksel (2004), Geotechnical characterization of lignite-bearing horizons in the Afsin-Elbistan lignite basin, SE Turkey, *Eng. Geol., 75*(2), 129–146.

van Asch, Th. W. J. (1984) Landslides: the deduction of strength parameters of material from equilibrium analysis, *Catena, 11*, 39–49.

van Asch, Th. W. J., and J. T. Buma (1997), Modelling groundwater fluctuations and the frequency of movement of landslide in the Terrs Noires Region of Barcelonnette (France), *Earth Surf. Process. Landforms, 22*, 131–141.

van Asch, Th. W. J., and V. H. Steijn (1991), Temporal patterns of mass movements in the French Alps, *Catena, 18*(5), 515–527.

Vandekerckhove, L., J. Poesen, D. Oostwoud Wijdenes, and G. Gyssels (2001), Short-term bank gully retreat rates in Mediterranean environments, *Catena, 44*(2), 133–161.

van Den Eeckhaut, M., J. Poesen, G. Verstraeten, V. Vanacker, J. Moeyersons, J. Nyssen, and L. P. H. van Beek (2005), The effectiveness of hillshade maps and expert knowledge in mapping in old deep-seated landslides, *Geomorphology, 67*, 351–363.

van der Hoven, S. J., D. K. Solomon, and G. R. Moline (2002), Numerical simulation of unsaturated flow along preferential pathways: implications for use of mass balance calculations for isotope storm hydrograph separation, *J. Hydrol., 268*, 214–233.

van Dijk, A. I. J. M. (2002), Water and sediment dynamics in bench-terraced agricultural steeplands in West Java, Indonesia. Ph.D. thesis, Free University, Amsterdam, The Netherlands.

VanDine, D. F. (1985), Debris flows and debris torrents in the Southern Canadian Cordillera, *Can. Geotech. J., 22*, 44–68.

VanDine, D. F., and M. Bovis (2002), History and goals of Canadian debris flow research, a review, *Nat. Hazards, 26*, 69–82.

van Genuchten, P. M. B., and H. de Rijke (1989), On pore water pressure variations causing slide velocities and accelerations observed in a seasonally active landslide, *Earth Surf. Process. Landforms, 14*, 577–586.

Van Noordwijk, M. (2000), Forest conversion and watershed functions in the humid tropics, *Proc. of IC-SEA/NIAES Workshop*, Bogor, Indonesia, pp. 1–10.

van Westen, C. J., and T.J. Terlien (1996), An approach towards deterministic landslide hazard analysis in GIS: a case study from Manizales (Columbia), *Earth Surf. Proc. Landforms, 21*, 853–868.

van Westen, C. J., R. Soeters, and N. Rengers (1993), Geographic information systems as applied to landslide hazard zonation, *Mapping Awareness and GIS in Europe, 7*(5), 9–13.

Vandre, B. C., and D. N. Swanston (1977), A stability evaluation of debris avalanches caused by blasting, *Bull. Assoc. Eng. Geol., 14*(4), 205–225.

Varnes, D. J. (1978), Slope movement types and processes, in *Landslide Analysis and Control*, edited by M. Clark, pp. 11–33, Special Rep. 176, Trans. Res. Board, National Academy of Science, National Res. Council, Washington, DC.

Varnes, D. J. (1981), Slope stability problems of Circum-Pacific Region as related to mineral and energy resources, *Am. Assoc. Pet. Geol. Studies Geol., 12*, 489–505.

Varnes, D. J. (1984), Landslide hazard zonation: a review of principles and practice, *Natural Hazard No. 3*, Commission on the Landslides of the IAEG, UNESCO, Paris.

Varnum, N. C., P. T. Tueller, and C. M. Skau (1991), A geographical information system to assess natural hazards in the east-central Sierra Nevada, *J. Imaging Technol., 17*(2), 57–61.

Verburg, P. H., T. A. Veldkamp, and J. Bouma (1999), Land use change under conditions of high population pressure: the case of Javam, *Global Environmental Change 9*, 303–312.

Vertessy, R. A., and H. Elsenbeer (1999), Distributed modeling of storm flow generation in an Amazonian rain forest catchment: Effects of model parameterization, *Water Resour. Res., 35*, 2173–2187.

Vick, S. G. (1996), Hydraulic tailings, in *Landslides—Investigation and Mitigation*, edited by A. K. Turner and R. L. Schuster, pp. 577–584, Special Report 247, Trans. Res. Board, National Res. Council, National Academic Press, Washington, DC.

Vieira, B. C., and N. F. Fernandes (2004), Landslides in Rio de Janeiro: the role played by variations in soil hydraulic conductivity, *Hydrol. Process., 18*, 791–805.

Voight, B. (1996), The management of volcano emergencies; Nevado Del Ruiz, in *Monitoring and Mitigation of Volcano Hazards*, edited by R. Scarpa and R. I. Tilling, pp. 719–769, Springer, Berlin.

Voight, B., and W. G. Pariseau (1978), Rockslides and avalanches: An introduction, in *Rockslides and Avalanches, I. Natural Phenomena, Dev. Geotech. Eng.*, vol. *14A*, edited by B. Voight, pp. 1–67, Elsevier, New York.

Voight, B., R. J. Janda, H. Glicken, and P. M. Douglas (1983), Nature and mechanics of the Mount St. Helens rock slide-avalanche of 18 May 1980, *Geotechnique, 33*, 243–273.

Vonder Linden, K. (1989), The Portuguese Bend Landslide, *Eng. Geol., 27*, 301–373.

Wafid, M. A., K. Sassa, H. Fukuoka, and G. Wang (2004), Evolution of shear-zone structure in undrained ring-shear tests, *Landslides, 1*(2) 101–112.

Waitt, R. B. (1979), Rockslide-avalanche across distributary of Cordilleran ice in Pasayten Valley, northern Washington, *Arct. Alp. Res., 11*(1), 33–40.

Wakatsuki, T., Y. Tanaka, and Y. Matsukura (2005), Soil slips on weathering-limited slopes underlain by coarse-grained granite or fine-grained gneiss near Seoul, Republic of Korea, *Catena, 60*(2), 181–203.

Waldron, L. J. (1977), The shear resistance of root-permeated homogeneous and stratified soil, *Soil Sci. Soc. Am. J., 41*(5), 843–849.

Waldron, L. J., and S. Dakessian (1981), Soil reinforcement by roots: calculation of increased soil shear resistance from root properties, *Soil Sci., 132*(6), 427–435.

Waldron, L. J., and S. Dakessian (1982), Effect of grass, legume and tree roots on soil shearing resistance, *Soil Sci. Soc. Am. J., 46*(5), 894–899.

Waldron, L. J., S. Dakessian, and J. A. Nemson (1983), Shear resistance enhancement of 1.22-meter diameter soil cross sections by pine and alfalfa roots, *Soil Sci. Soc. Am. J., 47*(1), 9–14.

Walkinshaw, J. (1992), Landslide correction costs on U.S. state highway systems, *Transport. Res. Record,* 1343, pp. 36–41.

Wang, F. W., K. Sassa, and H. Fukuoka (2000), Geotechnical simulation test for the Nikawa Landslide induced by January 17, 1995, Hyogoen–Nambu earthquake, *Soils and Foundations (Jpn. Geotech. Soc.), 40,* 35–46.

Wang, F. W., K. Sassa, and G. Wang (2002a), Mechanism of long-runout landslide triggered by the August 1998 heavy rainfall in Fukushima Prefecture, Japan, *Eng. Geol., 63,* 169–185.

Wang, G., and K. Sassa (2002), Post-failure mobility of saturated sands in undrained load-controlled ring shear tests, *Can. Geotech. J., 39,* 821–837.

Wang, G., K. Sassa, and H. Fukuoka (2003b), Downslope volume enlargement of a debris slide–debris flow in the 1999 Hiroshima, Japan, rainstorm, *Eng. Geol., 69,* 309–330.

Wang, W. L., and B. C. Yen (1974), Soil arching in slopes, *J. Geotech. Eng. Div. Am. Soc. Civ. Eng., 100* (GTI), 61–78.

Wang, W.-N., M. Chigira, and T. Furuya (2003a), Geological and geomorphological precursors of the Chiu-fen-erh-shan landslide triggered by the Chi-chi earthquake in central Taiwan, *Eng. Geol., 69,* 1–13.

Wang, Y., R. D. Summers, and R. J. Hofmeister (2002b), Landslide loss estimation, pilot project in Oregon, *Open-file-report O-02-05,* 22 p. Dept. of Geology and Mineral Industries, State of Oregon.

Warburton, J., J. Holden, and A. J. Mills (2004), Hydrological controls of surficial mass movement in peat, *Earth Sci. Rev., 67,* 139–156.

Ward, T. J., R. M. Li, and D. B. Simons (1981), Use of a mathematical model for estimating potential landslide sites in forested drainage basin, *IAHS Publ. 132,* 21–41.

Ward, T. J., R. M. Li, and D. B. Simons (1982), Mapping landslide hazards in forested watersheds, *J. Geotech. Eng. Div. Am. Soc. Civil Eng., 108* (GT2), 319–324.

Warkentin, B. P., and R. N. Yong (1962), Shear strength of montmorillonite and kaolinite related to interparticle forces, *Clays Clay Miner., 9,* 210–218.

Warrick, A. W., G. J. Mullen, and D. R. Nielsen (1977), Predictions of the soil water flux based upon field measured soil-water properties, *Soil Sci. Soc. Am., 41*(1), 14–19

Washburn, A. L. (1979), *Geocryology: a Survey of Periglacial Processes and Environment,* 401 p., Edward Arnold, London.

Washington Military Department (2003), Hazard profile—landslide, *Washington State Hazard Mitigation Plan,* 16 p.

Wasowski, J. (1998), Understanding rainfall-landslide relationships in man-modified environments: a case-history from Caramanico Terme, Italy, *Environ. Geol., 35*(2-3), 197-209.

Wasson, R. J. (1978), A debris flow at Reshūn, Pakistan Hindu Kush, *Geograf. Ann. 60A,* 151–159.

Wasson, R. J., and G. Hall (1982), A long record of mudslide movement at Waerenga-O-Kuri, New Zealand. *Z. Geomorph. N.F., 26,* 73–85.

Watanabe, K. (1988), Analysis of three-dimensional groundwater flow in the near-surface layer of a small watershed, *J. Hydrol., 102,* 287–300.

Watanabe, M., and K. Seno (1968), Landslides caused by heavy rainfalls, *Shin Sabo J. (J. Jpn. Soc. Erosion Contr. Eng.), 20*(4), 17–20 (in Japanese).

Water Induced Disaster Prevention Technical Centre (1993), *Annual Disaster Review,* 49 p., Water Induced Disaster Prevention Technical Centre, Khumaltar, Lalitpur, Nepal.

Watson, A. (1985), Soil erosion and vegetation damage near ski lifts at Cairn Gorm, Scotland, *Biol. Conserv.*, *33*, 363–381.

Watson, A. (2000), Wind-induced forces in the near-surface lateral roots of radiata pine, *For. Ecol. Manage.*, *135*, 133–142.

Watson, A., and C. O'Loughlin (1990), Structural root morphology and biomass of three age-classes of *pinus radiata*, *N.Z. J. For. Sci.*, *20*(1), 97–110.

Watson, A., M. Marden, and D. Rowan (1995), Tree species performance and slope stability, in *Vegetation and Slopes*, edited by D. H. Baker, pp. 161–171, Thomas Telford, London.

Watson, A., M. Marden, and D. Rowan (1997), Root-wood strength deterioration in Kanuka after clearfelling. *N. Z. J. For. Sci.*, *27*(2): 205–215.

Watson, A., C. Phillips, and M. Marden (1999), Root strength, growth, and rates of decay: root reinforcement changes of two species and their contribution to slope stability, *Plant and Soil*, *217*, 39–47.

Weaver, W. E., D. K. Hagans, and J. H. Popenoe (1995), Magnitude and causes of gully erosion in the lower Redwood Creek basin, Northwestern California, in Geomorphic Processes and Aquatic Habitat in the Redwood Creek basin, northwestern California, edited by K. M. Nolan and H. M. Marron, *Prof. Paper 1454*, pp. I1–I21, U.S. Geol. Surv.

Wegmann, K. W. (2003), Digital landslide inventory for the Cowlitz Country urban corridor-Kelso to Woodland (Coweeman River to Lewis River), Cowlitz Country, Washington, *Rep. of Investigation 34*, 20 p., Washington State Dept. of Natural Resour., Div. of Geology and Earth Resources, Olympia, WA.

Wei, Y., C. A. Lin, R. Benoit, and I. Zawadzki (1998), High resolution model simulation of precipitation and evaluation with Doppler radar observations, *Water Sci. Technol.*, *337*(11), 179–786.

Weidinger, J. T., J. Wang, and N. Ma (2002), The earthquake-triggered rock avalanche of Cui Hua, Qin Ling Mountains, P.R. of China—the benefits of a lake-damming prehistoric natural disaster, *Quaternary Int.*, *93/94*, 207–214.

Wells, W. G. (1987). The effects of fire on the generation of debris flows in southern California, in *Debris Flows/Avalanches: Processes, Recognition, Mitigation*, edited by J. E. Costa and G. F. Wieczorek, Rev. in Eng. Geol. 7, Geol. Soc. Am., Boulder, CO.

Wemple, B. C., J. A. Jones, and G. E. Grant (1996), Channel network extension by logging roads in two basins, western Cascades, Oregon, *Water Resour. Bull. 32*(6): 1195–1207.

Wemple, B. C., F. J. Swanson, and J. A. Jones (2001), Forest roads and geomorphic process interactions, Cascade Range, Oregon, *Earth Surf. Process. Landforms, 26*, 191–204.

Wen, B.-P., N. S. Duzgoren-Aydin, and A. Aydin (2004), Geochemical characteristics of the slip zones of a landslide in granitic saprolite, Hong Kong: implications for their development and microenvironments, *Environ. Geol.*, *47*(1), 140–154.

Wentworth, C. K. (1943), Soil avalanches on Oahu, Hawaii, *Geol. Soc, Am. Bull.*, *54*, 53–64.

Weyman, D. R. (1973), Measurements of the downslope flow of water in a soil, *J. Hydrol., 20*, 267–288.

Whipkey, R. Z. (1965), Sub-surface stormflow from forested slopes, *Bull. Int. Assoc. Sci. Hydrol.*, *10*, 74–85.

Whipkey, R. Z., and M. J. Kirkby (1978), Flow within the soil, in *Hillslope Hydrology*, edited by M. J. Kirkby, pp. 121–144, John Wiley & Sons, New York.

Whitaker, A., Y. Alila, J. Beckers, and D. Toews (2003), Application of the distributed hydrology soil vegetation model to redfish creek, British Columbia: model evaluation using internal catchment data, *Hydrol. Process.*, *17*(2), 199–224.

Whitehead, D., and P. G. Jarvis (1982), Coniferous forests and plantations, in *Water Deficits and Plant Growth*, vol. 6, edited by T. T. Kozlowski, pp, 49–152, Academic, New York.

Wieczorek, G. F. (1987), Effect of rainfall intensity and duration on debris flows in central Santa Cruz Mountains, California, *Rev. in Eng. Geol.*, vol. 7, pp. 93–104.

Wieczorek, G. F., and T. Glade (2005), Climatic factors influencing occurrence of debris flows, in *Debris-flow Hazards and Related Phenomena*, edited by M. Jakob and O. Hungr, pp. 325–362, Praxiz Springer, Berlin.

Wieczorek, G. F., S. Ellen, E. W. Lips, S. H. Cannon, and D. N. Short (1983), Potential for debris flow and debris flood along the Wasatch Front between Salt Lake City and Willard, Utah, and measured for their mitigation, *Open-File Report 83–635*, 46 p., U.S. Geol. Surv.

Wieczorek, G. F., M. C. Larsen, L. S. Eaton, B. A. Morgan, and J. L. Blair (2001), Debris-flow and flooding hazards associated with the December 1999 storm in coastal Venezuela and strategies for mitigation, *Open-File Report 01-0144*, U.S. Geol. Surv.

Wieczorek, G. F., E. W. Lips, and S. D. Ellen (1989), Debris flows and hyperconcentrated floods along the Wasatch Front, Utah, 1983 and 1984, *Bull. Assoc. Eng. Geol., 26*, 191–208.

Wieczorek, G. F., B. A. Morgan, and R. H. Campbell (2000), Debris-flow hazards in the Blue Ridge of Central Virginia, *Environ. Eng. Geosci., 6*(1), 3–23.

Wilford, D. J., M. E. Sakals, J. L. Innes, R. C. Sidle, and W. A. Bergerud (2004), Recognition of debris-flow hazard through watershed morphometrics, *Landslides, 1*, 61–66.

Wilkinson, P. L., M. G. Anderson, and D. M. Lloyd (2002), An integrated hydrological model for rain-induced landslide prediction, *Earth Surf. Process. Landforms, 27*, 1285–1297.

Williams, G. P. (1973), Changed spoil dump shape increases stability on the contour strip mines, presented at Research and Applied Technology Symp. on Mined-Land Reclamation, Pittsburgh, PA, March 7–8, 1973, pp. 243–249.

Wills, C. J., and T. P. McCrink (2002), Comparing landslide inventories: the map depends on the method, *Environ. Eng. Geosci., 8*(4), 279–293.

Wilson, C. J., and W. E. Dietrich (1987), The contribution of bedrock groundwater flow to storm runoff and high pore pressure development in hollows, *IAHS Publ. 165*, 49–60.

Wilson, C. M., and P. L. Smart (1984), Pipes and pipe flow process in and upland catchment, Wales, *Catena, 11*, 145–158.

Wilson, G. V., P. M. Jardine, R. J. Luxmoore, and J. R. Jones (1990), Hydrology of a forested hill-slope during storm events, *Geoderma, 46*, 119–138.

Wilson, S. D. (1974), Landslide instrumentation for the Minneapolis Freeway, *Transport. Res. Record 482*, 30–42.

Wohl, E. E., and P. P. Pearthree (1991), Debris flows as geomorphic agents in the Huachuca Mountains of southeastern Arizona, *Geomorphology, 4*, 273–292.

Wolfe, M. D., and J. W. Williams (1986), Rate of Landsliding as impacted by timber management activities in northwestern California, *Bull. Assoc. Eng. Geol., 23*(1), 53–60.

Wolle, C. M., and C. S. Carvalho (1994), Taludes naturias, in *Solos do Litoral de Sãn Paulo*, edited by F.F. Falconi and A. Negro, pp. 180–203, Associação Brasileira de Mecânica de Solos- Núcleo Regional de Sãn Paulo-ABMS, Sãn Paulo, Brazil.

Wong, H. L., and P. C. Jennings (1975), Effects of canyon topography on strong ground motion, *Seismol. Soc. Am. Bull., 65*(5), 1239–1257.

Wong, H. N., and K. K. S. Ho (1997), The 23 July 1994 landslide in Kwun Lung Lau, Hong Kong, *Can. Geotech. J., 34*, 825–840.

Woods, R., and L. Rowe (1996), The changing spatial variability of subsurface flow across a hill-slope, *J. Hydrol. N. Z., 35*, 51–86.

Wright, C., and A. Mella (1963), Modifications to the soil pattern of south central Chile resulting from seismic and associated phenomena during the period May to August 1960, *Bull. Seismol. Soc. Am., 53*, 1367–1402.

Wu, T. H. (1976), Investigation of landslides on Prince of Wales Island, Geotech. Eng. Rep. 5, 94 p., Civil Eng. Dept., Ohio State Univ., Columbus, Ohio.

Wu, T. H., and D. A. Sangrey (1978), Strength properties and their measurement, in *Landslides: Analysis and Control*, Special Rep. 176, pp. 139–154, Transport. Res. Board, National Academy of Science, National Res. Council, Washington, DC.

Wu, T. H., and D. N. Swanston (1980), Risk of landslides in shallow soils and its relation to clearcutting in southeastern Alaska, *Forest Science, 26*(3), 495–510.

Wu, T. H., W. P. McKinnel, and D. N. Swanston (1979), Strength of tree roots and landslides on Prince of Wales Island, Alaska, *Can. Geotech. J., 16*, 19–33.

Wu, T. H., W. H. Tang, and H. H. Einstein (1996), Landslide hazard and risk assessment, in *Landslides—Investigation and Mitigation*, edited by A. K. Turner and R. L. Schuster, pp. 102–118, Special Rep. 247, Transport. Res. Board, National Res. Council, National Academic Press, Washington, DC.

Wu, W. (1993), Distributed slope stability analysis in steep, forested basins, Ph.D. thesis, 148 p., Utah State Univ., Logan.

Wu, W., and R. C. Sidle (1995), A distributed slope stability model for steep forested hillslopes, *Water Resour. Res. 31*, 2097–2110.

Wu, W., and R. C. Sidle (1997), Application of a distributed shallow landslide analysis model (dSLAM) to managed forested catchments in coastal Oregon, *IAHS Publ. 245*, 213–221.

Wyllie, D. C., and N. I. Norrish (1996), Rock strength properties and their measurement, in *Landslides – Investigation and Mitigation*, edited by A. K. Turner and R. L. Schuster, pp. 474–504, Special Rep. 247, Trans. Res. Board, National Res. Council, National Academic Press, Washington, DC.

Wyss, M. (2001), Why is earthquake prediction research not progressing faster, *Tectonophysics, 338*, 217–223.

Wyss, M., and S. Matsumura (2002), Most likely locations of large earthquakes in the Kanto and Tokai areas, Japan, based on the local recurrence times, *Phys. Earth Planet. In., 131*, 173–184.

Wyss, W., and S. Yim (1996), Vulnerability and adaptability of Hong Kong to hazards under climatic change conditions, *Water Air Soil Poll., 92*, 181–190.

Yagi, N., A. Enoki, and R. Yatabe (1988), Prediction of failure of sandy slopes during rain, *Proc. Symp. on Landslide Mechanisms and Measures*, Shikoku Branch of Jpn. Geotech. Soc., pp. 37–42 (in Japanese).

Yamada, G. (1970), The analysis of the landslides along JNR. *Railway Tech. Res. Rep. No, 719*, 92 p., The Railway Tech. Res. Inst., Japanese National Railways (in Japanese).

Yamada, S. (1997), Seasonal variation in soil creep on a forested hillslope near Sapporo, Hokkaido, northern Japan, *Trans. Jpn. Geomorph. Union, 18*, 117–130.

Yamada, S. (1999), The role of soil creep and slope failure in the landscape evolution of a head water basin: field measurements in a zero order basin of northern Japan, *Geomorphology, 28*, 329–344.

Yamagishi, H., K. Amamiya, and K. Kurosawa (1999), Landslides induced by the Hokkaido Nansei-Oki Earthquake, 12 July, 1993, in *Landslides of the World*, edited by K. Sassa, pp. 211-214. Kyoto Univ. Press, Japan.

Yamamoto, H., H. Kadomura, R. Suzuki, and T. Imagawa (1980), Mudflows from a 1977-1978 tephra-covered watershed on Usu volcano, Hokkaido, Japan, *Trans. Jpn. Geomorph. Union, 1*, 73–88.

Yanase, H., H. Ochiai, and S. Matsuura (1985), A large-scale landslide on Mt. Ontake due to the Naganoken-Seibu earthquake, 1984, *Proc. 4th Int. Conf. and Field Workshop on Landslides*, Tokyo, pp. 323–328.

Yatabe, R., N. Yagi, and A. Enoki (1986), Investigation of method for predicting failure of sandy slopes during rain, *Soils and Foundations (Jpn. Geotech. Soc.), 376*, 297–305 (in Japanese).

Yates, S. R., A. W. Warrick, and D. O. Lomen (1985), Hillside seepage: an analytical solution to a nonlinear Dupuit-Forchheimer problem, *Water Resour. Res., 21*, 331–336.

Yatsu, E. (1966), *Rock Control in Geomorphology*, Sozosha, Tokyo, 133 p.

Yatsu, E. (1967), Some problems on mass movements, *Geograf. Ann., 49A* (2–4), 396–401.

Yatsu, E. (1988), *The Nature of Weathering, An Introduction*, Sozosha, Japan, 624 p.

Yee, C. S., and R. D. Harr (1977), Influence of soil aggregation on slope stability in the Oregon Coast Ranges, *Environ. Geol., 1*, 367-377.

Yeend, W. E. (1973), Slow-sliding slumps, Grand Mesa, Colorado, *The Mountain Geologist, 10*(1), 25–28.

Yesilnacar, E., and T. Topal (2005), Landslide susceptibility mapping: comparison of logistic regression and neural networks methods in a medium scale study, Hendek region (Turkey), *Eng. Geol., 79*, 251–266.

Yokota, S., and A. Iwamatsu (1999), Weathering distribution in a steep slope of soft pyroclastic rocks as an indicator of slope instability, *Eng. Geol., 55*, 57–68.

Yong, R. N., and B. P. Warkentin (1975), *Soil Properties and Behavior*, pp. 335–359, Elsevier, New York.

Yoon, G. G. (1991), Extent and economic significance of landslides in Korea, in *Landslides* edited by D. H. Bell, pp. 1071–1076, Balkema, Rotterdam, The Netherlands.

Yoshida, H., M. Nishim, and M. Nanbu (2001), Damage to residential fills in the 1995 Hanshin-Awaji Earthquake, *Nat. Hazards, 23*, 87–97.

Yoshimatsu, H. (1990), A large rock fall along a coastal highway in Japan, *Landslide News (Jpn. Landslide Soc.), 4*, 4–5.

Zaruba, Q., and V. Mencl (1969), *Landslides and their Control*, 214 p., Elsevier, New York.

Zehe, E., and H. Flühler (2001), Slope scale variation of flow patterns in soil profiles, *J. Hydrol., 247*, 116–132.

Zehe, E., T. Maurer, J. Ihringer, and E. Plate (2001), Modeling water flow and mass transport in a loess catchment, *Phys. Chem. Earth (B), 26*(7–8), 487–507.

Zêzere, J. L., A. B. Ferreira, and M. L. Rodrigues (1999), The role of conditioning and triggering factors in the occurrence of landslides: a case study in the area north of Lisbon (Portugal), *Geomorphology, 30*, 133–146.

Zheng, G., Y. Lang, B. Takano, M. Matsuo, A. Kuno, and H. Tsushima (2002), Iron speciation of sliding mud in Toyama Prefecture, Japan, *J. Asian Earth Sci., 20*, 955–963.

Zhou, Y. (1999), A case study on effects of Yunnan pine forest erosion control – with special reference to the alpine gorge, NW Yunnan, Southwest Jiaotong University Press, China, 194 p..

Ziemer, R. R. (1978), An apparatus to measure the crosscut shearing strength of roots, *Can. J. For. Res., 8*, 142–144.

Ziemer, R. R. (1981), Roots and the stability of forested slopes, *IAHS Publ. 132*, 343–361.

Ziemer, R. R., and D. N. Swanston (1977), Root strength changes after logging in southeast Alaska, *Res. Note PNW-306*, 10 p., For. Serv., U.S. Dep. of Agric., Portland, OR.

Zierholz, C., P. Hairsine, and F. Booker (1995), Runoff and soil erosion in the bushland following the Sydney bushfires, *Australian J. Soil Water Conserv., 8*(4), 28–37.

Zimbone, S. M., A. Vickers, R. P. C. Morgan, and P. Vella (1996), Field investigations of different techniques for measuring surface soil shear strength, *Soil Technol., 9*, 101–111.

Zimmermann, M., and W. Haeberli (1992), Climatic change and debris flow activity in high mountain areas—a case study in the Swiss Alps. *Catena Suppl., 22*, 59–72.

Zipper, C. E., W. E. Daniels, and J. C. Bell (1989), The practice of "approximate original contour" in the Central Appalachians. II. Economic and environmental consequences of an alternative, *Landscape Urban Plan., 18*, 139–152.

Index

Printed in the United States
By Bookmasters